Nonlinear Dispersive Equations

Existence and Stability of Solitary and Periodic Travelling Wave Solutions

A mi mujer Martha y mi hija Victoria Mel, por supuesto.

Contents

Preface xi

Part 1. History, Basic Models, and Travelling Waves 1

Chapter 1. Introduction and a Brief Review of the History 3

Chapter 2. Basic Models 17
- 2.1. Introduction 17
- 2.2. Models 17
- 2.3. Comments 22

Chapter 3. Solitary and Periodic Travelling Wave Solutions 25
- 3.1. Introduction 25
- 3.2. Travelling Wave Solutions 25
- 3.3. Examples 27
- 3.4. The Poisson Summation Theorem and Periodic Wave Solutions 39
- 3.5. Comments 42

Part 2. Well-Posedness and Stability Definition 47

Chapter 4. Initial Value Problem 49
- 4.1. Introduction 49
- 4.2. Some Results about Well-Posedness 49
- 4.3. Some Results about Global Well-Posedness 57
- 4.4. Comments 58

Chapter 5. Definition of Stability 61
- 5.1. Introduction 61
- 5.2. Orbital Stability 61
- 5.3. Comments 64

Part 3. Stability Theory 67

Chapter 6. Orbital Stability—the Classical Method 69
- 6.1. Introduction 69
- 6.2. Stability of Solitary Wave Solutions for the GKdV 70
- 6.3. "Stability of the Blow-up" for a Class of KdV Equations 81
- 6.4. Comments 87

Chapter 7. Grillakis-Shatah-Strauss's Stability Approach 91
- 7.1. Introduction 91

7.2. Geometric Overview of the Theory 91
7.3. Stability of Solitary Wave Solutions 93
7.4. Stability of Solitary Waves for KdV-Type Equations 98
7.5. On Albert-Bona's Spectrum Approach 99
7.6. Comments 100

Part 4. The Concentration-Compactness Principle in Stability Theory 103

Chapter 8. Existence and Stability of Solitary Waves for the GBO 105
8.1. Introduction 105
8.2. Solitary Waves for the GBO 107
8.3. Stability of Solitary Waves for the GBO Equations 119
8.4. Comments 124

Chapter 9. More about the Concentration-Compactness Principle 127
9.1. Introduction 127
9.2. Solitary Wave Solutions of Benjamin-Type Equations 127
9.3. Stability of Solitary Wave Solutions: the GKdV Equations 128
9.4. Stability of Solitary Wave Solutions: the Benjamin Equation 129
9.5. Stability of Solitary Wave Solutions: the Fourth-Order Equation 133
9.6. Stability of Solitary Wave Solutions: the GKP-I Equations 133
9.7. Comments 135

Chapter 10. Instability of Solitary Wave Solutions 137
10.1. Introduction 137
10.2. Instability of Solitary Wave Solutions: the GB Equations 139
10.3. Fifth-Order Korteweg-de Vries Equations 150
10.4. A Generalized Class of Benjamin Equations 152
10.5. Linear Instability and Nonlinear Instability 153
10.6. Comments 157

Part 5. Stability of Periodic Travelling Waves 159

Chapter 11. Stability of Cnoidal Waves 161
11.1. Introduction 161
11.2. Stability of Cnoidal Waves with Mean Zero for KdV Equation 164
11.3. Stability of Constant Solutions for the KdV Equation 174
11.4. Cnoidal Waves for the 1D Benney-Luke Equation 177
11.5. Angulo and Natali's Stability Approach 183
11.6. Comments 196

Part 6. APPENDICES 199

Appendix A. Sobolev Spaces and Elliptic Functions 201
A.1. Introduction 201
A.2. Lebesgue Space $L^p(\Omega)$ 201
A.3. The Fourier Transform in $L^1(\mathbb{R}^n)$ 201
A.4. The Fourier Transform in $L^2(\mathbb{R}^n)$ 202
A.5. Tempered Distributions 202

A.6. Sobolev Spaces 204
A.7. Sobolev Spaces of Periodic Type 206
A.8. The Symmetric Decreasing Rearrangement 207
A.9. The Jacobian Elliptic Functions 208

Appendix B. Operator Theory 211
B.1. Introduction 211
B.2. Closed Linear Operators: Basic Theory 211
B.3. Pseudo-Differential Operators and Their Spectrum 229
B.4. Spectrum of Linear Operators Associated to Solitary Waves 231
B.5. Sturm-Liouville Theory 237
B.6. Floquet Theory 240

Bibliography 245

Index 255

Preface

This book originated from lectures given by the author in January and February 2000 at IMPA's summer program, in the months of August to December 2002 at the State University of Campinas, and in the 24° Colóquio Brasileiro de Matemática, 2003, Brazil.

The intention of this book is to provide a self-contained presentation of classical and new methods in the mathematical studies of wave phenomena that are related to the existence and stability of travelling wave solutions (solitary and periodic waves) for nonlinear dispersive evolution equations. Although many results may be found in the existing literature, in this book we offer new results. This book has also been designed to be instructive as well as to be a new source of reference for students and for mature scientists interested in nonlinear wave phenomena. Simplicity and concrete applications are emphasized in order to make the material easily assimilated. Also, I hope that it inspires future developments in this important and useful subject.

The preparation of this book had partial support from *O Conselho Nacional de Desenvolvimento Científico e Tecnológico* (CNPq) and from *Coordenação de Aperfeiçoamento de Pessoal de Nível Superior* (CAPES), which support Brazilian research. Also, my appreciation goes to the Department of Mathematics of the State University of São Paulo, São Paulo, Brazil (where I am a professor) and to the Department of Mathematics of the University of California, Santa Barbara, where part of this book was finished.

I am indebted to many friends who gave me the initial inspiration for the treatment of this subject, the support, the encouragement, and suggestions to complete this book. I express my hearty thanks to Professors J. Albert, H. Biagioni, J. Bona, R. Iorio, F. Linares, and M. Scialom.

Last but not least gratitude goes to my wife, Martha, who was incredibly tolerant and cooperative during the evolution of this book, and also to my daughter, Victoria Mel, who gave me part of her valuable time to finish this manuscript.

Jaime Angulo Pava
State University of São Paulo
May 2009

Part 1

History, Basic Models, and Travelling Waves

Russell, John Scott (1808–882)

I believe I shall best introduce this phenomenon by describing the circumstances of my own first acquaintance with it. I was observing the motion of a boat which was rapidly drawn along a narrow channel by a pair of horses, when the boat suddenly stopped—not so the mass of water in the channel which it had put in motion; it accumulated round the prow of the vessel in a state of violent agitation, then suddenly leaving it behind, rolled forward with great velocity, assuming the form of a large solitary elevation, a rounded, smooth and well-defined heap of water, which continued its course along the channel apparently without change of form or diminution of speed. I followed it on horseback, and over-took it still rolling on at a rate of some eight or nine miles an hour, preserving its original figure some thirty feet long and a foot to a foot and a half in height. Its height gradually diminished, and after a chase of one or two miles I lost it in the winding of the channel.

In *Report on Waves*, Rep. 14th. Meet. Brit. Assoc. Adv. Sci., York, 1844, 311–390.

CHAPTER 1

Introduction and a Brief Review of the History

During the last thirty years the theory of stability of travelling wave solutions for nonlinear evolution equations has grown into a large field that attracts the attention of both mathematicians and physicists in view of its applications to real-world nonlinear models and of the novelty of the problems. The qualitative theory of nonlinear equations includes, in particular, investigations on the well-posedness of various problems for these equations, the behavior of solutions such as blowing-up, the existence and stability of solitary wave and periodic travelling wave solutions as well as properties of the dynamical system generated by these equations.

The purpose of this book is to give a self-contained presentation of some basic and detailed results concerning the existence and stability of travelling wave solutions. It is intended to be a new source for modern research dealing with nonlinear phenomena of dispersive type. The selection of the material is mainly related to the author's scientific interest. There are four main topics: the existence of solitary wave and periodic travelling wave solutions, the problems of the stability of these special kinds of solutions, the applicability of the *Concentration-Compactness Principle* in the study of the stability of solitary wave solutions of nonlinear dispersive equations, and the instability of solitary waves. A detailed and clear explanation of every concept and method introduced is given. The exposition is accompanied by a careful selection of modern examples. The book provides information that puts the reader at the forefront of current research.

Nonlinear evolution equations for modelling waves take into account both nonlinearity and dispersion effects. Their birth was the discovery of the solitary wave, or great wave of translation, observed on the Edinburgh-Glasgow canal in 1834 by J. Scott Russell. The story of the first encounter of Russell with the solitary wave was reported by him to the British Association in 1844 with the name *Report on Waves* [**246**]. Fascinated with this long water wave without a change in shape, which he called the "great wave of translation, or solitary wave", Russell made some laboratory experiments on this phenomenon, generating solitary waves by dropping a weight at one end of a water channel. He deduced empirically that the volume of water in the wave is equal to the volume displaced by the weight and that the speed c of the solitary wave and its maximum amplitude a above the free surface liquid of finite depth h satisfy the relation

$$c^2 = g(h + a),$$

where g is the acceleration due to gravity. His description of solitary waves contradicted the theories of water waves according to G. B. Airy and G. G. Stokes; they raised questions on the existence of Russell's solitary waves and conjectured that such waves cannot propagate in a liquid medium without a change of form. Despite the mathematical theory, the experimental evidence in favor of solitary waves was convincing. It was not until the 1870s that Russell's prediction was finally and independently confirmed by both J. Boussinesq (1871, [**70**]) and Lord Rayleigh (1876, [**240**]). Assuming that a solitary wave has a length much greater than the depth of the water, they derived from the equations of motion for an inviscid incompressible liquid that the wave height above the mean level h, $z = \psi(x,t)$, is given by

$$\psi(x,t) = a\text{sech}^2[\beta(x-ct)], \tag{1.1}$$

where $\beta^2 = 3a/[4h^2(h+a)]$ for any positive amplitude $a > 0$. Although these authors found the sech^2 solution, they did not write any equation for η that produces (1.1) as a solution. However, Boussinesq did a lot more ([**71**]) and discovered that if a water wave propagates along a flat-bottomed channel of undisturbed depth h and has large wavelength and small amplitude relative to h, then the elevation η of the water surface considered as a function of the coordinate x along the channel and the time t will approximately satisfy the nonlinear evolution equation

$$\eta_{tt} - gh\eta_{xx} - gh\left[\frac{3}{2h}\eta^2 + \frac{h^2}{3}\eta_{xx}\right]_{xx} = 0, \tag{1.2}$$

where g is the gravitational acceleration and $\sqrt{gh}$ is the speed of the shallow water waves. This equation is known as the *Boussinesq (bidirectional) equation.* Using this equation, he obtained an explicit representation of solitary waves travelling in both positive and negative x-directions, namely,

$$\eta_{\text{sol}}(x,t) = a\ \text{sech}^2\left[\sqrt{\frac{3a}{h^3}}(x \pm ct)\right]. \tag{1.3}$$

As we will see below, Boussinesq's inquiry about the behavior of the solutions of (1.2), with initial data being a slight perturbation of a solitary wave (1.3), was the genesis of the theory nowadays called the theory of stability of solitary wave solutions.

In 1895, D. J. Korteweg and G. de Vries [**168**] formulated a mathematical model which provided an explanation of the phenomenon observed by Russell (they were apparently unaware of the work of Boussinesq). They derived the now-famous equation for the propagation of waves in one direction on the surface of water of density ρ in the form

$$v_t = \frac{\sqrt{gh}}{h}\left[\left(\varepsilon + \frac{3}{2}v\right)v_x + \frac{1}{2}\sigma v_{xxx}\right] \tag{1.4}$$

where $v = v(x,t)$, x is a coordinate chosen to be moving with the wave, ε is a small parameter, and

$$\sigma = h\left(\frac{h^2}{3} - \frac{T}{g\rho}\right) \sim \frac{1}{3}h^3,$$

when the surface tension $T(\ll \frac{1}{3}g\rho h^2)$ is negligible. This is essentially the original form of the *Korteweg-de Vries equation.* We shall call it the KdV equation. We note that in the approximation used to derive this equation one considers long wave propagating in the direction of increasing x. Equation (1.4) is one of the simplest and most useful nonlinear model equations for solitary waves, and it represents the long-time evolution of wave phenomena in which *the steepening effect of the nonlinear term is counterbalanced by dispersion.*

In the first half of the twentieth century, solitary waves and related evolution equations were not a major topic of scientific conversation. Modern developments in the theory and applications of the KdV solitary waves began with the seminal work published as a Los Alamos Scientific Laboratory Report in 1955 by Enrico Fermi, John Pasta, and Stanislaw Ulam [**106**] on a numerical model of a discrete nonlinear mass-spring system. In 1914, Debye suggested that the finite thermal conductivity of an anharmonic lattice is due to the nonlinear forces in the springs. This suggestion led Fermi, Pasta, and Ulam to believe that a smooth initial state would eventually relax to an equipartition of energy among all modes because of nonlinearity. But their studies led to the remarkable conclusion that there is no equipartition of energy among the modes. What they found did not correspond well to heat conduction; it seems this simple mass-and-spring system features near recurrence of initial states, and not the kind of thermalization that one expects. Although all the energy was initially in the lowest modes, after flowing back and forth among various low-order modes, it eventually returns to the lowest mode, and the end states are a series of recurring states. This remarkable fact has become known as the Fermi-Pasta-Ulam (FPU) recurrence phenomenon. A Los Alamos report was duly constructed and the issue then lay dormant.

A few years later Gardner and Morikawa [**117**] studied the stability of a cold collisionless plasma as it arose in a putative description of nuclear fusion. Starting from the full magneto-hydrodynamic equations and making assumptions about the motion of the plasma, they derived the same equations as had Boussinesq and Korteweg-de Vries, although the physical context was different.

Afterwards, Fermi, Pasta, and Ulam and Gardner and Morikawa inspired Kruskal and Zabusky [**170**] to formulate a continuum model for the nonlinear mass-spring system to understand why recurrence occurred. The system of ordinary differential equations went over to a partial differential equation in this limit, and the equation in question was the Korteweg-de Vries equation. In fact, they considered the initial value problem for the KdV equation

$$u_t + uu_x + \delta u_{xxx} = 0, \tag{1.5}$$

where $\delta > 0$, with the initial condition

$$u(x,0) = \cos \pi x, \qquad 0 \le x \le 2,$$

and the periodic boundary conditions with period 2, so that $u(x,t) = u(x+2,t)$ for all t. Their numerical studies with $\sqrt{\delta} = 0.022$ produced a lot of new and interesting results. In fact, at later times the solutions develop a series of *eight* well-defined waves, each like sech^2 functions with the taller (faster) waves ever catching up and overtaking the shorter (slower) waves. These waves undergo nonlinear interaction, according to the KdV equation, and then emerge from the interaction without a change of form and amplitude, but with only a small change in their phases. So the most remarkable feature is that *these waves retain their identities after the nonlinear interaction.* Another surprising fact is that the initial profile reappears, very close to the FPU recurrence phenomenon. In view of their preservation of shape and their resemblance to the particle-like character of these waves, Kruskal and Zabusky called these solitary waves *solitons* (like photon, proton, etc.). Historically, Kruskal and Zabusky's paper in 1965 marked the birth of the new concept of the *soliton,* a name intended to signify particle-like quantities. Lax in 1968 [**179**] gave rigorous analytical proof that two distinct solitons are preserved through the nonlinear interaction governed by the KdV equation. These discoveries have led, in turn, to extensive theoretical, experimental, and computational studies over the last thirthy-five years.

For instance, we have the ingenious method for finding the exact solution of the KdV equation formulated by Gardner, Greene, Kruskal, and Miura in 1974 [**116**], which is known as the *Method of Inverse Scattering* or the *Inverse Scattering Transform.* This novel method has been generalized to solve several other nonlinear equations (see [**2**] and [**3**]).

Solitary waves in water were first observed scientifically by Russell [**246**], and in his study of these wave forms, he found that individually they appear to be stable states of motion. Propagating along a uniform canal, a solitary wave displays a remarkable property of permanence, such as to give an observer immediate confidence in its stability. Afterward, in the seminal work of Boussinesq, [**71**], after Boussinesq had deduced Russell's empirical great wave of translation (1.3), a new question arose with regard to the stability of these solitary wave solutions by the flow generated by equation (1.2). In fact, Boussinesq proposed to show that the solitary waves (1.3) are stable in the sense that a slight perturbation of a solitary wave will continue to resemble a solitary wave all of the time, rather than evolving into some other wave form. Such a result in fact can be an explanation of why solitary waves are so easily produced and observed in experiments.

Boussinesq obtained three invariant physical quantities, E, F, and V, related to equation (1.2), defined by

$$E(\eta) = \int_{-\infty}^{\infty} \left[\eta_x^2(x) - \frac{3}{h}\eta^3(x)\right] dx, \qquad F(\eta) = \int_{-\infty}^{\infty} \eta^2(x)\, dx,$$
$$V(\eta) = \int_{-\infty}^{\infty} \eta(x)\, dx. \tag{1.6}$$

If $\eta = \eta(x,t)$ is a suitable smooth solution of the nonlinear equation (1.2), then $E(\eta(x,t))$, $F(\eta(x,t))$, and $V(\eta(x,t))$ are independent of the temporal variable t. Evidently, F and V represent the energy and the volume, respectively. The quantity E was named by Boussinesq as the *moment of instability.* Boussinesq noticed that the constraint variational problem

$$\delta E = 0 \qquad \text{for } F \text{ fixed} \tag{1.7}$$

is solved by the solitary wave solution (1.3) with a specific value of a. Boussinesq also realized that the extremal property of solitary waves is related to stability, and so he tentatively explained the permanence of these waves. He asserted that, within the class of wave forms whose energy has a given value, those profiles which correspond to the greatest moments of instability will differ the most from solitary wave profiles, while the minimum value of the moment of instability within this class is attained at a solitary wave profile. So by considering η to be close to η_{sol}, *Boussinesq assumed* that $E(\eta_{\text{sol}})$ is the absolute minimum for a given $F(u) = F(\eta_{\text{sol}})$, so that

$$\Delta E(\eta, \eta_{\text{sol}}) \equiv E(\eta) - E(\eta_{\text{sol}}) \geqq 0.$$

But $F(\eta)$ and $\Delta E(\eta, \eta_{\text{sol}})$ are both independent of time if η is a solution. So, if an initial wave form is initially close to η_{sol}, by continuity, its evolution is subject to the constraint that $\Delta E(\eta(t), \eta_{\text{sol}})$ keeps the same small value as in the beginning. Boussinesq then suggested that any wave initially close to a solitary wave remains so for all time. However, while he was on the right track, by modern standards his overall argument for the stability of solitary waves contains some gaps: first, he only proved that $E(\eta_{\text{sol}})$ is stationary, not that it is a minimum. Second, even though $\Delta E(\eta(t), \eta_{\text{sol}})$ was nonnegative and small, it is not evident that this is an effective measure of the difference between the functions $\eta(t)$ and η_{sol} (see Chapter 6).

The first rigorous proof of stability of solitary wave solutions appeared only a century later, in Benjamin's article [**42**] on solitary wave solutions of the KdV equation (1.5) with $\delta = 1$ (obviously this normalization does not affect in any sense the theory). If we consider a wave of the form $u(x,t) = \phi_c(x - ct)$ and substitute it in (1.5) with the boundary condition that $\phi_c(\xi) \to 0$ as $|\xi| \to \infty$, we see that the wave profile ϕ_c satisfies the nonlinear differential equation

$$-\phi_c''(\xi) + c\phi_c(\xi) - \frac{1}{2}\phi_c^2(\xi) = 0 \tag{1.8}$$

where $\xi = x - ct$. It is well known that, via integration, the wave profile ϕ_c is given by

$$\phi_c(\xi) = 3c\,\mathrm{sech}^2\Big(\frac{1}{2}\sqrt{c}\,\xi\Big). \tag{1.9}$$

Making an elaboration of Boussinesq's original ideas, Benjamin showed that the solitary wave solutions (1.9) are stable by the flow generated by the KdV equation for every $c > 0$. The arguments given by Benjamin were also based on the functionals E ($h = 1$) and F, which are also invariants of motion for the KdV equation. In a simple analysis, we cannot expect to prove stability of solitary wave solutions with respect to the metric in $H^s(\mathbb{R})$, $s \geqq 0$ (see Chapter 5). So, one way of avoiding this difficulty is to give up knowing where the solution is in exchange for knowing its *shape* very well. This discrimination leads to a new measure of distance and a new notion of stability, which Benjamin called *stability in shape* (or orbital stability). This notion in a few words tell us that ϕ is stable whenever an initial wave form ψ is sufficiently near to ϕ in the $H^1(\mathbb{R})$-norm and then for each instant t there is a translation, $\gamma(t)$, such that the shape of the function $x \to u(x + \gamma(t), t)$ will resemble and remain close to ϕ in the $H^1(\mathbb{R})$-norm. The heart of Benjamin's analysis was to show that if $\psi \in H^1(\mathbb{R})$ is sufficiently close to ϕ_c in $H^1(\mathbb{R})$-norm and $F(\psi) = F(\phi_c)$, then

$$E(\psi) - E(\phi_c) \geq A \inf_{y\in\mathbb{R}} \|\psi - \phi_c(\cdot + y)\|_1^2 \tag{1.10}$$

where A denotes a positive constant which is independent of ψ. Note that inequality (1.10) shows that ϕ_c is a local minimizer of the functional E, but it does not show Boussinesq's assertion (in the case of equation (1.5)) that ϕ_c is a minimizer of E over the set of all admissible functions ψ satisfying $F(\psi) = F(\phi_c)$. As we will see later, this assertion was only established years later as a consequence of the *Concentration-Compactness Method* (Lions [**193**], [**194**]) and from the pioneering work of Cazenave and Lions [**82**] about the orbital stability of standing waves for some nonlinear Schrödinger equations.

The local analysis given in [**42**] is centered on the quadratic form $\langle \mathcal{L}\varphi, \varphi\rangle$ generated by the closed, unbounded, self-adjoint operator on $L^2(\mathbb{R})$

$$\mathcal{L} = -\frac{d^2}{dx^2} + c - \phi_c. \tag{1.11}$$

Upper and lower bounds on the quadratic term are the key ingredients to obtain inequality (1.10) and so establish stability. Upper bounds are straightforward in this case, and the crux of the matter is effective lower bounds. At this point, a careful spectral theory for the operator $\mathcal{L}$ arises as a crucial piece of information for completing the stability. The basic properties about $\mathcal{L}$ are

(1.12) *$\mathcal{L}$ has a single negative eigenvalue which is simple, the zero eigenvalue is simple with eigenfunction ϕ_c', and the remainder of the spectrum is positive and bounded away from zero.*

So, since ϕ_c in (1.9) has a single maximum and is an even function monotonically decreasing to zero at infinity, the Sturm-Liouville theory immediately implies the required conditions in (1.12). So, inequality (1.10) and other ingredients imply that by suitably translating the solitary wave profile ϕ_c, one can nearly match it to the solution u and so the assertion that the solitary wave is stable in shape is a consequence.

For several years after these initial advances there were no new developments in the stability theory of solitary wave solutions for nonlinear evolution equations. This is owing, in part, to the overall complexity of Benjamin's theory and, in part, to the difficulty of establishing the crucial information (1.12).

The next advance appears to have been made by Bennet *et al.* [**49**] in their approach to the stability of the Benjamin-Ono equation's solitary wave (see (3.5) and (3.6)). Their theory still required a complete spectral analysis. Weinstein in 1986–1987 ([**274**], [**276**]) simplified the overall argument demonstrating stability, clarifying and sharpening what was required from the spectral analysis, and very considerably extending the range of the theory's applicability. A large number of the extensive collection of cases that fall within the theory produced the development of sharper conditions for the stability and instability of general travelling wave solutions for nonlinear evolution equations, such as the general stability theory established by Grillakis, Shatah, and Strauss in [**123**] and [**124**]. This theory was set for abstract Hamiltonian equations (or systems) of the form

$$u_t = JE'(u(t)) \tag{1.13}$$

where J should be an onto skew-symmetric linear operator. In [**123**] a sufficient condition of stability was established, one which will imply the crucial inequality (1.10). In the case of the KdV equation, this condition is given by the strict convexity property of the real-valued function

$$d(c) = E(\phi_c) + cF(\phi_c) \tag{1.14}$$

for $c \in (0, \infty)$. This condition is the same as the one given by Shatah in 1983 ([**251**]) and by Weinstein in 1987 ([**274**]) in their analyses of the stability of ground states for certain nonlinear evolution equations.

Instability conditions were also considered in the Grillakis *et al.* abstract results [**123**], in which the assumption on J is crucial. So, this result does not apply directly when J is a differential operator such as ∂_x. In this situation an improvement of the theory was given by Bona, Souganidis, and Strauss [**64**] and Souganidis and Strauss [**255**], by taking into consideration general one-dimensional evolution equations of the forms

$$u_t + f(u)_x - Mu_x = 0 \quad \text{and} \quad M_0 u_t + f(u)_x = 0,$$

where M and M_0 are pseudo-differential operators of order $\mu \geqq 1$. So, in the presence of the right spectral information (1.12), a solitary wave that is not proved to be stable by the theory is in fact unstable. Namely, the basic function d satisfies the property that $d''(c) < 0$.

Condition (1.12) is a crucial piece of information to be verified in order to obtain a stability or instability theory. In general, it is not easy to prove this condition. In particular the fact that zero is a simple eigenvalue is a delicate business. A more accurate study about this point in the theory has been developed by Albert in [**6**] and Albert and Bona in [**10**]. In their works novel sufficient conditions for obtaining the needed spectral information were established (see Chapter 7).

We have seen that the stability of solitary wave solutions, in a general form, is based on conditions that prove an inequality of the form (1.10). This means that we must show that the solitary wave solution is a local constrained minimizer of a Hamiltonian functional, and the procedure for this is carried out basically studying specific spectral properties of a linear operator obtained by linearizing the solitary wave equation. In practice this spectral analysis is particularly difficult to carry out (see Chapter 7). To avoid these difficulties, an alternative method of proving stability of solitary waves, which does not rely on local analysis, was developed by Cazenave and Lions in [**82**] using Lions's method of *Concentration-Compactness*. In this method, instead of starting with a given solitary wave and trying to prove that it reaches a local minimum for a constrained variational problem, one starts with the constrained variational problem and looks for global minimizers. When the method works and the functionals involved in the variational analysis are conserved quantities for the equation in question, one shows not only the existence of global minimizers which are solitary wave solutions, but also that the set of global minimizers is a stable set for the flow generated by the evolution equation under discussion. This means that a solution which starts near the set of minimizers will remain near it for all time (see Chapters 8 and 9).

In the last couple of decades, a series of applications and variants of the Concentration-Compactness Method was obtained for a great range of dispersive evolution equations. We cite for instance the works of Weinstein [**276**], Albert, Bona, and Saut [**14**], Kuznetsov [**172**], Kuznetsov, Rubenchik, and Zakharov [**173**], Albert [**8**], de Bouard and Saut [**95**], Kichenassamy [**164**], Lopes [**195**], [**197**], [**198**], Angulo [**21**], and Albert and Angulo [**9**].

The next development in the study of the applicability of the Concentration-Compactness Method in the existence and stability of solitary wave solutions was achieved by Levandosky [**182**]. Here Levandosky studied the stability of a fourth-order wave equation (see Chapter 8, Section 8.3) and observed that the method can still be used in the stability theory if the functionals involved in the variational problem are not conserved quantities.

Overall, the study of the qualitative properties of solitary wave solutions and its influence on the development of the theory of nonlinear evolution equations have produced a large number of papers in the past decade such that it now becomes difficult to trace the developments. In particular, a very substantial contribution relative to the large-time asymptotic behavior of KdV-type equations has been established in the works of Pego and Weinstein [**235**], [**236**], Bona and Soyeur [**63**], Laedke, Blaha, Spatschek, and Kuznetsov [**176**], Merle [**220**], Martel and Merle [**210**], [**211**], [**212**], [**214**], [**216**], [**217**], Martel [**208**], [**209**], and Merle and Vega [**221**].

Now, the situation regarding the existence and stability of periodic travelling wave solutions is very different from that for solitary wave solutions; because of these progressive wave trains have received comparatively little attention. A first study of these wave forms was made by Benjamin in [**45**]. This study focused on the periodic steady solutions called *cnoidal waves*, which were found initially by Korteweg-de Vries in [**168**] for the KdV equation (1.5) ($\delta = 1$). Benjamin put forward an approach to proving the stability of cnoidal waves of the form

$$\varphi(\xi) = \beta_2 + (\beta_3 - \beta_2)\mathrm{cn}^2\left(\sqrt{\frac{\beta_3 - \beta_1}{12}}\xi; k\right), \tag{1.15}$$

but he did not provide a detailed justification of his assertions, and several aspects of his argument seemed problematic. Recently, Angulo, Bona, and Scialom in [**28**] gave a complete theory of the stability of cnoidal waves for the KdV equation (see also [**26**]). To obtain this result, they used the theory of elliptic integrals and the modern theory of stability of Grillakis, Shatah, and Strauss [**123**] which they adapted to the periodic context.

Afterwards, Angulo in [**25**] found new formulas of periodic travelling wave solutions of the focusing nonlinear Schrödinger equation

$$iu_t + u_{xx} + |u|^2 u = 0$$

based on the Jacobian elliptic function of type *cnoidal and dnoidal* and established a stability and instability study of the wave form of dnoidal type (see also Gallay and Hărăguş [**113**], [**114**]). An similar study was also obtained for the models Modified Korteweg-de Vries and Hirota-Satsuma in [**25**], [**26**], [**27**].

It is remarkable to see that in all these works it was necessary to use the method of quadrature to obtain the explicit profile of the periodic travelling wave solutions, in other words, to put our *differential equation* in the form

$$[\psi']^2 = F(\psi).$$

Moreover, the necessary spectral information for studying stability or instability was obtained via an elaborated spectral theory for the periodic eigenvalue problem associated with the Jacobi form of the *Lamé's equation*, namely,

$$\begin{cases} \dfrac{d^2}{dx^2}\Psi + [\rho - n(n+1)k^2 sn^2(x;k)]\Psi = 0, \\ \Psi(0) = \Psi(2K(k)), \quad \Psi'(0) = \Psi'(2K(k)), \end{cases}$$

when $n \in \mathbb{N}$, $sn(\cdot;k)$ represents the Jacobian elliptic function snoidal, $k \in (0,1)$, and K is the complete elliptic integral of the first kind:

$$K(k) = \int_0^1 \frac{dt}{\sqrt{(1-t^2)(1-k^2t^2)}}.$$

We note that Gardner in [**115**] provided a theory for determining that the large wavelength periodic waves are *linearly unstable* whenever the limiting homoclinic wave (solitary wave) is unstable. He applied his theory to diverse types of nonlinear evolutions equations in one space variable. In the case of the generalized KdV equations

$$u_t + u^p u_x + u_{xxx} = 0,$$

with $p \in \mathbb{N}$, if we assume that this equation produces a family of large wavelength periodic waves U^α such that the period $2T_\alpha$ tends to infinity as α tends to zero, then this family is unstable whenever $p > 4$ and $\alpha > 0$ is sufficiently small.

Recently Angulo and Natali in [**37**] developed a novel theory for studying the existence and stability of periodic travelling waves based on the theory of totally positive operators and the *Poisson Summation Theorem.* Their theory was based on the study of the general equation of KdV-type:

$$u_t + u^p u_x - M u_x = 0,$$

where $p \geq 1$ is an integer and M is a differential or pseudo-differential operator in the framework of periodic functions. In other words, M is defined as a Fourier multiplier operator by

$$\widehat{Mg}(k) = \alpha(k)\widehat{g}(k), \quad k \in \mathbb{Z},$$

where α is assumed to be a measurable, locally bounded, even function on $\mathbb{R}$, satisfying the condition

$$A_1|k|^{m_1} \leq \alpha(k) \leq A_2(1 + |k|)^{m_2}$$

for $m_1 \leq m_2$, $|k| \geq k_0$, $\alpha(k) > b$ for all $k \in \mathbb{Z}$, and $A_i > 0$. The travelling wave solutions

$$u(x,t) = \varphi_c(x - ct)$$

have a profile φ_c being a real-valued smooth periodic function with an *a priori* fundamental period $2L$, $L > 0$. Hence substituting the form of u given above and integrating once (assuming the constant of integration to be zero), one obtains that $\varphi = \varphi_c$ is a solution of the equation

$$(M + c)\varphi_c - \frac{1}{p+1}\varphi_c^{p+1} = 0. \tag{1.16}$$

So we find the associated linear, closed, unbounded, self-adjoint operator $\mathcal{L}$: $D(\mathcal{L}) \to L^2_{\text{per}}([-L, L])$ defined on a dense subspace of $L^2_{\text{per}}([-L, L])$ by

$$\mathcal{L}_M u = (M + c)u - \varphi_c^p u. \tag{1.17}$$

The problem about the existence of a nontrivial smooth curve of periodic solutions for (1.16) above presents new and delicate issues that need to be handled. The possibility of explicitly finding solutions will naturally depend on the form of M. If it is a differential operator of the form $M = -\partial_x^2$, the use of the quadrature method and the theory of elliptic functions has been shown to be a main tool, and so the solutions will depend on a general form of the Jacobian elliptic functions of the *snoidal, cnoidal,* and *dnoidal* types (see **[25]**, **[26]**, **[27]**, **[45]**). Since the period of these functions depends on the complete elliptic integral of the first kind $K(k)$, we have that the modulus k will depend on the velocity c and therefore we have that *a priori* the period of φ_c will depend on c. So, the required smooth branch of periodic travelling wave solutions with a fixed minimal period has been obtained in many cases by using the Implicit Function Theorem. We note that this procedure based on the quadrature method obviously in general does not work if M is a pseudo-differential operator (see Albert **[7]** in which it is shown that the travelling wave equations associated to the Intermediate Long Wave (ILW) and the Benjamin-Ono (BO) equations can be reduced to a quadrature form).

In Angulo and Natali **[37]** a different approach was established to obtain explicit solutions of (1.16) for a specific form of M and values of p. This approach is based on the classical *Poisson Summation Theorem* **[258]**. There are at least two important advantages to this new approach. The first one is that it can be used for obtaining solutions when M is a pseudo-differential operator, for example in the case of the Benjamin-Ono equation. The other one is related to computing the expression

$$\frac{d}{dc}\int \varphi_c^2(x)dx. \tag{1.18}$$

In general, determining the sign of the expression in (1.18) is very difficult in the periodic case. As has been shown in the literature, the use of nontrivial identities for the complete elliptic integrals of the first and the second kinds sometimes comes on the scene as a fundamental piece in the analysis, and so verifying this property can become a challenge. As we will see, the verification of (1.18) can be very easily obtained by using a combination of the Poisson Summation Theorem and the *Parseval Theorem.*

With regard to the spectral conditions associated with (1.17), the problem of obtaining (1.12) is very delicate. One of the most remarkable results in the theory of stability of solitary wave solutions was given by Albert and by Albert and Bona in [**6**] and [**10**], in which sufficient conditions were given to obtain the properties in (1.12) associated with $\mathcal{L}_M$. The advantage of Albert's approach is that it does not require an explicit computation of the spectrum of the linear operator (1.17), since it is obtained exclusively from positivity properties of the Fourier transform of the solitary wave in question. The Angulo-Natali theory establishes an extension of the theory in [**6**] and [**10**] for the case of positive even periodic travelling wave solutions. The periodic problem has new points not encountered when considering issues related to the solitary waves. The analysis in [**37**] also relies upon the theory of totally positive operators and so the class $PF(2)$ defined by Karlin in [**149**] is used.

The theory in [**37**] leads to a significant simplification of some recent proofs of stability of periodic travelling wave solutions of KdV-type equations, such as in the case of the Korteweg-de Vries and the modified Korteweg-de Vries equations, since in those cases the verification of the spectral conditions requires the determination of the instability intervals associated with the Lamé equation above and of an explicit formula of at least the first three eigenvalues ρ. Our analysis does not require this information.

The theory in [**37**] has established two very interesting and novel results. The first one is about the stability of the periodic travelling wave solutions found by Benjamin in [**41**] for the Benjamin-Ono equation

$$u_t + uu_x - \mathcal{H}u_{xx} = 0, \tag{1.19}$$

where $\mathcal{H}$ denotes the periodic Hilbert transform defined by

$$\mathcal{H}f(x) = \frac{1}{2L}\,\text{p.v.}\int_{-L}^{L} \text{cotg}\Big[\frac{\pi(x-y)}{2L}\Big] f(y)\,dy.$$

The associated periodic waves for (1.19) with a minimal period $2L$ are given for $c > \frac{\pi}{L}$ as

$$\varphi_c(x) = \frac{2\pi}{L}\frac{\sinh(\gamma)}{\cosh(\gamma) - \cos\left(\frac{\pi x}{L}\right)},$$

such that $\gamma > 0$ satisfies $\tanh(\gamma) = \dfrac{\pi}{cL}$. The second result is the existence and stability/instability of positive periodic travelling waves for the critical KdV equation

$$u_t + 5u^4u_x + u_{xxx} = 0 \tag{1.20}$$

and the critical nonlinear Schrödinger equation

$$iu_t + u_{xx} + |u|^4 u = 0.$$

More exactly, in [**38**] there is shown the existence of a smooth curve $c \in \left(\frac{\pi^2}{L^2}, +\infty\right) \to \varphi_c$ of periodic travelling wave solutions for (1.20) where the profile φ_c is given by

$$\varphi_c(z) = \frac{\sqrt{\eta_3}\, dn(\frac{4}{3g}z; k)}{\sqrt{1 + \beta^2 sn^2(\frac{4}{3g}z; k)}}$$

with η_3, g, and k smooth functions depending on c. Moreover, there is a unique (threshold) value c^* of the speed-wave c of φ_c, which separates two different global scenarios of the evolution of a localized initial perturbation of φ_c: for $c < c^*$ we have nonlinear stability and for $c > c^*$ we have nonlinear instability.

We note that several instability analyses for evolution partial differential equations found in the literature have been carried out in a general setting, in which the periodic waves have a distinguished asymptotic limit (see Gardner in [**115**] and Sandstede and Scheel in [**247**]). The results presented in [**37**] do not fall into this class. More precisely, the instability result for the critical KdV and the critical Schrödinger equations is not obtained from an instability linearized analysis or from an analysis of the spectra of large wavelength periodic waves which follow the unstable solitary wave. We believe that detecting instabilities via the last approach is a difficult task. Moreover, from the instability results obtained in Gardner [**115**] it is possible to say that the approaches presented for the critical KdV equation complete the picture about the existence of a branch of periodic travelling wave solutions, which are unstable for the generalized Korteweg-de Vries equation provided $p \geqq 4$.

It is important to note that the stability results about periodic waves are obtained in general by periodic initial disturbance having the same minimal period of the periodic travelling wave solutions being studied. The results of stability in $H^1_{\text{per}}([0, L])$ for the orbit generated by the cnoidal waves φ in (1.15),

$$\Omega_\varphi \equiv \{\varphi(\cdot + s)\}_{s\in\mathbb{R}},$$

for the flow of the periodic KdV equation, are obtained for initial disturbances of φ having the same period L. It is a conjecture on the part of Benjamin [**45**] that cnoidal waves of minimal period L are unstable by perturbations, for example, of period $2L$. Some evidence in favor of this scenario was recently shown by Angulo [**25**] in the case of the nonlinear Schrödinger equation

$$iu_t + u_{xx} + |u|^2 u = 0.$$

Now we describe how this book has been divided. The following chapters will present many ideas which may determine or at least suggest whether solitary wave solutions or periodic travelling wave solutions for a given nonlinear dispersive equation are or are not stable.

The second chapter provides a set of equations which give the reader a clear understanding of the applicability and physical relevance of nonlinear dispersive evolution equations. In the next chapters, many of these models have been chosen to exemplify the theories of stability or instability. Chapter 3, we consider the principal issue of our study, namely, travelling wave solutions. Here we give a good set of explicit formulas for solitary waves and periodic travelling waves for nonlinear equations of interest.

Chapter 4 contains results on the well-posedness of the Cauchy problem and initial boundary value problem belonging to some equations put forward in Chapter 2. In particular we present an idea of the proof of the theory of local well-posedness in H^s_{per}, $s \geq 0$, for the KdV equaton in the Bourgain spaces. Chapter 5 establishes the definition of stability, which is understood in the Liapunov sense.

Chapter 6 develops the theory of stability of solitary wave solutions for the Generalized Korteweg-de Vries equations. In this case we apply in detail the germinal ideas of Benjamin, Bona, and Weinstein. We have called this approach the "classical method". Also, we include an application of this local method by obtaining a theory of "*stability of the blow-up*" for a class of Kortweg-de Vries equations. The chapter finishes with a small review of improvements in the theory of solitary wave solutions and solitons for KdV-type equations.

Chapter 7 is devoted to exemplifying the Grillakis, Shatah, and Strauss stability theory. Special attention is given to model equations that arise in the study of long wave propagation. The chapter finishes with some comments and other applications of their theory. It includes the basic theorems of the theory of Albert and Bona, with regard to sufficient conditions for the required spectral information in the stability theory of solitary wave solutions.

In Chapter 8, we consider the Concentration-Compactness Principle and its applicability to the study of the existence and stability of solitary wave solutions of nonlinear evolution equations. Special attention is given to the existence and stability of solitary wave solutions for the Generalized Benjamin-Ono equations. Using the theory of symmetric decreasing rearrangements of functions and Levandosky's approach, we present a *new proof of the stability of Benjamin-Ono's solitary waves.* In Chapter 9, we consider other applications of the Concentration-Compactness principle. In particular, we review the stability theory for Kadomtsev-Petviashvili equations.

Chapter 10 deals with the instability of solitary wave solutions. Special attention is devoted to the solitary wave solutions for Generalized Benjamin equations. Here the solitary wave solutions in consideration are those obtained via a variational approach, and the method to be established does not depend on spectral or convexity conditions. The chapter contains other applications of the theory. In particular, we review the instability theory for fifth-order Korteweg-de Vries equations. We also show a linearized instability theory for weakly coupled Korteweg-de Vries systems.

Chapter 11 is dedicated exclusively to the study of the existence and stability of periodic travelling wave solutions. Here we present an extension of the Grillakis, Shatah, and Strauss theory to the case when the *travelling wave solutions do not have the property of being critical points.* We apply it to the stability of cnoidal wave solutions with mean zero with regard to the flow of the KdV and Benney-Luke equations. We also establish the basic theory of Angulo and Natali and apply it to obtain the existence and stability of positive cnoidal wave solutions for the Korteweg-de Vries and Benjamin-Ono equation. We also apply it to show the stability/instability of a smooth curve of periodic travelling wave solutions for the critical Korteweg-de Vries and the critical nonlinear Schrödinger equations.

We finish these notes with two appendices. The first one establishes the basic tools of the Fourier transform, the space of tempered distributions, the theory of Sobolev spaces $H^s(\mathbb{R}^n)$ of type $L^2(\mathbb{R})$, as well as the periodic Sobolev spaces on $\mathbb{R}$.

Also, we pull together some facts about the symmetric decreasing rearrangement of a function in $\mathbb{R}$ and the theory of Jacobian elliptic functions. The second appendix contains a detailed and clear explanation of the theory of closed linear operators in Hilbert spaces that is used in this work. Special attention is given to the spectral theory required in the stability approach. A careful treatment is given to the Sturm-Liouville theory on $\mathbb{R}$ of second-order differential operators and to the Floquet theory associated with the Lamé equation.

The literature concerned with the subject of this book is immense. Therefore here we cite those papers most closely connected with the text of the book without claiming completeness. An updated and vast bibliography is included to stimulate new interest in future study and research.

CHAPTER 2

Basic Models

2.1. Introduction

The central equations of study in this book are model equations for waves which take account of both nonlinearity and dispersion effects. For example, by modelling equations for waves in dispersion media, we begin by considering a body of water of finite depth under the influence of gravity, bounded below by an impermeable surface. Ignoring the effects of viscosity and assuming that the flow is incompressible and irrotational, the motion is taken to be governed by the Euler equations together with suitable boundary conditions on the rigid surface and on the water-air interface. So, by making assumptions and approximations about the physical interest in question, we can obtain a set of model dispersive equations formally valid for the description of waves propagating in just one direction for small amplitude long wavelength motion, as in the case of the classical model of the Boussinesq equation (1.2) and the case of the Korteweg-de Vries equation (1.5). We refer the reader to the notes by Bona in [**59**] where one can find a pedagogical exposition about the modelling and an analysis of nonlinear wave phenomena. We also refer the reader to Billingham and King's book [**57**], Whitham's book [**277**], or the apt reference for the modelling of other nonlinear evolution equations which will be established below.

2.2. Models

In a general form the equations that interest us in this book will be those that arise in the modelling of the propagation of long waves, one-dimensional equations of the form

$$u_t + f(u)_x - Mu_x = 0 \tag{2.1}$$

where $u = u(x,t)$ is a real-valued function for $x, t \in \mathbb{R}$, $f : \mathbb{R} \to \mathbb{R}$ is a super-linear $C^\infty(\mathbb{R})$ function, typically a polynomial with $f(0) = 0$, and M is the dispersion operator defined as a Fourier multiplier,

$$\widehat{Mu}(\xi) = \alpha(\xi)\widehat{u}(\xi) \tag{2.2}$$

where $\widehat{g}$ denotes the Fourier transform of g with respect to the spatial variable (see Appendix A). The symbol α of M typically denotes a continuous even real-valued function, which satisfies (for example) the conditions

$$a_1|\xi|^{\beta_1} \leqq \alpha(\xi) \leqq a_2(1+|\xi|)^{\beta_2} \qquad \text{for } \xi \in \mathbb{R} \tag{2.3}$$

where $a_1, a_2 > 0$ and $\beta_2 \geqq \beta_1 \geqq 1$.

Some equations that exemplify the general model (2.1), whether by their relevance in the world of nonlinear dispersive evolution equations, by their own right or by relevance in studies of research today, are the following.

EXAMPLE 2.1 (The Korteweg-de Vries Equation; KdV equation henceforth). Consider $f(u) = \frac{1}{2}u^2$ and $\alpha(\xi) = \xi^2$. The linear differential operator M is given by $M = -\partial_x^2$ and (2.1) is transformed into

$$u_t + uu_x + u_{xxx} = 0. \tag{2.4}$$

As is well known, this model appears in different physical contexts; for example in the unidirectional propagation of shallow water waves on the surface of a body of water, it governs the behavior of small amplitude waves of long wavelength, when they are compared with the water depth. Other physical scenarios where this equation appears are in longitudinal waves propagating in a one-dimensional lattice of equal masses coupled by nonlinear springs (the Fermi-Pasta-Ulam problem [**106**] in Kruskal and Zabusky [**170**]), ion-acoustic waves in a cold plasma (Taniuti and Washini [**264**]), rotating flow in a tube (Leibovich [**181**]), and longitudinal dispersive waves in elastic rods (Nariboli [**227**]). To see more applications, we refer the reader to Dodd *et al.* [**98**], Jeffrey and Kakutani [**145**], and Miura [**223**].

Associated with the KdV equation, we have the Generalized Korteweg-de Vries equations (GKdV equations henceforth),

$$u_t + u^p u_x + u_{xxx} = 0 \tag{2.5}$$

with $p \in \mathbb{N}$.

EXAMPLE 2.2 (The Benjamin-Ono Equation; BO equation henceforth). Consider $f(u) = \frac{1}{2}u^2$ and $\alpha(\xi) = |\xi|$. The linear differential operator M is given by $M = \mathcal{H}\partial_x$, where $\mathcal{H}$ represents the Hilbert transform defined by

$$\mathcal{H}f(x) = \frac{1}{\pi}\,\text{p.v.}\int \frac{f(y)}{x-y}\,dy = \frac{1}{\pi}\lim_{\epsilon\to 0}\int_{|y-x|\geq\epsilon} \frac{f(y)}{x-y}\,dy.$$

Via the Fourier transform we have that $\widehat{\mathcal{H}f}(\xi) = -i\text{sgn}(\xi)\widehat{f}(\xi)$ (see Appendix A). Therefore (2.1) is transformed into

$$u_t + uu_x - \mathcal{H}u_{xx} = 0. \tag{2.6}$$

This equation was derived by Benjamin in [**42**] as a model for the unidirectional propagation of internal waves of finite amplitude in a stratified fluid of great depth. In this particular case, one has an incompressible nondiffusive heterogeneous fluid for which the density is nonconstant only within a layer whose thickness is much smaller than the total depth.

Associated with the BO equation, we also have the Generalized Benjamin-Ono equations (GBO equations henceforth)

$$u_t + u^p u_x - \mathcal{H}u_{xx} = 0 \tag{2.7}$$

with $p \in \mathbb{N}$.

EXAMPLE 2.3 (The Intermediate Long-Wave Equation; ILW equation henceforth). Consider $f(u) = \frac{1}{2}u^2$ and

$$\alpha_H(\xi) = \xi\coth(\xi H) - \frac{1}{H}, \qquad H \in (0,\infty).$$

By denoting M_H the pseudo-differential operator generated by α_H, equation (2.1) is transformed into the family of nonlinear evolution equations,

$$u_t + uu_x - M_H u_x = 0. \tag{2.8}$$

In this case we have a model for nonlinear dispersive waves on the interface between two fluids of different positive densities contained at rest in a long channel with a horizontal top and bottom, the lighter fluid forming a horizontal layer above a layer of the same depth of the heavier fluid (see Joseph [**146**]).

An examination of the symbols of the dispersion operators associated with the KdV equation, BO equation and ILW equation reveals a close relationship between these equations; namely, when H is near 0, we obtain the symbol $\alpha_H(\xi)$ approximates to ξ^2 (at least for small values of ξ), while $\alpha_H(\xi) \to |\xi|$ as $H \to \infty$. This indicates that the ILW equation interpolates between the KdV and BO equations in the interval $0 < H < \infty$, approximating one or the other equation near the endpoints of the interval. This relationship is also suggested by the nature of the physical phenomena which the equations are intended to model; see Kubota, Ko, and Dobbs [**171**] and Albert, Bona, and Saut [**14**].

EXAMPLE 2.4 (The Benjamin Equation). Consider $f(u) = u^2$ and $\alpha(\xi) = \xi^2 - l|\xi|$, where $l \in \mathbb{R}$. The linear differential operator M is given by $M = -\partial_x^2 - l\mathcal{H}\partial_x$ and (2.1) is transformed into the pseudo-differential equation

$$u_t + 2uu_x + l\mathcal{H}u_{xx} + u_{xxx} = 0. \tag{2.9}$$

Benjamin, in [**43**] and [**44**], introduced this novel model which describes the unidirectional propagation of internal waves. This model governs approximately the evolution of waves on the interface of a two-fluid system in which surface tension effects can not ignored. The incompressible system at rest consists of a layer of depth h_1 of light fluid of density ρ_1 bounded above by a rigid plane and resting upon a layer of heavier fluid of density $\rho_2 > \rho_1$ of depth h_2 (very deep relative to the upper layer h_1), also resting on a rigid plane. We note that because of the density difference, waves can propagate along the interface between the two fluids (see Albert, Bona, and Restrepo [**13**]).

Associated with the Benjamin equation, we have the Generalized Benjamin equations (GB equations henceforth)

$$u_t + u^p u_x + u_{xxx} + l\mathcal{H}u_{xx} = 0 \tag{2.10}$$

with $p \in \mathbb{N}$.

EXAMPLE 2.5 (Korteweg-de Vries-type Equations). Consider $f(u) = \frac{u^{p+1}}{p+1}$ and $\alpha(\xi) = |\xi|^\beta$. The linear differential operator M is given by $M = D^\beta$, and so (2.1) is transformed in the pseudo-differential equation

$$u_t + u^p u_x - D^\beta u_x = 0.$$

This general equation is an extension of the GKdV and GBO equations.

Other nonlinear models that appear in the literature and that do not have the general form (2.1), but for which the theory to be developed in this book can be applied, are the following:

EXAMPLE 2.6 (The Nonlinear Schrödinger Equation; NLS henceforth). For $u = u(x,t) \in \mathbb{C}$, $x \in \mathbb{R}^n$, and $t \in \mathbb{R}$ we have the equation

$$iu_t + \Delta u + |u|^2 u = 0. \tag{2.11}$$

This important equation appears in various domains of the Mathematical Physics field, for example in nonrelativistic quantum mechanics ([**127**]), in the study of propagation laser beams ([**158**]), and in the propagation of a plane electromagnetic wave in a nonlinear medium ([**2**]) (see also Billingham and King's book [**57**]).

Associated with the NLS equation, we also have the following generalization:

$$iu_t + \Delta u + |u|^p u = 0, \tag{2.12}$$

with $p > 0$. We note that equation (2.11) is also called the focusing NLS.

EXAMPLE 2.7 (The Kadomtsev-Petviashvili Equation; KP-I henceforth).

$$(u_t + (u^2)_x + u_{xxx})_x = u_{yy} \tag{2.13}$$

where $u = u(x, y, t) \in \mathbb{R}$ and $x, y, t \in \mathbb{R}$. This model describes the propagation of weakly nonlinear dispersive long waves on the surface of a fluid, when the wave motion is essentially one-directional with weak transverse effects along the y-axis ([**147**]).

Associated with the KP-I equation, we also have the Generalized Kadomtsev-Petviashvili equations (GKP-I equations henceforth),

$$(u_t + (u^{m+1})_x + u_{xxx})_x = u_{yy} \tag{2.14}$$

where $m = m_1/m_2$, m_1 is an even integer, and m_2 is an odd integer. When $m = 2$, this equation is a model for describing the evolution of sound waves in antiferromagnetics ([**104**]). To some extent, the GKP-I equations can be viewed as a two-dimensional analog of the GKdV equations.

EXAMPLE 2.8 (Interaction Equation between a Short and Long Wave; S-BO henceforth).

$$\begin{cases} iu_t + u_{xx} = \alpha uv, \\ v_t + \gamma \mathcal{H} v_{xx} = \beta(|u|^2)_x \end{cases} \tag{2.15}$$

where $u = u(x, t) \in \mathbb{C}$, $v = v(x, t) \in \mathbb{R}$, $x, t \in \mathbb{R}$, $\mathcal{H}$ is the Hilbert transform, and α, β, γ are real constants. This model describes the interaction between long waves and short waves under a weakly coupled nonlinearity in a two-layer fluid and under the setting of deep and shallow flows. Here the fluid depth of the lower layer is sufficiently large in comparison with the wavelength of the internal wave, and the fluids have different densities ([**112**]). We note that the short wave term is described by u and the long wave by v.

EXAMPLE 2.9 (The Schrödinger-Korteweg-de Vries Systems; S-KdV henceforth).

$$\begin{cases} i(u_t + c_1 u_x) + \delta_1 u_{xx} = \alpha uv, \\ v_t + c_2 v_x + \delta_2 v_{xxx} + \gamma (v^2)_x = \beta(|u|^2)_x \end{cases} \tag{2.16}$$

where $u = u(x, t) \in \mathbb{C}$, $v = v(x, t) \in \mathbb{R}$, $x, t \in \mathbb{R}$, and $c_i, \delta_i, \alpha, \beta, \gamma$ are real. Systems of the form (2.16) appear as models for interactions between long and short waves in a variety of physical settings. For example, Kawahara, Sugimoto, and Kakutani [**156**] derived (2.16) as a model for the interaction between long gravity waves and capillary waves on the surface of shallow water, in the case when the group velocity of the capillary wave coincides with the velocity of the long wave. In Makhankov [**207**], Nishikawa, Hojo, Mima, and Ikesi [**229**], a system of equations is derived for resonant ion-sound/Langmuir wave interactions in plasmas, which reduces to (2.16) under the assumption that the ion-sound wave is unidirectional. Similarly, one can obtain (2.16) as the unidirectional reduction of a model for the resonant interaction of acoustic and optical modes in a diatomic lattice (see Yajima and Satsuma [**278**]).

EXAMPLE 2.10 (General Systems of Long Wave Equations). Many systems of equations which have been derived as models for the interaction of two (or more) long waves, each of which corresponds to a different underlying mode or vertical structure, take the vectorial form

$$\mathbf{h}_t + D^{-1}(\nabla N(\mathbf{h}) - Q\mathbf{h})_x = 0, \tag{2.17}$$

in which $\mathbf{h}$ is an $\mathbb{R}^n$-valued function of the variables x and t. The operators D, ∇N, and Q in (2.17) are defined as follows:

- D is an $n \times n$ diagonal matrix with positive entries along the diagonal.
- ∇N is the gradient of a function $N : \mathbb{R}^n \to \mathbb{R}$.
- The dispersion operator Q is a matrix Fourier multiplier operator defined by
$$\widehat{Q\mathbf{h}}(\xi) = A(\xi)\widehat{\mathbf{h}}(\xi)$$
for $\xi \in \mathbb{R}$, where $A(\xi)$, the symbol of Q, is for each $\xi \in \mathbb{R}$ a symmetric $n \times n$ matrix with real entries, and $A(\xi)$ satisfies $A(-\xi) = A(\xi)$ for each $\xi \in \mathbb{R}$. Further, there exist positive constants C_1, C_2 and a number $\theta \geqq \frac{1}{2}$ such that
$$C_1|k|^{2\theta}|\mathbf{v}|^2 \leqq \langle A(\xi)\mathbf{v}, \mathbf{v}\rangle \leqq C_2|k|^{2\theta}|\mathbf{v}|^2$$
for all vectors $\mathbf{v}$ in $\mathbb{R}^n$ and all sufficiently large values of $|k|$.

A specific model in the form (2.17) is the Liu, Kubota, and Ko system. This system was derived in [**191**] to model the interaction between a disturbance $h_1(x,t)$ located at an upper pycnocline and another disturbance $h_2(x,t)$ located at a lower pycnocline in a three-layer fluid. It can be written as

$$\begin{cases} h_{1t} - c_1 h_{1x} + \alpha_1 h_1 h_{1x} - \gamma_1 (M_1 h_1)_x - \gamma_2[(M_2 h_1)_x - (Sh_2)_x] = 0, \\ h_{2t} - c_2 h_{2x} + \alpha_2 h_2 h_{2x} - \gamma_3 (M_3 h_2)_x - \gamma_4[(M_2 h_2)_x - (Sh_1)_x] = 0. \end{cases}$$

Here c_i, α_i, γ_i are real constants, with γ_i positive for $i = 1, 2, 3, 4$. The operators M_i are Fourier multiplier operators defined by

$$\widehat{M_i f}(k) = m_i(k)\widehat{f}(k),$$

where

$$m_i(k) = k\coth(kH_i) - \frac{1}{H_i}$$

for $i = 1, 2, 3$, with H_1, H_2, H_3 being positive constants related to the depths of the three fluid layers. The operator S is also a Fourier multiplier operator,

$$\widehat{Sf}(k) = n(k)\widehat{f}(k),$$

where

$$n(k) = \frac{k}{\sin(kH_2)}.$$

EXAMPLE 2.11 (Regularized Models for Long Waves). For M being the dispersion operator defined in (2.2), these regularized models have the form

$$u_t + u_x + u^p u_x - M u_t = 0, \tag{2.18}$$

where $u = u(x,t)$ is an $\mathbb{R}$-valued function for $x, t \in \mathbb{R}$ and $p \geqq 1$ is an integer. Specific models in the form (2.18) are the reguralized Benjamin-Bona-Mahony equation [**47**] (the BBM equation)

$$u_t + u_x + uu_x - u_{xxt} = 0,$$

and the reguralized Benjamin-Ono equation

$$u_t + u_x + uu_x - \mathcal{H}u_{xt} = 0.$$

2.3. Comments

There are many nonlinear evolution equations or systems in one and two spatial dimension such that their study of existence and stability of travelling wave solutions could be obtained via the stability theory to be established in the next chapters.

For example, we have the second equation in the modified Korteweg-de Vries ($p = 2$ in (2.5)) and BO hierarchies respectively:

(1) (The fifth-order modified Korteweg-de Vries equation)

$$u_t - u_{xxxxx} - 30u^4u_x - 10u^2u_{xxx} - 10(u_x)^3 - 40uu_xu_{xx} = 0.$$

(2) (The third-order Benjamin-Ono equation)

$$u_t - 4u_{xxx} + 3u^2u_x - 3(u\mathcal{H}u_x)_x - 3\mathcal{H}(uu_x)_x = 0.$$

In water waves theory we can consider the following systems:

(1) (The Gear and Grimshaw system) For real variables $u = u(x,t)$, $v = v(x,t)$, and $a_i, b_i \in \mathbb{R}$ with $b_1, b_2 > 0$,

$$\begin{cases} u_t + uu_x + u_{xxx} + a_3v_{xxx} + a_1vv_x + a_2(uv)_x = 0, \\ b_1v_t + rv_x + vv_x + v_{xxx} + b_2a_3u_{xxx} + b_2a_2uu_x + b_2a_1(uv)_x = 0. \end{cases}$$

(2) (Boussinesq systems) For real variables $\eta = \eta(x,t)$, $u = u(x,t)$

$$\begin{cases} \eta_t + u_x + (u\eta)_x + au_{xxx} - b\eta_{xxt} = 0, \\ u_t + \eta_x + uu_x + c\eta_{xxx} - du_{xxt} = 0, \end{cases}$$

where for $\theta \in [0,1]$ the constants a, b, c, d satisfy the constraints

$$a + b = \frac{1}{2}(\theta^2 - \frac{1}{3}), \quad c + d = \frac{1}{2}(1 - \theta^2) \geqq 0, \quad a + b + c + d = \frac{1}{3}.$$

We also have the following interesting models for $u = u(x,t)$ being a complex-valued function:

(1) (The complex modified Korteweg-de Vries equation)

$$u_t + 3|u|^2u_x + u_{xxx} = 0.$$

(2) (The derivative nonlinear Schrödinger equation)

$$iu_t + u_{xx} \pm 2i(|u|^2u)_x = 0.$$

In nonlinear optical theory we have the following three coupled nonlinear Schrödinger systems:

(1) For complex variables $w = w(x,t)$, $v = v(x,t)$, and $r, s, \sigma, \alpha \in \mathbb{R}$

$$\begin{cases} iw_t + rw_{xx} - \theta w + \bar{w}v = 0, \\ i\sigma v_t + sv_{xx} - \alpha v + \dfrac{1}{2}w^2 = 0. \end{cases}$$

(2) For $u(x,t)$ and $w(x,t)$ complex-valued functions and $\sigma, \alpha \in \mathbb{R}$

$$\begin{cases} iu_t + u_{xx} - u + \Big(\frac{1}{9}|u|^2 + 2|w|^2\Big)u + \frac{1}{3}\overline{u}^2 w = 0, \\ i\sigma w_t + w_{xx} - \alpha w + \Big(9|w|^2 + 2|u|^2\Big)w + \frac{1}{9}u^3 = 0. \end{cases}$$

(3) (Nonlinear Schrödinger-Boussinesq equation) For $u = u(x,t) \in \mathbb{C}$, $n = n(x,t) \in \mathbb{R}$, $\phi = \phi(x,t) \in \mathbb{R}$, $\alpha, \beta \in \mathbb{R}$, and $p \in \mathbb{N}$

$$\begin{cases} iu_t + u_{xx} - nu - \beta|u|^2 u = 0, \\ n_t = \triangle\phi, \\ \phi_t = n + \alpha n^{p+1} - \triangle n + |u|^2. \end{cases}$$

In two dimension, we also have

(1) (The Davey-Stewartson system) For $u = u(x,y,t) \in \mathbb{C}$, $\phi = \phi(x,y) \in \mathbb{R}$, and $c_i \in \mathbb{R}$

$$\begin{cases} iu_t + c_0 u_{xx} + u_{yy} = c_1|u|^2 u + c_2 u\phi_x, \\ \phi_{xx} + c_3\phi_{yy} = (|u|^2)_x. \end{cases}$$

(2) (Zakharov system) For $u = u(x,y,t) \in \mathbb{C}$ and $n = n(x,y,t) \in \mathbb{R}$

$$\begin{cases} iu_t + \triangle u = un, \\ n_{tt} - n_{xx} = -\triangle(|u|^2). \end{cases}$$

In three dimension we can consider

(1) (The Gross-Pitaevskii equation) For $u = u(x,y,z,t) \in \mathbb{C}$

$$iu_t + \triangle u + V(x,y,z)u - |u|^2 u = 0.$$

(2) (The Kadomtsev-Petviashvili equation) For $u = u(x,y,z,t) \in \mathbb{R}$

$$\begin{cases} u_t + uu_x + u_{xxx} + av_y + bw_z = 0, \\ v_x = u_y, \\ w_x = u_z, \end{cases}$$

where the constants a, b measure the transverse dispersion effects.

(3) (Vector nonlinear Schrödinger systems) For $E = E(x,t) : \mathbb{R}^3 \times \mathbb{R} \to \mathbb{C}^3$

$$iE_t + \nabla(\nabla \cdot E) - \alpha^2 \nabla \times \nabla \times E + |E|^{2\sigma} E = 0,$$

where $\alpha^2 \in \mathbb{R}$ and $\sigma > 0$.

CHAPTER 3

Solitary and Periodic Travelling Wave Solutions

3.1. Introduction

Russell's original experiments showed the solitary wave to be a very stable aspect of surface wave motion with constant speed and without change of shape. Similar observations of other physical systems in more recent years have confirmed that when such waves exist, they appear to be very persistent and to play a distinguished role in the time-evolution of the equation in question. In this book our main interest will be to look at some fundamental qualitative properties of the solutions of nonlinear wave equations of the specific form $u(x,t) = \phi_c(x - ct)$ where $x, t \in \mathbb{R}$ and c is a specific wave speed. In general these solutions are called *travelling wave solutions*, and by fixing boundary conditions, we can obtain the so-called solitary wave solutions, or periodic travelling waves, or kink solutions (also known as bore solutions). Besides the basic properties of these solutions, we are interested here in their existence and stability or instability by the flow in question (see the coming chapters).

3.2. Travelling Wave Solutions

In this section we consider the principal issue of our study, and the spirit and heart of this book: travelling wave solutions. One fundamental mathematical representation of a wave moving in a one-dimensional medium is given by functions of two variables $u(x,t)$ of the form

$$u(x,t) = \phi(x - ct) \tag{3.1}$$

where ϕ is a function of one variable and c is a nonzero constant. The animation of such a function begins with the graph of the initial profile $u(x,0) = \phi(x)$. If c is positive, then the profile of $u(x,t) = \phi(x - ct)$ at a large time t is exactly a translation of the initial profile by an amount ct in the positive x-direction. Such a function represents a disturbance moving with constant speed c. Similarly, $u(x,t) = \phi(x - ct)$ with $c < 0$, representing a disturbance moving in the negative x-direction with speed $|c|$. In either case, *the profile at each time t does not get distorted and remains a recognizable feature of a wave as it is translated along the x-axis.*

Waves represented by functions of the form (3.1) are called **travelling waves**. Two basic features of any travelling wave are the underlying profile shape defined by ϕ and the speed $|c|$ at which the profile is translated along the x-axis.

A **travelling wave solution** of a partial differential equation is a solution of the differential equation which has the form of a travelling wave $u(x,t) = \phi(x - ct)$ where c is the wave speed and $\xi = x - ct$ is the characteristic variable. Finding travelling wave solutions generally begins by assuming the form (3.1) and then determining which function ϕ and constant c yield a solution to the differential equation. The determination of ϕ and c, as we will be see below, essentially depends on the boundary conditions given to ϕ. In general, ϕ will be a function depending of c and so we will denote this dependence by writing ϕ_c.

We can divide the family of travelling wave solutions associated with a given partial differential equation into three basic forms:

DEFINITION 3.1. The solitary wave solution: This is the name given to a travelling wave solution when $\phi : \mathbb{R} \to \mathbb{R}$ satisfies the boundary conditions

$$\lim_{\xi \to \pm\infty} \phi^{(n)}(\xi) = 0$$

for all $n \in \mathbb{N}$.

DEFINITION 3.2. The periodic travelling wave solution: This is the name given to a travelling wave solution when $\phi : \mathbb{R} \to \mathbb{R}$ is a periodic function with a given period L such that $\phi^{(n)}(\xi)$ exist for all $\xi \in \mathbb{R}$, with the boundary periodic condition on $[0, L]$

$$\phi^{(n)}(0) = \phi^{(n)}(L), \qquad \text{for all } n \in \mathbb{N}.$$

DEFINITION 3.3. The kink travelling wave solution: This is the name given to a travelling wave solution when $\phi : \mathbb{R} \to \mathbb{R}$ satisfies the boundary conditions

$$k_l = \lim_{\xi \to -\infty} \phi(\xi) \qquad \text{and} \qquad k_r = \lim_{\xi \to +\infty} \phi(\xi)$$

and $\infty > k_l > k_r > -\infty$ or $k_l < k_r$. The function ϕ sometimes can also satisfy the additional asymptotic condition

$$\lim_{\xi \to \pm\infty} \phi^{(j)}(\xi) = 0, \qquad j = 1, 2, 3, \dots .$$

The kink travelling wave solution is called, depending of the physical framework, a *bore* or a *wave front*.

In the case of a specific nonlinear evolution equation which is completely integrable, for example the KdV equation (see Ablowitz and Clarkson [**2**], Drazin [**99**]), the solitary wave solutions are called **solitons**. In general it is not obvious that travelling wave solutions exist for a specific model. Moreover, when a specific theory shows only the existence of solutions, it is not easy to find an explicit formula for them.

In this book our main interest will be the qualitative properties of the *solitary wave and periodic travelling wave solutions*. We note that the wave speed c is generally positive and it is given by the specific form of the models with which we will work. Sometimes, depending on the model, some extra conditions are necessary, for example, the case in the Benjamin equation (2.9) (see (3.10)).

Next, we present some explicit examples of travelling wave solutions associated with the general equation (2.1). In fact, if we substitute (3.1) in (2.1) and integrate once, we obtain that ϕ satisfies

$$M\phi(\xi) + c\phi(\xi) - f(\phi(\xi)) = A_\phi, \tag{3.2}$$

where A_ϕ is a real constant and $\xi = x - ct$. Initially we consider the case of solitary wave solutions. Therefore, we need to have that $A_\phi = 0$. In general is not easy to show a formula that represents a solitary wave solution. Sometimes the method of quadrature works, but in general there is not a procedure. The problem can become difficult especially when we work with a nonlocal operator, such as $\mathcal{H}\partial_x$.

3.3. Examples

EXAMPLE 3.4 (Solitary wave solutions of the GKdV equations). In this case equation (3.2) takes the form

$$-\phi''(\xi) + c\phi(\xi) - \frac{1}{p+1}\phi^{p+1}(\xi) = 0 \tag{3.3}$$

where " $'$ " $= \frac{d}{d\xi}$. So via the quadrature method, in other words, by writing equation (3.3) in the form

$$(\phi')^2 = F(\phi),$$

we obtain via integration that

$$\phi_c(\xi) = \Big[\frac{(p+1)(p+2)c}{2}\Big]^{1/p} \operatorname{sech}^{2/p}\Big(\frac{p\sqrt{c}}{2}\xi\Big) \tag{3.4}$$

satisfies (3.3) for every $\xi \in \mathbb{R}$.

We note that as a general pattern, if we show the existence of a solution ϕ of (3.2), then we have really found infinite solutions, because $\phi(\cdot + y)$ (any translation of ϕ) will also be a solution of (3.2). In the case of equation (3.3), it can be shown that the unique positive solution (modulo translations and signal) is exactly that given by (3.4). We refer the reader to Berestycki and Lions [**54**], which presents an elementary proof of this fact.

EXAMPLE 3.5 (Solitary wave solutions of the GBO equations). In this case equation (3.2) takes the form

$$\mathcal{H}\phi'(\xi) + c\phi(\xi) - \frac{1}{p+1}\phi^{p+1}(\xi) = 0. \tag{3.5}$$

Explicit solutions to equation (3.5) are known only when $p = 1$, which gives the BO equation. This solution was found by Benjamin in 1967 ([**41**]) and has the algebraic form

$$\phi_c(\xi) = \frac{4c}{1 + c^2\xi^2}. \tag{3.6}$$

The uniqueness problem (modulo translations) of this solution was shown by Amick and Toland in 1990 ([**18**]). Albert in [**7**] recovered Amick and Toland's result on the uniqueness of the BO solitary waves (3.6), which makes no use of complex analysis. Albert's uniqueness result is deduced directly from two properties of the dispersion operator $\frac{d}{dx}\mathcal{H}$: namely, the positivity of the resolvent $(\frac{d}{dx}\mathcal{H} + c)^{-1}$ and the product identity

$$f \cdot g + \mathcal{H}(f \cdot \mathcal{H}g + g \cdot \mathcal{H}f) - \mathcal{H}f \cdot \mathcal{H}g = 0.$$

These properties lead to the conclusion that every L^2 solution to equation (3.5) with $p = 1$ is also a solution of the *ordinary differential equation*

$$c\phi - \frac{1}{2}\phi^2 = \Big(\frac{\phi'}{\phi}\Big)',$$

which can be write in the quadrature form

$$(\phi')^2 = \phi^2(D + 2c\phi - \frac{1}{2}\phi^2), \tag{3.7}$$

where D is the constant of integration.

The existence of smooth solutions to (3.5) with a general $p \in \mathbb{N}$ can be shown via an analysis of variational type, more precisely via the *Concentration-Compactness*

Method developed by P.-L. Lions [**193**], [**194**]. In Chapter 8 of this book, we establish a theory of existence of even solutions to (3.5) for every $c > 0$, such that they are strictly positive decreasing functions of $|\xi|$ belonging to $H^\infty(\mathbb{R})$.

EXAMPLE 3.6 (Solitary wave solutions of the ILW equation). In this case a family of solitary wave solutions was discovered by Joseph in 1977 [**146**]. For a given $H \in (0, \infty)$ and $c > 0$, if $a \in (0, \frac{\pi}{2}H)$ and $b \in (0, \infty)$ are determined in terms of c and H by the equations

$$\begin{cases} 1 - \dfrac{aH}{\tan(aH)} + aH\tan(aH) = (c-1)H \quad and \\ aH\tan(aH) = \dfrac{1}{2}bH, \end{cases}$$

then the solutions are given by

$$\phi_{c,H}(\xi) = \frac{b}{\cosh^2(a\xi) + \frac{b^2}{16a^2}\sinh^2(a\xi)}. \tag{3.8}$$

The uniqueness problems of these solutions have been proved by Alber and Toland in [**16**]. Albert in [**7**] recovered Albert and Toland's result on the uniqueness of the ILW solitary waves (3.8), which makes no use of complex analysis. Albert's uniqueness result is deduced directly from the conclusion that every L^2 solution to equation

$$M_H\phi + c\phi = \frac{1}{2}\phi^2$$

is also a solution of the *ordinary differential equation*

$$c\phi - \frac{1}{2}\phi^2 = \left(\frac{\phi'}{\phi}\right)',$$

which can be write in the quadrature form

$$(\phi')^2 = \phi^2(D + 2c\phi - \frac{1}{2}\phi^2), \tag{3.9}$$

where D is the constant of integration being necessarily positive.

As we saw in Example 2.3 there is a close relationship between the ILW equation and the KdV and BO equations. A similar relationship can also be seen in the framework of solitary wave solutions. In fact, it is possible to show that if we choose $b = 4$ in (3.8), then

$$\begin{array}{ll} \lim_{H\to\infty} \phi_{1,H}(\xi) = \phi_{1,BO}(\xi) & \text{uniformly for } \xi \in \mathbb{R}, \\ \lim_{H\to 0} \phi_{\frac{4}{3},H}(\gamma\xi) = \phi_{\frac{4}{3},KdV}(\xi) & \text{uniformly for } \xi \in \mathbb{R}, \end{array}$$

where $\gamma = \sqrt{\frac{H}{3}}$, and $\phi_{c,KdV}$, $\phi_{c,BO}$ represent, respectively, the solitary wave solutions of the KdV equation and the BO equation with wave speed c (equation (3.4) with $p = 1$ and (3.6)).

EXAMPLE 3.7 (Solitary wave solutions of the Benjamin equation). In this case equation (3.2) takes the form

$$\phi''(\xi) + l\mathcal{H}\phi'(\xi) + \phi^2(\xi) = c\phi(\xi). \tag{3.10}$$

The existence of explicit solutions to equation (3.10) with $l \neq 0$ is not known yet. The problem of existence of solitary wave solutions was pioneered by Benjamin in [**43**] and [**44**], where, using a Leray-Schauder degree theory, he showed the existence of even solutions for (3.10) if $\gamma = \frac{1}{2\sqrt{c}} l \in (0,1)$, namely, if we have that $c > \frac{1}{4}l^2$. Later, Albert, Bona, and Restrepo, in [**13**], by using the Implicit Function Theorem and the existence of solutions to (3.10) when $l = 0$ (i.e., the KdV solitary wave solutions), proved the existence of a continuous curve of even solitary wave solutions to (3.10) if l is sufficiently near zero. Afterwards, Angulo in [**21**], by using the *Concentration-Compactness Method*, established a theory of existence of solitary wave solutions with γ close to 1 (see Section 9.4). Finally, Bona and Chen in [**84**] proved the existence of solutions to the generalized Benjamin equation (2.10), based also on the *Concentration-Compactness Method* with $p \geqq 1$ and $c > \frac{1}{4}l^2$ (see Chapter 9).

Other types of travelling wave solutions also appear when we consider evolution equations in higher dimensions or systems. For example, in the case of the NLS equation (2.12) the travelling wave solutions to be considered are solutions u of the form

$$u(x,t) = e^{i\omega t}\phi_{\omega,\theta}(x - \theta t) \tag{3.11}$$

where $\phi_{\omega,\theta} : \mathbb{R}^n \to \mathbb{C}$, $|\phi_{\omega,\theta}(\xi)| \to 0$ as $|\xi| \to \infty$, ω is a scalar, and $\theta \in \mathbb{R}^n$. Of special interest in applications are the solutions with $\theta = 0$, $\omega > 0$, and $\phi_{\omega,\theta}$ real. Such solutions

$$u(x,t) = e^{i\omega t}\phi_\omega(x) \tag{3.12}$$

are called *standing wave* solutions or are often referred to as *bound states.* Standing wave solutions have been studied intensively in the last years (Berestycki and Lions [**54**], Berestycki, Gallouet, and Kavian [**52**], Berestycki and Lions [**53**], Cazenave [**79**], Berestycki, Lions, and Peletier [**55**], Strauss [**261**]). Also of special interest in many physical situations are the so-called *ground states.* These wave forms ϕ_ω, which are analogous to the solitary wave solutions of the KdV equation, are positive solutions which belong to $H^1(\mathbb{R}^n)$.

EXAMPLE 3.8 (Ground state solution of the NLS equation). If we consider the form (3.12) in (2.12), then ϕ_ω must satisfy the elliptic equation

$$\Delta\phi_\omega - \omega\phi_\omega + |\phi_\omega|^p\phi_\omega = 0. \tag{3.13}$$

It is well known that by choosing $\omega > 0$ one can find a solution $\phi_\omega \in H^1(\mathbb{R}^n)$ of (3.13) with $0 < p < \frac{4}{n-2}$ ($p > 0$, if $n = 1$ or 2) which is positive and radially symmetric and for which $e^{\varepsilon|x|}\phi_\omega \in L^\infty(\mathbb{R})$ with an $\varepsilon > 0$ ([**53**], [**79**], [**261**]). If we consider $n = 1$, then equation (3.13) is reduced to a differential equation similar to equation (3.3) and therefore a simple quadrature analysis tell us that all the solutions are of the hyperbolic-secant type. If we take $n \geqq 2$, an explicit form of ϕ_ω is not known yet. The uniqueness problem of solutions for (3.13) is a difficult issue. The uniqueness of positive solutions, modulo translations and rotations, has been established by Kwong in [**174**].

By considering the case of the KP-I equation (2.13), the travelling wave solutions to be considered are solutions of the form

$$u(x,y,t) = \phi(x - ct, y) \tag{3.14}$$

where $c > 0$ and $\phi : \mathbb{R}^2 \to \mathbb{R}$ satisfies $|\nabla\phi(\xi,y)| \to 0$ as $|\xi|^2 + |y|^2 \to \infty$. In this case, solutions of the form (3.14) are also called solitary wave solutions.

EXAMPLE 3.9 (Solitary wave solutions of the KP-I equation). If we substitute the form of u given by (3.14) in (2.13), we find that ϕ must satisfy the differential equation

$$\partial_\xi^2(-c\phi + \phi^2 + \partial_\xi^2\phi) = \partial_y^2\phi. \tag{3.15}$$

An explicit solution to (3.15) is known and is called a *lump solution*, namely

$$\phi(\xi, y) = 8\,\frac{c - \frac{\xi^2}{3} + \frac{y^2}{3c}}{(c + \frac{\xi^2}{3} + \frac{y^2}{3c})^2}\,. \tag{3.16}$$

Others results about the existence of solutions (but not about explicit solutions) to equation (3.15) have been obtained by de Bouard and Saut in [**95**] and [**96**], using the *Concentration-Compactness Method.* A result about the uniqueness of solutions to equation (3.15) has not been found, so we do not know if the solitary wave solutions found by de Bouard and Saut are lump solutions. With regard to the existence of solitary wave solutions of the GKP-I equations (2.14), we known solutions of the form (3.14) exist for $c > 0$ and $0 < m < 4$, [**95**], [**192**] (see also Chapter 9).

Finally, another type of travelling wave solutions appears when we consider the systems (2.15) and (2.16). In this case the travelling wave solutions have the form

$$\begin{cases} u(x,t) = e^{i\omega t}e^{ic(x-ct)/2}\phi(x-ct), \\ v(x,t) = \psi(x-ct), \end{cases} \tag{3.17}$$

where $\phi, \psi : \mathbb{R} \to \mathbb{R}$ and $c, \omega \in \mathbb{R}$. When ϕ, ψ are functions which vanish at infinity in some sense (usually ϕ, ψ are in $H^1(\mathbb{R})$), we call the solutions in (3.17) *bound state* solutions.

EXAMPLE 3.10 (Bound state solutions of equation (2.15)). If we substitute the form of u and v given by (3.17) into system (2.15), we obtain after some calculations that (ϕ, ψ) satisfies the pseudo-differential system

$$\begin{cases} \phi'' - \sigma\phi = \alpha\,\psi\phi, \\ \gamma\mathcal{H}\psi' - c\psi = \beta\phi^2, \end{cases} \tag{3.18}$$

where $\sigma = \omega - \frac{c^2}{4}$. If we consider α, β, σ positive constants and $\gamma < 0$ in (3.18), then it is possible to show the existence of even solutions ϕ, ψ either via the Implicit Function Theorem or via the *Concentration-Compactness Method* (see Angulo and Montenegro [**34**], [**35**]). The existence of explicit solutions to equation (3.18) is not known yet, except in the case $\gamma = 0$. When $\gamma = 0$, (3.18) reduces to equation (3.3) with $p = 2$ and different constant coefficients, so solutions of the form (3.4) are found.

EXAMPLE 3.11 (Bound state solutions of the Schrödinger-Korteweg-de Vries systems). By substituting (u, v) of the form (3.17) in (2.16) and by considering without loss of generality $c_1 = c_2 = 0$, $\delta_1 = 1$, we obtain the system of differential equations for (ϕ, ψ)

$$\begin{cases} \phi'' - \sigma\phi = \alpha\,\psi\phi, \\ \delta_2\psi'' - c\psi + \gamma\psi^2 = \beta\phi^2, \end{cases} \tag{3.19}$$

where $\sigma = \omega - \frac{c^2}{4}$. Many results about existence of solutions to (3.19) have been obtained; for instance, Funakoshi and Oikawa in [**112**] list the following explicit one-parameter families of solutions to (3.19). For $\gamma \leqq 2$, define

$$\begin{cases} \phi(x) = \pm 6B^2\sqrt{2-3q}\,\operatorname{sech}^2(Bx), \\ \psi(x) = 6B^2\operatorname{sech}^2(Bx) \end{cases} \tag{3.20}$$

where $B > 0$ is arbitrary. Then (ϕ, ψ) satisfy (3.19) with $\sigma = 4B^2$, $c = 8B^2$, $\alpha = \beta = -1$, and $\delta_2 = 2$. On the other hand, if $\gamma \geqq 2$, then we have that

$$\begin{cases} \phi(x) = \pm 6B^2\sqrt{3q-2}\,\operatorname{sech}(Bx)\tanh(Bx), \\ \psi(x) = 6B^2\operatorname{sech}^2(Bx) \end{cases} \tag{3.21}$$

is a solution of (3.19) with $\sigma = B^2$, $c = 2B^2(3\gamma - 2)$, $\alpha = \beta = -1$, and $\delta_2 = 2$. When $\gamma = 2$, of course, these solutions coincide with the obvious solution given by $\phi = 0$ and $\psi(x) = (12B^2/\gamma)\operatorname{sech}(Bx)$, which satisfies (3.19) with $c = 8B^2$, $\alpha = \beta = -1$, and $\delta_2 = 2$, for all $\gamma \neq 0$. Now, in [**83**], L. Chen considered (3.19) in the special case when $\gamma = 6$, $\alpha = \beta = -1$, $\delta_2 = 2$, and he found a two-parameter family of explicit solutions given by

$$\begin{cases} \phi(x) = \pm\sqrt{2B^2(c-8B^2)}\,\operatorname{sech}(Bx), \\ \psi(x) = 2B^2\operatorname{sech}^2(Bx) \end{cases}$$

where $B^2 = \sigma$, $c > 0$, and $\sigma \in (0, c/8)$ are arbitrary. More recently, Angulo in [**22**] has shown the existence of other solutions for (3.19) via an analysis variational. Finally, in Albert and Angulo [**9**], a theory of existence of ground-state solutions (ϕ and ψ positive functions in (3.17)) has been shown for a certain range of values of γ.

EXAMPLE 3.12 (Cnoidal wave solutions of the KdV equation). Next we will obtain an "explicit" family of periodic travelling wave solutions to the KdV equation called *cnoidal waves*. So, we are looking for solutions in the form $u(x,t) = \phi_c(x-ct)$ with $c > 0$. Substituting into the KdV equation and integrating once, we obtain the second-order nonlinear ordinary differential equation for ϕ_c,

$$\phi_c'' + \frac{1}{2}\phi_c^2 - c\phi_c = A_{\phi_c} \tag{3.22}$$

where A_{ϕ_c} is an integration constant. Multiplying the result by ϕ_c, a second integration is possible, yielding the first-order equation (in a quadrature form)

$$\begin{aligned} [\phi_c']^2 &= \frac{1}{3}[-\phi_c^3 + 3c\phi_c^2 + 6A_{\phi_c}\phi_c + 6B_{\phi_c}] \equiv \frac{1}{3}F_{\phi_c}(\phi_c) \\ &= \frac{1}{3}(\phi_c - \beta_1)(\phi_c - \beta_2)(\beta_3 - \phi_c) \end{aligned} \tag{3.23}$$

where A_{ϕ_c}, B_{ϕ_c} are integration constants and $\beta_1, \beta_2, \beta_3$ are the zeros of the polynomial $F_{\phi_c}(t) = -t^3 + 3ct^2 + 6A_{\phi_c}t + 6B_{\phi_c}$. So, we need the β_i's to satisfy the relations

$$\begin{cases} 3c = \beta_1 + \beta_2 + \beta_3, \\ A_{\phi_c} = -\frac{1}{6}(\beta_1\beta_2 + \beta_2\beta_3 + \beta_1\beta_3), \\ B_{\phi_c} = \frac{1}{6}\,\beta_1\beta_2\beta_3. \end{cases} \tag{3.24}$$

If F_{ϕ_c} has only one real root β_1, say, then $\phi'(z)$ can vanish only when $\phi(z) = \beta$. This means the maximum value ϕ takes on in its period domain is the same as its minimum value there, and so ϕ is constant. Since we are interested in nontrivial periodic solutions, the β_i's are real and satisfy $c^2 + 2A_{\phi_c} > 0$. Hence we can assume that they are ordered in the form $\beta_1 < \beta_2 < \beta_3$. In this case from (3.23) we need to have $\beta_2 \leqq \phi \leqq \beta_3$. We now define $\varphi = \phi/\beta_3$. Then (3.23) becomes

$$(\varphi')^2 = \frac{\beta_3}{3}(\varphi - \eta_1)(\varphi - \eta_2)(1 - \varphi)$$

where $\eta_i = \beta_i/\beta_3$, $i = 1, 2$. The variable φ lies in the interval $[\eta_2, 1]$. By translation of the spatial coordinates, we may locate a maximum value of φ at $z = 0$. As the only critical points of φ for values of φ in $[\eta_2, 1]$ occur when $\varphi = \eta_2 < 1$ and when $\varphi = 1$, it must be the case that $\varphi(0) = 1$. One checks that $\varphi'' > 0$ when $\varphi = \eta_2$ and $\varphi'' < 0$ when $\varphi = 1$. Thus, it is clear that our putative periodic solution must oscillate monotonically between the values $\varphi = \eta_2$ and $\varphi = 1$. A simple analysis would now allow us to conclude such periodic solutions exist, but we are pursuing a specific formula, not just existence.

Changing variables again by letting

$$\varphi = 1 + (\eta_2 - 1)\sin^2\psi,$$

we see that $\psi(0) = 0$ and

$$(\psi')^2 = \frac{\beta_3}{12}(1 - \eta_1)\Big[1 - \Big(\frac{1 - \eta_2}{1 - \eta_1}\Big)\sin^2\psi\Big].$$

In order to write this last relation in standard form, we define

$$k^2 = \frac{1 - \eta_2}{1 - \eta_1}, \qquad l = \frac{\beta_3}{12}(1 - \eta_1).$$

It follows that $0 < k^2 < 1$ and $l > 0$. Then we obtain that

$$F(\psi(\xi); k) = \int_0^{\psi(\xi)} \frac{dt}{\sqrt{1 - k^2\sin^2 t}} = \sqrt{l}\,\xi.$$

Therefore from the definition of the Jacobian elliptic function $y = \mathrm{sn}(u; k)$, we can write the last equality as

$$\sin\psi = \mathrm{sn}(\sqrt{l}\,\xi; k),$$

and hence $\varphi = 1 + (\eta_2 - 1)\mathrm{sn}^2(\sqrt{l}\,\xi; k)$. Using the relation $\mathrm{sn}^2 u + \mathrm{cn}^2 u = 1$, we finally arrive at the conventional form of the so-called *cnoidal wave* solutions,

$$\phi_c(\xi) \equiv \phi_c(\xi; \beta_1, \beta_2, \beta_3) = \beta_2 + (\beta_3 - \beta_2)\,\mathrm{cn}^2\Big[\sqrt{\frac{\beta_3 - \beta_1}{12}}\,\xi; k\Big] \tag{3.25}$$

with

$$k^2 = \frac{\beta_3 - \beta_2}{\beta_3 - \beta_1}, \quad \beta_1 + \beta_2 + \beta_3 = 3c, \quad \beta_1 < \beta_2 < \beta_3.$$

Now, since $\mathrm{cn}(u + 2K) = -\mathrm{cn}(u)$, it follows that cn^2 has fundamental period $2K$ and therefore the cnoidal wave solution ϕ_c in (3.25) has the fundamental period (wavelength) T_{ϕ_c} given by

$$T_{\phi_c} = \frac{4\sqrt{3}}{\sqrt{\beta_3 - \beta_1}}\,K(k). \tag{3.26}$$

An important feature of the fundamental period T_{ϕ_c} is that *it depends on the speed* c. In fact, by using (3.26), we get the inequality

$$T_{\phi_c}^2 > \frac{(2\pi)^2}{\sqrt{c^2 + 2A_{\phi_c}}}. \tag{3.27}$$

We note that equation (3.25) contains a huge number of periodic travelling wave solutions for the KdV equation, which are obtained basically by varying the modulus k . Moreover, formula (3.25) contains, at least formally, two basic solutions of the KdV equation which are obtained as an approximation of periodic solutions. In fact, by considering β_1 and β_2 going toward zero through positive values, $k^2 \to 1^-$ and $\beta_3 \to 3c^-$. The elliptic functions and their periods also simplify in this limit, with $\mathrm{sn}(u;1^-) \sim \tanh u$ and $K(k) \to +\infty$ as $k \to 1^-$. The cnoidal wave loses its periodicity in this limit and we obtain a wave form with a single hump and with "infinity period" of the form

$$\phi_c(\xi;0,0,3c) = 3c\,\mathrm{sech}^2\Big(\frac{\sqrt{c}}{2}\xi\Big),$$

which is exactly the classical solitary wave solution of the KdV equation with speed c (see (3.4)). Next, if we consider the limiting case $\beta_3 - \beta_2 \ll 1$, we obtain that k^2 is small and so $\mathrm{cn}(u;0^+) \sim \cos\, u$ and $K(0^+) \sim \pi/2$. At leading order

$$T_{\phi_c} = \frac{2\pi\sqrt{3}}{\sqrt{\beta_3 - \beta_1}}, \qquad 3c = \beta_1 + 2\beta_2.$$

This leads to the expression

$$\phi_c(\xi) = \beta_2 + \frac{1}{2}(\beta_3 - \beta_2) + \frac{1}{2}(\beta_3 - \beta_2)\cos\Big(\frac{\pi}{T_{\phi_c}}\xi\Big).$$

We note that this harmonic wave is the expected solution of the linearized problem in (3.22).

Finally, we note that in the regimen of the cnoidal wave solutions in which the individual peaks are tall and narrow and overlap very slightly, an obvious approximation of the cnoidal wave is an infinite series of spatially repeated solitary waves. In fact, one can put solitary waves of identical shape centered on $\xi = n$ for $n = 0, \pm1, \pm2, \ldots$ and obtain formally the periodic function

$$\varphi(\xi) = 3\omega \sum_{n=-\infty}^{\infty} \mathrm{sech}^2\Big(\frac{\sqrt{\omega}}{2}(\xi + 2nl)\Big),$$

which is a $2l$ periodic solution. So we have the natural question: Is this periodic function an *exact* solution of (3.22) for some specific integration constant A_φ and speed c? In fact, we will see below that the *Poisson Summation Theorem* can show us that the infinite series of spatially repeated solitary waves is in fact a periodic travelling wave solution for the KdV, with a specific wave-speed c which depends of ω.

EXAMPLE 3.13 (Dnoidal wave solutions of the mKdV equation). Now we are interested in the periodic travelling wave solutions of the following modified KdV equation:

$$u_t + 3u^2u_x + u_{xxx} = 0. \tag{3.28}$$

We are looking for periodic solutions of the form $u(x,t) = \phi_c(x-ct)$; it follows from this specific form of u and from (3.28) that ϕ_c must satisfy

$$\phi_c'' + \phi_c^3 - c\phi_c = A_{\phi_c}$$

where A_{ϕ_c} is an integration constant. Here, we are interested in the case $A_{\phi_c} = 0$. It follows that ϕ_c satisfies the first-order equation (in quadrature form)

$$[\phi_c']^2 = \frac{1}{2}[-\phi_c^4 + 2c\phi_c^2 + 4B_{\phi_c}] = \frac{1}{2}(\eta_1^2 - \phi_c^2)(\phi_c^2 - \eta_2^2) \tag{3.29}$$

where B_{ϕ_c} is an integration constant and η_1, η_2 are the positive zeros of the polynomial $F_{\phi_c}(t) = -t^4 + 2ct^2 + 4B_{\phi_c}$. We assume without loss of generality that $\eta_1 > \eta_2 > 0$. So, if we are looking for positive solutions, then we need to have $\eta_2 \leqq \phi_c \leqq \eta_1$ and the η_i's to satisfy

$$\begin{cases} 2c = \eta_1^2 + \eta_2^2, \\ 4B_{\phi_c} = -\eta_1^2\eta_2^2. \end{cases} \tag{3.30}$$

We define $\varphi = \phi_c/\eta_1$ and $k^2 = (\eta_1^2 - \eta_2^2)/\eta_1^2$; then (3.29) becomes

$$(\varphi')^2 = \frac{\eta_1^2}{2}(1-\varphi^2)(\varphi^2 - 1 + k^2).$$

Defining an additional variable ψ via the relation $\varphi^2 = 1 - k^2\sin^2\psi$, we see that

$$(\psi')^2 = \frac{\eta_1^2}{2}(1 - k^2\sin^2\psi),$$

so we obtain for $l = \eta_1/\sqrt{2}$ that

$$\int_0^{\psi(\xi)} \frac{dt}{\sqrt{1-k^2\sin^2 t}} = l\,\xi.$$

Therefore, from the definition of the Jacobian elliptic function $y = \mathrm{sn}(u;k)$, we obtain from the last equality that $\sin\psi = \mathrm{sn}(l\xi;k)$, and hence

$$\varphi(\xi) = \sqrt{1 - k^2\mathrm{sn}^2(l\xi;k)} = \mathrm{dn}(l\xi;k).$$

Then by returning to the initial variable, we obtain the so-called *dnoidal wave* solutions:

$$\phi_c(\xi) \equiv \phi_c(\xi;\eta_1,\eta_2) = \eta_1\,\mathrm{dn}\Big(\frac{\eta_1}{\sqrt{2}}\,\xi;k\Big) \tag{3.31}$$

with

$$k^2 = \frac{\eta_1^2 - \eta_2^2}{\eta_1^2}, \qquad \eta_1^2 + \eta_2^2 = 2c, \qquad 0 < \eta_2 < \eta_1. \tag{3.32}$$

Since $\mathrm{dn}(u + 2K) = \mathrm{dn}\ u$, it follows that the dnoidal wave solution ϕ_c in (3.31) has the fundamental period (wavelength) T_{ϕ_c}, given by

$$T_{\phi_c} \equiv \frac{2\sqrt{2}}{\eta_1} K(k). \tag{3.33}$$

By using (3.32), we see that given a fixed wave speed $c > 0$, then $0 < \eta_2 < \sqrt{c} < \eta_1 < \sqrt{2c}$. Hence we can consider (3.33) as a function of η_2, namely

$$T_{\phi_c}(\eta_2) = \frac{2\sqrt{2}}{\sqrt{2c - \eta_2^2}} K(k(\eta_2)) \quad \text{with} \quad k^2(\eta_2) = \frac{2c - 2\eta_2^2}{2c - \eta_2^2}. \tag{3.34}$$

We are now in a position to examine in more detail the wave form (3.31) and the wavelength (3.33). Initially, we will see that

$$T_{\phi_c} > \sqrt{2}\pi/\sqrt{c}. \tag{3.35}$$

If $\eta_2 \to 0$, then $k(\eta_2) \to 1^-$, and so $K(k(\eta_2)) \to +\infty$. Therefore, $T_{\phi_c}(\eta_2) \to +\infty$ as $\eta_2 \to 0$. Now, if $\eta_2 \to \sqrt{c}$, then $k(\eta_2) \to 0$, and so $K(k(\eta_2)) \to \pi/2$. Therefore, $T_{\phi_c}(\eta_2) \to \pi\sqrt{2}/\sqrt{c}$ as $\eta_2 \to \sqrt{c}$. Finally, since $\eta_2 \to T_{\phi_c}(\eta_2)$ is a strictly decreasing function, we obtain (3.35).

By looking at (3.35), we see the following interesting feature about the wavelength of dnoidal wave solutions for the mKdV. If we fix $c > 0$ and a given wavelength $L > 0$, then we can see that if the mKdV equation has a dnoidal wave solution of the form (3.31) with fundamental period L, then it is necessary that $L > \pi\sqrt{2}/\sqrt{c}$.

Equation (3.31) contains a huge number of periodic travelling wave solutions for the mKdV equation, all of which are positive. Moreover, formula (3.31) contains, at least formally, two basic solutions of the mKdV equation obtained as an approximation of periodic solutions. In fact, by considering η_2 as it tends to zero, we see $\eta_1^2 \to 2c^-$ and $k \to 1^-$. The Jacobian elliptic function dn and its period also simplify at this limit, with $\mathrm{dn}(u;1^-) \sim \mathrm{sech}u$ and $K(k) \to +\infty$ as $k \to 1^-$. The dnoidal wave loses its periodicity at this limit and we obtain a wave form with a single hump and with "infinity period" of the form

$$\phi_c(\xi;\sqrt{2c},0) = \sqrt{2c}\,\mathrm{sech}(\sqrt{c}\,\xi),$$

which is exactly the classical solitary wave solutions of the mKdV equation with speed c. Next, if we consider the limiting case $\eta_2 \to \eta_1$, we obtain that $\eta_1^2 \sim c^-$, k^2 is small, and therefore $\mathrm{dn}(u;0^+) \sim 1$. At this limit we obtain the constant wave form

$$\phi_c(\xi;\sqrt{c},\sqrt{c}) = \sqrt{c}.$$

We note that this constant wave form is the expected constant solution of equation (3.29).

EXAMPLE 3.14 (Cnoidal wave solutions of the mKdV equation). Next we show the existence of other periodic solutions associated to equation (3.28). We note that by obtaining the dnoidal wave solutions (3.31), we have assumed in (3.29) that the polynomial $F_{\phi_\omega}(t) = -t^4 + 2ct^2 + 4B_{\phi_c}$ has the real zeros $-\eta_1, \eta_1, -\eta_2, \eta_2$ (note that F_{ϕ_c} is even). Now, we will consider the case when we have a pure imaginary root and two roots are real. So, we need to have

$$[\phi_c']^2 = \frac{1}{2}(a^2+\phi_c^2)(b^2-\phi_c^2). \tag{3.36}$$

By supposing $b > 0$, we need to have that

$$-b \leqq \phi_c \leqq b$$

and

$$\begin{cases} 2c = b^2 - a^2, \\ 4B_{\phi_c} = a^2b^2 > 0. \end{cases} \tag{3.37}$$

Define $\chi = \phi_c/b$ (suppose $\chi(0) = 0$) and $k^2 = b^2/(a^2+b^2)$; then (3.36) becomes

$$(\chi')^2 = \frac{b^2}{2}(1-\chi^2)\Big(\frac{a^2}{b^2}+\chi^2\Big).$$

Now, defining ψ as $\chi^2 = 1 - \sin^2\psi$, we get that

$$(\psi')^2 = \frac{a^2+b^2}{2}(1 - k^2\sin^2\psi).$$

Then for $\beta = \sqrt{\frac{a^2+b^2}{2}}$, we obtain via integration and the definition of the Jacobian elliptic function sn that $\sin\psi(\xi) = \mathrm{sn}(\beta\xi; k)$. Therefore, $\chi^2 = 1 - \mathrm{sn}^2(\beta\xi; k) = \mathrm{cn}^2(\beta\xi; k)$. So, we obtain that

$$\phi_c(\xi) = b\,\mathrm{cn}(\beta\xi; k) \tag{3.38}$$

is a solution of (3.36) with

$$k^2 = \frac{b^2}{a^2+b^2} \quad \text{and} \quad \beta = \sqrt{\frac{a^2+b^2}{2}}.$$

Since $\mathrm{cn}(\cdot; k)$ has fundamental period $4K(k)$, it follows that the cnoidal wave solution (3.36) has fundamental period (wavelength), T_{ϕ_c}, given by

$$T_{\phi_c} \equiv \frac{4\sqrt{2}}{\sqrt{a^2+b^2}}\,K(k). \tag{3.39}$$

We note that (3.38) represents a nonpositive periodic solution of (3.28). Moreover from (3.37), c can be positive or negative.

Next, by using the Implicit Function Theorem, we can obtain the following smooth families of cnoidal waves with a minimal period L:

(1) for every $c > 0$ there is a unique $b = b(c)$ such that $b \in (\sqrt{2c}, +\infty)$ and $c \in (0, +\infty) \to b(c)$ is a strictly increasing smooth function such that

$$\frac{4K(k)}{\sqrt{b^2 - c}} = L.$$

Here the modulo $k = k(c)$ satisfies $k^2 = b^2/(2b^2 - 2c)$ and $k'(c) > 0$. Moreover, the cnoidal wave

$$\phi_{c,1}(\xi) = b\,\mathrm{cn}\Big(\sqrt{b^2 - c}\,\xi; k\Big)$$

is a solution of (3.28) with fundamental period L, and the mapping

$$c \in (0, +\infty) \to \phi_{c,1} \in H^1_{\mathrm{per}}([0, L])$$

is a smooth function.

(2) For every $c \in (-\frac{4\pi^2}{L^2}, 0)$ there is a unique $a = a(c)$ such that $a \in (\sqrt{-2c}, +\infty)$ and $c \in (-\frac{4\pi^2}{L^2}, 0) \to a(c)$ is a strictly decreasing smooth function such that

$$\frac{4K(k)}{\sqrt{a^2 + c}} = L.$$

Here the modulo $k = k(c)$ satisfies $k^2 = (a^2+2c)/(2a^2+2c)$ and $k'(c) > 0$. Moreover, the cnoidal wave

$$\phi_{c,2}(\xi) = \sqrt{a^2 + 2c}\,\mathrm{cn}\Big(\sqrt{a^2 + c}\,\xi; k\Big)$$

is a solution of (3.28) with fundamental period L, and the mapping

$$c \in \Big(-\frac{4\pi^2}{L^2}, 0\Big) \to \phi_{c,2} \in H^1_{\mathrm{per}}([0, L])$$

is a smooth function.

EXAMPLE 3.15 (Periodic standing wave solutions of the NLS equation). Now we are interested in periodic travelling wave solutions of the NLS (2.11) in one dimension. The wave has the form

$$u(x,t) = e^{i\omega t}\phi_\omega(x) \tag{3.40}$$

where $\phi_\omega : \mathbb{R} \to \mathbb{R}$ is a periodic function. We will call these solutions *periodic standing waves.* So, from (2.11) we know ϕ_ω must satisfy

$$\phi_\omega'' + \phi_\omega^3 - \omega\phi_\omega = 0,$$

which is exactly the same equation satisfied by the periodic travelling wave solutions associated with the mKdV equation (3.28). Thus ϕ_ω can have the form (3.31) or (3.38).

EXAMPLE 3.16 (Periodic solutions of the Critical Korteweg-de Vries equation). Now we will show the existence of a family of periodic travelling wave solutions to the critical KdV equation written in the form

$$u_t + 5u^4u_x + u_{xxx} = 0.$$

These are solutions of the form $u(x,t) = \varphi(x-ct)$ with φ satisfying

$$-c\varphi_c + \varphi_c^5 + \varphi_c'' = 0. \tag{3.41}$$

In fact, by multiplying equation (3.41) by φ_c' and integrating once, we obtain

$$[\varphi_c']^2 = \frac{1}{3}(-\varphi_c^6 + 3c\varphi_c^2 + 6B_{\varphi_c}), \tag{3.42}$$

where B_{φ_c} is a nonzero constant of integration. Next, we consider the ansatz $\varphi_c = \psi_c^{\frac{1}{2}}$ (that is, we are considering positive solutions). Replacing φ_c in (3.42), we obtain the equation

$$[\psi_c']^2 = \frac{4}{3}(-\psi_c^4 + 3c\psi_c^2 + 6B_{\psi_c}\psi_c). \tag{3.43}$$

The formula (3.43) can be rewritten in quadrature form as

$$[\psi_c'(\xi)]^2 = \frac{4}{3}F_\psi(\psi_c(\xi)), \tag{3.44}$$

with $F(t) := F_\psi(t) = -t^4 + 3ct^2 + 6B_{\psi_c}t$. Let us consider η_1, η_2, and η_3, the nonzero roots of the polynomial F_ψ such that $F(t) = t(t-\eta_1)(t-\eta_2)(\eta_3 - t)$. Because of equation (3.44), we must have

$$\begin{cases} \eta_1 + \eta_2 + \eta_3 = 0, \\ \eta_1\eta_2 + \eta_2\eta_3 + \eta_1\eta_3 = -3c, \\ \eta_1\eta_2\eta_3 = 6B_{\varphi_c}. \end{cases}$$

The first of the relationships above shows us that the roots η_i, $i = 1,2,3$, should satisfy $\eta_1 < 0 < \eta_2 < \eta_3$. Therefore

$$\eta_2 \le \psi_c \le \eta_3.$$

Consider a wave-speed $c > 0$ arbitrary but fixed. We are looking for non-constant periodic solutions ψ_c such that its maximum value and minimum value on its period domain $[0, L]$ are given by $\psi_c(0) = \eta_3$ and $\psi_c(\sigma) = \eta_2$ for some $\sigma \in (0, L)$, respectively. The differential equation (3.44) allows us to use Leibnitz's rule and to conclude that

$$\int_{\psi(\xi)}^{\eta_3} \frac{dt}{\sqrt{t(\eta_3 - t)(t-\eta_2)(t-\eta_1)}} = \frac{2}{\sqrt{3g}}(\xi + M_c), \tag{3.45}$$

where M_c is a constant of integration and g is defined below. Since (3.41) is invariant under translations, we can conclude from Byrd and Friedman [**75**] (formula 257.00) that

$$\psi_c(\xi) = \frac{\eta_3(\eta_2-\eta_1) + \eta_1(\eta_3-\eta_2)\mathrm{sn}^2\left(\frac{2}{\sqrt{3g}}\xi; k\right)}{(\eta_2-\eta_1) + (\eta_3-\eta_2)\mathrm{sn}^2\left(\frac{2}{\sqrt{3g}}\xi; k\right)}, \tag{3.46}$$

where

$$g = \frac{2}{\sqrt{\eta_3(\eta_2-\eta_1)}}, \qquad k^2 = \frac{-\eta_1(\eta_3-\eta_2)}{\eta_3(\eta_2-\eta_1)}.$$

So, using the expression for k, we arrive at the following compact form of ψ_c:

$$\psi_c(\xi) = \eta_3\left[\frac{\mathrm{dn}^2\left(\frac{2}{\sqrt{3g}}\xi; k\right)}{1+\beta^2\mathrm{sn}^2\left(\frac{2}{\sqrt{3g}}\xi; k\right)}\right], \tag{3.47}$$

with $\beta^2 = -\eta_3 k^2/\eta_1 > 0$ and where we have denoted only for convenience $\alpha^2 = -\beta^2$.

Now, the periodic travelling wave solution $\psi = \psi_c$ obtained above has a fundamental period given by

$$T_\psi = \sqrt{3g}K(k_\psi) = \frac{2\sqrt{3}K(k_\psi)}{\sqrt{\eta_3(\eta_2-\eta_1)}}, \tag{3.48}$$

where

$$k_\psi^2 = \frac{-\eta_1(\eta_3-\eta_2)}{\eta_3(\eta_2-\eta_1)}.$$

By fixing $c \in \mathbb{R}$, we can determine that η_2 and η_3 satisfy the equation

$$\eta_2^2 + \eta_3^2 + \eta_2\eta_3 = 3c. \tag{3.49}$$

Therefore, from (3.48) and the fact that $\eta_2, \eta_3 > 0$ we get $c > 0$ and $\eta_1 < 0 < \eta_2 < \sqrt{c} < \eta_3 < \sqrt{3c}$. Moreover, it is possible to obtain the fact that T_ψ can be seen as a function of η_3 in the form

$$T_\psi(\eta_3) = \frac{2\sqrt{3}K(k)}{(12c\eta_3^2 - 3\eta_3^4)^{\frac{1}{4}}},$$

with k given by formula (3.51) below.

Next we gather some important information deduced from relation (3.48). First, for $\eta_3 \to \sqrt{c}$ we have $\eta_2 \to \sqrt{c}$ and then $k \to 0$. Therefore, $T_\psi \to \pi/\sqrt{c}$. On the other hand, since $\eta_3 \to \sqrt{3c}$, we have $\eta_2 \to 0$, so we can conclude that $k \to 1$ and thus $T_\psi \to \infty$. Since $\eta \in (\sqrt{c}, \sqrt{3c}) \mapsto T_\psi(\eta)$ is an increasing function, we have the main relation

$$T_\psi > \pi/\sqrt{c}.$$

Next from (3.49) we express k, η_1, and η_2 as functions of η_3 and c. In fact, we have

$$\eta_1 = \frac{-\eta_3 - \sqrt{12c - 3\eta_3^2}}{2} \quad \text{and} \quad \eta_2 = \frac{-\eta_3 + \sqrt{12c - 3\eta_3^2}}{2}. \tag{3.50}$$

Note that $12c - 3\eta_3^2 > 0$ because $\eta_3 < \sqrt{3c}$. So we see from the definition of k^2 that

$$k^2 = \frac{3\eta_3^2 + \sqrt{12c\eta_3^2 - 3\eta_3^4} - 6c}{2\sqrt{12c\eta_3^2 - 3\eta_3^4}}. \tag{3.51}$$

How, we can formally see the asymptotic behavior of the solutions ψ_c. We start with the constant solutions: it follows from (3.51) that if $\eta_3 \to \sqrt{c}$ (c fixed), then $k \to 0$ and so $\psi_c(z) \to \eta_3 = \sqrt{c}$, which gives us the nontrivial constant solution $\varphi_c \equiv c^{1/4}$ for the critical KdV. On the other hand, if $\eta_3 \to \sqrt{3c}$, then $k \to 1$, and so

$$\psi_c(\xi) \to f(\xi) = \sqrt{3c}\,\text{sech}(2\sqrt{c}\,\xi).$$

Then $\sqrt{f}$ is the solitary wave solution associated to the critical KdV equation above. Note that via the relation $\phi(\xi) \equiv 5^{1/4}\sqrt{f(\xi)}$ we get (3.4) with $p = 4$.

3.4. The Poisson Summation Theorem and Periodic Wave Solutions

In this section we will see an application of the *Poisson Summation Theorem* for obtaining periodic travelling wave solutions. We start with the classical theorem.

THEOREM 3.17 (Poisson Summation Theorem). *Let* $\widehat{f}(x) = \int_{-\infty}^{+\infty} f(y)e^{-2\pi ixy}dy$ *and* $f(y) = \int_{-\infty}^{\infty} \widehat{f}(x)e^{2\pi ixy}dx$ *satisfying*

$$|f(y)| \le \frac{A}{(1+|y|)^{1+\delta}} \qquad \textit{and} \qquad |\widehat{f}(x)| \le \frac{A}{(1+|x|)^{1+\delta}},$$

where $\delta > 0$ *and* $A > 0$ *(* $\widehat{f}$ *and* f *can be assumed to be continuous functions). Thus*

$$\sum_{n=-\infty}^{+\infty} f(x+2Ln) = \frac{1}{2L}\sum_{n=-\infty}^{+\infty} \widehat{f}\left(\frac{n}{2L}\right)e^{\frac{\pi inx}{L}}, \qquad L > 0.$$

In particular, when $x = 0$, *we have*

$$\sum_{n=-\infty}^{+\infty} f(2Ln) = \frac{1}{2L}\sum_{n=-\infty}^{+\infty} \widehat{f}\left(\frac{n}{2L}\right).$$

The four series above converge absolutely.

PROOF. See [**258**]. ☐

EXAMPLE 3.18 (Periodic travelling wave solutions of the BO equation). In this case, we will show that the BO equation (2.6) has explicit periodic travelling wave solutions of the form $u(x,t) = \phi_c(x - ct)$ for an arbitrary fundamental period $2l$ and wave speed $c > \frac{\pi}{l}$. In fact, in [**41**] Benjamin obtained that for

$$\begin{cases} \beta = 2c\tanh\gamma \qquad \text{and} \\ \tanh\gamma = \dfrac{1}{cl}\,\pi, \end{cases} \tag{3.52}$$

the function

$$\phi_c(\xi) = \frac{\beta \sinh \gamma}{\cosh\Big(\gamma\Big) - \cos\Big(\frac{\pi x}{l}\Big)} \tag{3.53}$$

satisfies the pseudo-differential equation

$$\mathcal{H}\phi_c' + c\phi_c - \frac{1}{2}\phi_c^2 = 0. \tag{3.54}$$

We note that the solitary wave solutions (3.6) associated with the BO equation may be obtained from (3.53). In fact, if we take the limit $l \to \infty$, then $\gamma \to 0$ and $\beta \to 0$ in such a way that $l\gamma \to c\pi$, $l\beta \to 2\pi$. Therefore, equation (3.53) gives us the form (3.6) at the limit.

Next, we will present a different method for obtaining the explicit solution (3.53) by using the Poisson Summation Theorem. The equation

$$\mathcal{H}\phi_\omega' + \omega\phi_\omega - \frac{1}{2}\phi_\omega^2 = 0$$

determines the solitary wave solutions

$$\phi_\omega(x) = \frac{4\omega}{1+\omega^2 x^2}, \quad \omega > 0,$$

and their Fourier transform is given by

$$\widehat{\phi_\omega}(x) = 4\pi e^{\frac{-2\pi}{\omega}|x|}.$$

Then, by the Poisson Summation Theorem, we obtain the following periodic function, ψ_ω:

$$\begin{aligned}\psi_\omega(x) &\equiv \sum_{n=-\infty}^{+\infty} \phi_\omega(x+2ln) = \frac{2\pi}{l}\sum_{n=-\infty}^{+\infty} e^{\frac{-\pi|n|}{\omega l}} e^{\frac{\pi i n x}{l}} \\ &= \frac{2\pi}{l}\sum_{n=0}^{+\infty} \varepsilon_n e^{\frac{-\pi n}{\omega l}} \cos\Big(\frac{n\pi x}{l}\Big) = \frac{2\pi}{l}\mathrm{Re}\left[\coth\Big(\frac{\frac{\pi}{\omega l}}{2} + \frac{i\pi x}{2l}\Big)\right] \\ &= \frac{2\pi}{l}\left(\frac{\sinh\big(\frac{\pi}{\omega l}\big)}{\cosh\big(\frac{\pi}{\omega l}\big) - \cos\big(\frac{\pi x}{l}\big)}\right),\end{aligned} \tag{3.55}$$

where

$$\varepsilon_n = \begin{cases} 1, & \text{if} \quad n = 0, \\ 2, & \text{if} \quad n = 1, 2, 3, \ldots. \end{cases}$$

Now, let ϕ_c, with $c \in \mathbb{R}$, be a smooth periodic solution of equation (3.54). Then ϕ_c can be expressed as a Fourier series:

$$\phi_c(x) = \sum_{n=-\infty}^{+\infty} a_n e^{\frac{in\pi x}{l}}. \tag{3.56}$$

Substituting the expression above into the BO equation, we get

$$\left[\frac{\pi|n|}{l} + c\right] a_n = \frac{1}{2}\sum_{m=-\infty}^{+\infty} a_{n-m}a_m.$$

Now, from (3.55) we consider

$$a_n \equiv \frac{2\pi}{l} e^{-\gamma|n|},$$

$n \in \mathbb{Z}$, $\gamma \in \mathbb{R}$. Substituting a_n into the last identity, we have

$$\sum_{m=-\infty}^{+\infty} a_{n-m}a_m = \frac{4\pi^2}{l^2}e^{-\gamma|n|}\left[|n| + 1 + 2\sum_{k=1}^{+\infty} e^{-2\gamma k}\right] = \frac{4\pi^2}{l^2}e^{-\gamma|n|}(|n| + \coth\gamma).$$

Then, we conclude that

$$c + \frac{\pi|n|}{l} = \frac{2\pi}{l}\cdot\frac{1}{2}(|n| + \coth\gamma). \tag{3.57}$$

We denote $\gamma = \dfrac{\pi}{\omega l}$ and consider $c > \dfrac{\pi}{l}$. Then, if we choose $\omega = \omega(c) > 0$ such that

$$\tanh(\gamma) = \frac{\pi}{cl}, \tag{3.58}$$

we see from (3.55) that $\psi_{\omega(c)} = \phi_c$. Therefore we obtain that ϕ_c has the form (3.53) with $\gamma > 0$ satisfying (3.58). Moreover, since we have that $\phi_c > 0$ and that $\gamma := \gamma(c) = \tanh^{-1}\left(\frac{\pi}{cl}\right)$ is a differentiable function for $c > \frac{\pi}{l}$, it follows that we obtain the following smooth curve of positive periodic travelling waves for the BO equation:

$$c \in \left(\frac{\pi}{l}, +\infty\right) \mapsto \phi_c \in H^n_{\text{per}}([0, 2l]).$$

EXAMPLE 3.19 (Periodic solutions of the mKdV Equation). Next, we will establish another way to obtain the existence of periodic travelling wave solutions for the mKdV equation of the form $u(x,t) = \varphi(x - ct) := \varphi_c(\xi)$, where $\xi = x - ct$, $c \in \mathbb{R}$, and its profile of period L. The equation which determines the periodic travelling wave solutions is

$$\varphi_c'' + \varphi_c^3 - c\varphi_c = 0. \tag{3.59}$$

Next, we obtain an explicit solution for (3.59) using the Poisson Summation Theorem. Indeed for $\omega > 0$, the solitary wave solution for the mKdV equation on $\mathbb{R}$ is

$$\phi_\omega(x) = \sqrt{2\omega}\,\text{sech}(\sqrt{\omega}x), \qquad x \in \mathbb{R}.$$

Its Fourier transform is $\widehat{\phi_\omega}(x) = \sqrt{2}\pi\text{sech}\left(\frac{\pi x}{2\sqrt{\omega}}\right)$, and where $\omega > 0$, it will be chosen later. From the Poisson Summation Theorem we obtain the following periodic function of period L:

$$\psi_\omega(\xi) = \frac{\sqrt{2}\pi}{L}\sum_{n=0}^{\infty}\varepsilon_n\text{sech}\left(\frac{\pi n}{2\sqrt{\omega}L}\right)\cos\left(\frac{2\pi n\xi}{L}\right), \tag{3.60}$$

where

$$\varepsilon_n = \begin{cases} 1, & n = 0, \\ 2, & n = 1, 2, 3, ... \end{cases}$$

On the other hand, it considers the Fourier expansion of the Jacobi elliptic function *dnoidal*, dn, of period L (see [**75**], [**205**]),

$$\frac{2K}{L}\text{dn}\left(\frac{2K\xi}{L}; k\right) = \frac{\pi}{L} + \frac{4\pi}{L}\sum_{n=1}^{+\infty}\frac{q^n}{1 + q^{2n}}\cos\left(\frac{2n\pi\xi}{L}\right),$$

where $K = K(k)$ is the complete elliptic integral of the first kind and q, known as "nome", is such that

$$q = e^{\left(-\frac{\pi K'}{K}\right)}.$$

Here, $K'(k) = K(\sqrt{1-k^2})$. We can conclude that

$$\frac{q^n}{1+q^{2n}} = \frac{1}{2}\text{sech}\left(\frac{n\pi K'}{K}\right).$$

Therefore,

$$\frac{2K}{L}\text{dn}\left(\frac{2K\xi}{L};k\right) = \frac{\pi}{L} + \frac{2\pi}{L}\sum_{n=1}^{+\infty}\text{sech}\left(\frac{n\pi K'}{K}\right)\cos\left(\frac{2n\pi\xi}{L}\right).$$

Because of the shape of the series that determines ψ_ω given above, let

$$\varphi_c(\xi) = \eta\text{dn}\left(\frac{\eta\xi}{\sqrt{2}};k\right)$$

be a periodic solution of period L for (3.59), with $\eta > 0$ and $k \in (0,1)$ fixed. Then, the following identities should be satisfied:

$$c = \frac{\eta^2}{2}(1+k'^2) \quad \text{and} \quad \eta = \frac{2\sqrt{2}K(k)}{L} \tag{3.61}$$

with $k'^2 = 1-k^2$. Thus, for $k \in (0,1)$ we should have that $\eta \in (\sqrt{c}, \sqrt{2c})$, and from the asymptotic properties of K, that $c > \dfrac{2\pi^2}{L^2}$. Indeed, for $k \in (0,1)$ fixed, η is immediately defined from (3.61) and then

$$c = \frac{4K^2(k)}{L^2}(2-k^2) > \frac{2\pi^2}{L^2}, \tag{3.62}$$

since $k \to K^2(k)(2-k^2)$ is a strictly increasing function. So, with $k \in (0,1)$ and c defined in (3.62), we define $\omega = \omega(c)$ as

$$\omega = \frac{c}{16(2-k^2)K'^2(k)}.$$

Therefore from (3.60) it follows that $\varphi_c \equiv \psi_{\omega(c)}$ is a solution of (3.59).

3.5. Comments

The Poisson Summation Theorem is a strong tool in the construction of periodic travelling wave solutions as those shown in the examples of Section 3.4. This approach also gives information about the convergence of periodic waves to solitary waves in the long-wave limit. Indeed, suppose that ϕ_c is a specific solitary wave solution with wave speed given by c and such that for $m \in \mathbb{N}$, $\phi_c \in H^m_{\text{per}}([-L,L])$ for $L > 0$. Consider the periodic function $\psi(x) = P_L(\phi_c)(x)$ where

$$P_L(\phi_c)(x) = \frac{1}{2L}\sum_{n=-\infty}^{+\infty}\widehat{\phi_c}\left(\frac{n}{2L}\right)e^{\frac{\pi i n x}{L}}.$$

Then it is possible to show that $P_L(\phi_c) \in H^m_{\text{per}}([-L,L])$ and that for every $\epsilon > 0$ there is an $L_\epsilon > 0$ sufficiently large, such that when both $P_L(\phi_c)$ and ϕ_c are restricted to the interval $(-L,L)$,

$$\|P_L(\phi_c) - \phi_c\|_{H^m_{\text{per}}([-L,L])} < \epsilon \quad \text{for all } L \geqq L_\epsilon.$$

Now, at the time of writing this book, I did not find any precise information about the problem of uniqueness (modulo translations) of periodic travelling wave solutions associated to nonlinear dispersive equations. Examples 3.13 and 3.14 above, associated to the existence of periodic solutions for the mKdV equation, tell us that some restrictions on the signal of the solutions needed to be given, for getting some uniqueness results. I feel that this problem can be done with a moderate amount of effort.

For the power of the Poisson Summation Theorem in the theory of N-polycnoidal waves, we refer the reader to the works of J. P. Boyd [**72**], [**73**] and references therein.

Next we give an example of nonexistence of standing wave type solutions for the following nonlocal nonlinear Schrödinger equation (see Pelinovsky and Grimshaw [**237**]):

$$\partial_t u = -i\alpha\partial_x^2 u + \beta u\partial_x(|u|^2) - i\beta u\mathcal{T}_h\partial_x(|u|^2) + i\gamma|u|^2 u, \tag{3.63}$$

where $u = u(x,t) \in \mathbb{C}$, $x,\ t \in \mathbb{R}$, $\mathcal{T}_h$ is the nonlocal operator

$$\mathcal{T}_h u(x) = \frac{1}{2h}\text{p.v.}\int_{-\infty}^{\infty} \coth\left(\frac{\pi(y-x)}{2h}\right)u(y)dy,$$

where $\alpha > 0$, $\beta \geq 0$, and $\gamma \geq 0$ are real constants, and $h \in (0,+\infty)$. Here $\mathcal{T}_h \longrightarrow \mathcal{H}$ when $h \longrightarrow +\infty$, where $\mathcal{H}$ is the Hilbert transform. Next, by normalizing equation (3.63) with $\alpha = \beta = 1$, we obtain that it is gauge equivalent to the equation

$$\partial_t v = -i\partial_x^2 v + \frac{i}{4}|v|^4 v - iv\mathcal{T}_h\partial_x(|v|^2) + i\gamma|v|^2 v, \tag{3.64}$$

where

$$v(x,t) = u(x,t)e^{\frac{i}{2}\rho(x,t)} \tag{3.65}$$

and

$$\rho(x,t) = \int_{-\infty}^{x} |u(y,t)|^2 dy. \tag{3.66}$$

Standing waves for (3.64) are solutions of the form

$$v(x,t) = e^{i\omega t}\varphi(x), \tag{3.67}$$

where $x,t \in \mathbb{R}$, ω is a real constant which represents the speed of rotation, and φ is a smooth real function. Next, we obtain the following theorem (see Angulo and Moura [**36**]):

THEOREM 3.20. *Let $\gamma \geqq 0$ in (3.64). Then there is no solution for equation (3.64) in the form $v(x,t) = e^{i\omega t}\varphi(x)$, for φ being a real smooth function satisfying the property $x\varphi'(x) \longrightarrow 0$ if $|x| \longrightarrow \infty$, and when the following conditions are satisfied:*

(1) *$h < +\infty$ and $\omega \leq 0$,*
(2) *with $\mathcal{H}$ substituting $\mathcal{T}_h$ (that is, $h \longrightarrow +\infty$) and with $\omega \in \mathbb{R}$.*

The Hilbert transform satisfies the following property.

LEMMA 3.21. *For any function $f \in \mathcal{S}(\mathbb{R})$, we have that*

$$\int_{-\infty}^{\infty} xf'\mathcal{H}(f')dx = 0.$$

PROOF. Since $\mathcal{H}$ is a skew-adjoint operator, it follows that

$$\int_{-\infty}^{\infty} xf'\mathcal{H}(f')dx = -\int_{-\infty}^{\infty} \mathcal{H}(xf')f'dx. \tag{3.68}$$

By definition of the Hilbert transform, we have

$$\begin{aligned}\mathcal{H}(xf') &= \text{p.v.}\frac{1}{\pi}\int_{-\infty}^{\infty}\frac{yf'(y)}{y-x}dy \\ &= \text{p.v.}\frac{1}{\pi}\int_{-\infty}^{\infty}\frac{(y-x)f'(y)}{y-x}dy + x\,\text{p.v.}\frac{1}{\pi}\int_{-\infty}^{\infty}\frac{f'(y)}{y-x}dy \;=\; x\mathcal{H}(f').\end{aligned}\tag{3.69}$$

Now it is enough to substitute identity (3.69) in (3.68). □

Next we prove Theorem 3.20.

PROOF OF THEOREM 3.20. Substituting (3.67) in equation (3.64), we obtain

$$-\varphi'' - \omega\varphi + \frac{1}{4}\varphi^5 + \gamma\varphi^3 - \varphi\mathcal{T}_h(\varphi^2)' = 0.\tag{3.70}$$

Multiplying equation (3.70) by φ and integrating by parts, we arrive at the identity

$$\begin{aligned}\int_{-\infty}^{\infty}(\varphi')^2dx &= \omega\int_{-\infty}^{\infty}\varphi^2dx - \frac{1}{4}\int_{-\infty}^{\infty}\varphi^6dx \\ &\quad - \gamma\int_{-\infty}^{\infty}\varphi^4dx + \int_{-\infty}^{\infty}\varphi^2\mathcal{T}_h(\varphi^2)'dx.\end{aligned}\tag{3.71}$$

Multiplying equation (3.70) by $x\varphi'$ and integrating properly by parts, we see that φ must obey

$$\begin{aligned}&\int_{-\infty}^{\infty}(\varphi')^2dx + \omega\int_{-\infty}^{\infty}\varphi^2dx - \frac{1}{12}\int_{-\infty}^{\infty}\varphi^6dx \\ &\qquad - \frac{\gamma}{2}\int_{-\infty}^{\infty}\varphi^4dx - \int_{-\infty}^{\infty}x(\varphi^2)'\mathcal{T}_h(\varphi^2)'dx = 0.\end{aligned}\tag{3.72}$$

From identity (3.72) we have that

$$\begin{aligned}0 < \int_{-\infty}^{\infty}(\varphi')^2dx &= -\omega\int_{-\infty}^{\infty}\varphi^2dx + \frac{1}{12}\int_{-\infty}^{\infty}\varphi^6dx + \frac{\gamma}{2}\int_{-\infty}^{\infty}\varphi^4dx \\ &\quad + \int_{-\infty}^{\infty}x(\varphi^2)'\mathcal{T}_h(\varphi^2)'dx.\end{aligned}\tag{3.73}$$

Now we apply (3.73) to (3.71) and arrive at the following expression:

$$\begin{aligned}\omega\int_{-\infty}^{\infty}\varphi^2dx + \frac{1}{2}\int_{-\infty}^{\infty}\varphi^2\mathcal{T}_h(\varphi^2)'dx &= \frac{1}{6}\int_{-\infty}^{\infty}\varphi^6dx + \frac{3\gamma}{4}\int_{-\infty}^{\infty}\varphi^4dx \\ &\quad + \frac{1}{2}\int_{-\infty}^{\infty}x(\varphi^2)'\mathcal{T}_h(\varphi^2)'dx.\end{aligned}\tag{3.74}$$

By Parseval's identity we see that

$$\int_{-\infty}^{\infty}\varphi^2\mathcal{T}_h(\varphi^2)'dx < 0,\tag{3.75}$$

for both cases $h < +\infty$ and $h \longrightarrow +\infty$.

(1) Since $\gamma \geq 0$ and by hypothesis $\omega \leq 0$, it follows from (3.74) and (3.75) that

$$\int_{-\infty}^{\infty}x(\varphi^2)'\mathcal{T}_h(\varphi^2)'dx < 0.\tag{3.76}$$

Identity (3.73) provides

$$\omega \int_{-\infty}^{\infty} \varphi^2 dx < \frac{1}{12} \int_{-\infty}^{\infty} \varphi^6 dx + \frac{\gamma}{2} \int_{-\infty}^{\infty} \varphi^4 dx + \int_{-\infty}^{\infty} x(\varphi^2)' \mathfrak{T}_h (\varphi^2)' dx. \tag{3.77}$$

From (3.74), (3.76), and (3.77) we obtain

$$\begin{aligned} \omega \int_{-\infty}^{\infty} \varphi^2 dx + \frac{1}{2} \int_{-\infty}^{\infty} \varphi^2 \mathfrak{T}_h (\varphi^2)' dx > \omega \int_{-\infty}^{\infty} \varphi^2 dx + \frac{1}{12} \int_{-\infty}^{\infty} \varphi^6 dx \\ + \frac{\gamma}{4} \int_{-\infty}^{\infty} \varphi^4 dx - \frac{1}{2} \int_{-\infty}^{\infty} x(\varphi^2)' \mathfrak{T}_h (\varphi^2)' dx. \end{aligned} \tag{3.78}$$

Thus, from (3.75) and (3.78), we obtain

$$\begin{aligned} 0 &> \frac{1}{2} \int_{-\infty}^{\infty} \varphi^2 \mathfrak{T}_h (\varphi^2)' dx \\ &> \frac{1}{12} \int_{-\infty}^{\infty} \varphi^6 dx + \frac{\gamma}{4} \int_{-\infty}^{\infty} \varphi^4 dx - \frac{1}{2} \int_{-\infty}^{\infty} x(\varphi^2)' \mathfrak{T}_h (\varphi^2)' dx > 0, \end{aligned} \tag{3.79}$$

which is a contradiction. Therefore (1) is proven.

(2) In this case, we have the Hilbert transform instead of the operator $\mathfrak{T}_h$ in equation (3.64). From Lemma 3.21 and identity (3.72) we have that

$$0 < \int_{-\infty}^{\infty} (\varphi')^2 dx = -\omega \int_{-\infty}^{\infty} \varphi^2 + \frac{1}{12} \int_{-\infty}^{\infty} \varphi^6 dx + \frac{\gamma}{2} \int_{-\infty}^{\infty} \varphi^4 dx. \tag{3.80}$$

Now we apply (3.80) to (3.71) and arrive at the following expression:

$$\begin{aligned} \omega \int_{-\infty}^{\infty} \varphi^2 dx + \frac{1}{2} \int_{-\infty}^{\infty} \varphi^2 \mathfrak{T}_h (\varphi^2)' dx = \frac{1}{12} \int_{-\infty}^{\infty} \varphi^6 dx + \frac{\gamma}{2} \int_{-\infty}^{\infty} \varphi^4 dx + \frac{1}{12} \int_{-\infty}^{\infty} \varphi^6 dx \\ + \frac{\gamma}{4} \int_{-\infty}^{\infty} \varphi^4 dx > \omega \int_{-\infty}^{\infty} \varphi^2 dx + \frac{1}{12} \int_{-\infty}^{\infty} \varphi^6 dx + \frac{\gamma}{4} \int_{-\infty}^{\infty} \varphi^4 dx, \end{aligned} \tag{3.81}$$

so from (3.81) we obtain an inequality similar to inequality (3.79):

$$0 > \frac{1}{2} \int_{-\infty}^{\infty} \varphi^2 \mathfrak{T}_h (\varphi^2)' dx > \frac{1}{12} \int_{-\infty}^{\infty} \varphi^6 dx + \frac{\gamma}{4} \int_{-\infty}^{\infty} \varphi^4 dx > 0,$$

and again we have a contradiction. Therefore (2) is proven. □

At this time, we do not know how to prove case (1) of Theorem 3.20 when $\omega > 0$.

Part 2

Well-Posedness and Stability Definition

Fourier, Jean Baptiste Joseph (1768–1830)

Mathematics compares the most diverse phenomena and discovers the secret analogies that unite them.

Hadamard, Jacques (1865–1963)

During the celebration of his seventieth birthday, he said:

There is no sentiment more comforting for me today than to see around me the young researchers whom I was able, and still hope to encourage in their first steps in research, as well as their elders, these disciples who are my friends and by whose efforts certain aspects of science have evolved to the point of being sometimes unrecognisable to me. Nothing can be more precious for a scientist than to feel himself passed by on the same road he started to build....

In Maz'ya and Shaposhnikova [**219**], p. 196.

CHAPTER 4

Initial Value Problem

4.1. Introduction

In this chapter we initially consider several results of the well-posedness in the sense of Hadamard of the initial value problem associated with the evolution equations put forward in Chapter 2. Our exposition in this case has the purpose of establishing the results which will be used in the next chapters. In particular we present the basic ideas for showing that the periodic initial value problem associated with the KdV equation is globally well-posed in the Sobolev space $H^s_{\text{per}}([0, 2\pi])$, $s \geqq 0$.

For comprehensive information about the latest results, conjectures, bibliography, concepts, and other material on the local and global well-posedness problems (and related questions) for nonlinear dispersive and wave equations, we suggest the readers go to the DispersiveWiki project: *http://tosio.math.utoronto.ca/wiki.*

4.2. Some Results about Well-Posedness

In this section, some results belonging to the well-posedness of the Cauchy problem

$$\begin{cases} u_t + f(u)_x - Mu_x = 0, & x, t \in \mathbb{R}, \\ u(x, 0) = u_0(x) \end{cases} \tag{4.1}$$

will be established. The conditions about M and f are the same as those given in Chapter 2 (see (2.2) and (2.3)).

Let $(X, \|\cdot\|_X)$ be a Banach reflexive space which we will suppose is continuously embedded in the basic Hilbert space $L^2(\mathbb{R})$ (or periodic $L^2_{\text{per}}([-L, L])$). We also suppose there exits another Banach reflexive space $(Y, \|\cdot\|_Y)$ which is continuously embedded in X ($\|u\|_X \leq \|u\|_Y$ for all $u \in Y$). Then we have the following definition:

DEFINITION 4.1 (Well-Posedness). We say that the initial value problem (4.1) is *locally well-posed* (l.w.p.) in Y if the following conditions are satisfied:

(1) For each $u_0 \in Y$ there is a $T > 0$ and a unique solution u of (4.1) such that $u \in C([-T, T]; Y)$ and $u(0) = u_0$.

(2) For each $T_0 \in (0, T)$ the mapping

$$u_0 \in Y \longmapsto u \in C([-T_0, T_0]; Y)$$

is continuous. Here u is the solution of (4.1) associated with the initial data u_0. In other words, the solution depends continuously on the initial data.

We say that the initial valued problem (4.1) is globally well-posed (g.w.p.) in Y if we can choose $T = +\infty$.

In this book the type of spaces X which will be used are the classical Sobolev spaces $H^s(\mathbb{R})$ (or spaces of periodic type) which satisfy the relation $H^r(\mathbb{R}) \subseteq H^s(\mathbb{R})$ for $r \geqq s$ (see Appendix A).

To establish a result of well-posedness for a given equation, for example (4.1), the method generally used is based on solving the corresponding integral equation associated with (4.1), namely the Duhamel formula:

$$u(t) = W(t)u_0 - \int_0^t W(t-\tau)[f(u)_x](\tau)\, d\tau,$$

where $\{W(t) : t \in (-\infty, \infty)\}$ denotes the group describing the solution of the linear problem associated with (4.1):

$$\begin{cases} v_t - Mv_x = 0, & x, t \in \mathbb{R}, \\ v(x,0) = f(x), \end{cases}$$

in other words, $v(t) = W(t)f = e^{\partial_x M t} f$. The solution $u(t)$ is obtained by the contraction mapping principle applied in a suitable function space, determined by the specific form of the operator $\partial_x M$ and the nonlinearity $f(u)$.

The following theorem of well-posedness associated with (4.1) will be sufficient for our purpose. Here we will consider $f(u) = \frac{1}{p+1}u^{p+1}$, $p \geqq 1$, $p \in \mathbb{N}$. We note that there is extensive literature on the subject of l.w.p. and we will review it briefly.

THEOREM 4.2. *Let $p \in \mathbb{N}$ and suppose $s > \frac{3}{2}$. Then the initial value problem* (4.1) *is locally well-posed in $H^s(\mathbb{R})$. Moreover, if $\beta_1 = \beta_2$ in* (2.3), *$p < 2\beta_1$, and $\beta > 3$, the initial value problem* (4.1) *is globally well-posed in $H^s(\mathbb{R})$.*

The proof of l.w.p. in Theorem 4.2, which will not be given here, makes use only of standard techniques, which will not be given here. We refer the reader to the theories established in Kato [**150**], [**152**], Iorio [**136**], and Abdelouhad, Bona, Felland, and Saut [**1**].

We note that by using the Sobolev Embedding Theorem, $H^r(\mathbb{R}) \hookrightarrow C_0(\mathbb{R})$ with $r > \frac{1}{2}$, it follows that if $s > \beta_1 + \frac{3}{2}$, then the solution obtained in Theorem 4.2 is classical in the sense that each term in equation (4.1) is a continuous function and (4.1) is satisfied pointwise everywhere for $t \in [-T, T]$. The result of g.w.p. is deduced from the l.w.p. theory by using the conservation laws described in (4.10) for equation (2.1).

Many recent results have been obtained with regard to the l.w.p. theory associated with the following particular equation in (4.1):

$$u_t + u^p u_x - D^\beta u_x = 0, \tag{4.2}$$

where the homogeneous linear operator $M = D^\beta$ is defined by $\widehat{D^\beta f}(\xi) = |\xi|^\beta \widehat{f}(\xi)$ and $\beta \geqq 1$. Equation (4.2) contains many of the models established in the previous chapter. Indeed, by choosing $\beta = 2$, we obtain the GKdV equations (2.5), and for $\beta = 1$, we get the GBO equations (2.7).

A large number of results on the best Sobolev exponent s in $H^s(\mathbb{R})$ for which the initial value problem associated with the GKdV and BO equations is locally well-posed have been obtained in the last fifteen years, for example those by Bourgain, Colliander, Keel, Kenig, Molinet, Ponce, Staffilani, Takaoka, Tao, Tzvetkov, and Vega.

One example that relates to our interest in the theory of stability is the following result in [**31**] associated with equation (4.2).

THEOREM 4.3. *Let $p \in \mathbb{N}$ and $\beta \geqq 2$. Then equation* (4.2) *is locally well-posed in $H^s(\mathbb{R})$ with $s \geqq \beta/2$.*

The best results known, in $H^s(\mathbb{R})$, for the case of the GKdV equations may be summarized in the following theorem.

THEOREM 4.4. *Let $p \in \mathbb{N}$. Then we have the following results of well-posedness for the GKdV equations* (2.5).

(1) *For $p = 1$,* (2.5) *is g.w.p. if $s > -\frac{3}{4}$* [**88**].
(2) *For $p = 2$,* (2.5) *is l.w.p. if $s \geqq \frac{1}{4}$* [**160**] *and g.w.p. if $s > \frac{1}{4}$* [**88**].
(3) *For $p = 3$,* (2.5) *is g.w.p. if $s \geqq -\frac{1}{6}$* [**265**] *and g.w.p. if $s \geqq 0$* [**125**].
(4) *For $p > 4$,* (2.5) *is l.w.p. if $s > s_p = \frac{1}{2} - \frac{2}{p}$* [**160**].
(5) *For $p = 4$,* (2.5) *is l.w.p. if $s \geqq 0$* [**160**] *and g.w.p. in $H^1(\mathbb{R})$ if*

$$\|u_0\| < \|\phi_c\|,$$

where ϕ_c is the solitary wave solution (3.4) *with an arbitrary c (see Weinstein* [**272**] *or Theorem 6.11 of this book).*
(6) *For $p > 4$,* (2.5) *is g.w.p. in $H^1(\mathbb{R})$ if $\|u_0\|_1$ is small* [**31**].

REMARK 4.5. Merle, in [**220**], has shown that for $p = 4$ it is possible to find solutions of (2.5) that blow up in a finite or infinite time in the $H^1(\mathbb{R})$-norm. Recently, in [**110**], global existence was shown for data $u_0 \in H^s(\mathbb{R})$, $s > 3/4$, with $\|u_0\| < \|\phi_c\|$.

With regard to the best known results in the case of the GBO equations we have the following theorem (see earlier works [**136**] and [**239**]):

THEOREM 4.6. *Let $p \in \mathbb{N}$. Then we have the following results of well-posedness for the GBO equations* (2.7).

(1) *For $p = 1$,* (2.7) *is g.w.p. if $s \geqq 0$* [**224**].
(2) *For $p = 2$,* (2.7) *is g.w.p. for large data in H^1* [**159**] *and l.w.p. for small data in H^s, $s > 1/2$* [**225**].
(3) *For $p = 3$,* (2.7) *is l.w.p. if $s > 5/6$* [**159**].
(4) *For $p \geqq 4$,* (2.7) *is l.w.p. if $s \geqq 3/4$ and g.w.p. if $s \geqq 1$* [**161**].

The value $p = 2\beta$, $\beta \geqq 1$, in (4.2) is called the value critical to equation (4.2), which we will denote by p_{crit}. As we will see in the next chapters, this name arises from the property that for all $p < p_{\text{crit}}$ the solitary wave solutions determined by (4.2) are orbitally stable and for $p > p_{\text{crit}}$ they are unstable (see [**64**] or Chapters 7 and 9 of this book). In the case of the GKdV equations (2.5), we can consider $p \geqq p_{\text{crit}} = 4$ (Martel and Merle [**212**]). We also note that it is possible to show that if $\beta \geqq 2$ and $p = p_{\text{crit}}$ in (4.2), then we have a g.w.p. theory in $H^{\beta/2}(\mathbb{R})$ if $\|u_0\| < \|\phi_c\|$, where ϕ_c is the solitary wave solution determined by (4.2) (see [**30**] or Chapter 6, Theorem 6.11, of this book).

Now, since in this book we will show some results about the stability of periodic travelling wave solutions for the KdV equation, we establish a result of well-posedness in the periodic case for the KdV equation (see the earlier works [**162**] and [**65**]).

THEOREM 4.7. *Let $L > 0$ be fixed. Then we have that the periodic initial value problem associated with the KdV equation* (2.4) *is l.w.p. in $H^s_{\text{per}}([0, L])$, $s \geqq -\frac{1}{2}$.*

We will give an idea of the proof of l.w.p. for the KdV equation in H^s_{per}, $s \geqq 0$ (see [**162**], [**88**], and [**148**] for the general case). We start with the definition of the Bourgain spaces in the periodic case; without loss of generality we take $L = 2\pi$.

- **Bourgain spaces:** We consider the Schwartz space of semiperiodic type $\mathcal{S}_{\text{per},2\pi}$ such that $f \in \mathcal{S}_{\text{per},2\pi} \Leftrightarrow f \in C^\infty(\mathbb{R}^2 : \mathbb{C})$ and satisfies the conditions

$$f(x + 2\pi, t) = f(x, t),$$
$$\sup_{(x,t)\in\mathbb{R}^2} |t^{k_1} \partial_t^{k_2} \partial_x^{k_3} f(x, t)| < +\infty, \quad k_i \in \mathbb{N}.$$

The *space-time Fourier transform* $\hat{f} = \hat{f}(k, \tau)$ for $k \in \mathbb{Z}$ and $\tau \in \mathbb{R}$ is defined as

$$\hat{f}(k, \tau) = c \int_{-\infty}^{\infty} \int_0^L f(x, t) e^{-ikx} e^{-i\tau t} dx dt.$$

So, we have that $\hat{f}(k, \tau) = [\widehat{f}^{(x)}(k)]^{(t)}\ (\tau)$. This transform is inverted by

$$f(x, t) = c \sum_{k\in\mathbb{Z}} \Big[\int_{-\infty}^{\infty} \hat{f}(k, \tau) e^{i\tau t} d\tau \Big] e^{ikx}.$$

It follows that $\widehat{f(t)}^{(x)}(k) = c \int_{-\infty}^{\infty} \hat{f}(k, \tau) e^{i\tau t} d\tau$.

DEFINITION 4.8. For $s, b \in \mathbb{R}$ we define the Bourgain spaces $X_{s,b}$ to be the closure of $\mathcal{S}_{\text{per},2\pi}$ with regard to the norm

$$\|u\|_{X_{s,b}} = \Big(\sum_{n\in\mathbb{Z}} (1 + |n|)^{2s} \int_{-\infty}^{\infty} (1 + |\tau - n^3|)^{2b} |\widehat{u}(n, \tau)|^2 d\tau \Big)^{1/2}.$$

So, if we denote $\langle \sigma \rangle = 1 + |\sigma|$, then

$$\|u\|_{X_{s,b}} = \|\langle n \rangle^s \langle \tau - n^3 \rangle^b \widehat{u}(n, \tau)\|_{\ell^2(L^2_\tau)}.$$

Also, we define the space Y_s, $s \in \mathbb{R}$, as the closure of the Schwartz functions $\mathcal{S}_{\text{per},2\pi}$, with respect to the norm

$$\|u\|_{Y_s} = \|u\|_{X_{s,\frac{1}{2}}} + \Big[\sum_{n\in\mathbb{Z}} (1 + |n|)^{2s} \Big(\int_{-\infty}^{\infty} |\widehat{u}(n, \tau)|\, d\tau \Big)^2 \Big]^{1/2}. \tag{4.3}$$

So, $\|u\|_{Y_s} = \|u\|_{X_{s,\frac{1}{2}}} + \|\langle n \rangle^s \widehat{u}(n, \tau)\|_{\ell^2(L^1_\tau)}$. We similarly introduce the space Z_s defined via the norm

$$\|u\|_{Z_s} = \|u\|_{X_{s,-\frac{1}{2}}} + \Big[\sum_{n\in\mathbb{Z}} (1 + |n|)^{2s} \Big(\int_{-\infty}^{\infty} \frac{|\widehat{u}(n, \tau)|}{1 + |\tau - n^3|}\, d\tau \Big)^2 \Big]^{1/2}.$$

Next we look at some basic properties of the space $X_{s,b}$.

- **Basic Bourgain space embedding:** Here we consider some embedding relations for the spaces $X_{s,b}$ and Y_s.

(1) For $s \geqq s_0$ and $b \geqq b_0$ we have $X_{s,b} \subset X_{s_0,b_0}$.
(2) If $2 \leq p < \infty$ and $b \geqq \frac{1}{2} - \frac{1}{p}$, then $X_{s,b} \hookrightarrow L^p_t H^s_{\text{per}}$.
(3) If $2 \leq p, q < \infty$, $b \geqq \frac{1}{2} - \frac{1}{p}$, and $s \geqq \frac{1}{2} - \frac{1}{q}$, then $X_{s,b} \hookrightarrow L^p_t L^q_x$.
(4) If $1 < p \leq 2$ and $b \leq \frac{1}{2} - \frac{1}{p}$, then $L^p_t H^s_{\text{per}} \hookrightarrow X_{s,b}$.

(5) For $b > 1/2$, $X_{s,b} \subset C(\mathbb{R}_t; H^s_{\rm per})$. In fact, from the classical Sobolev embedding $L^\infty(\mathbb{R}_t) \hookrightarrow H^b_t(\mathbb{R})$ it follows that

$$\begin{aligned}\sum_{n\in\mathbb{Z}} \langle n\rangle^{2s} |\widehat{f}^{(x)}(n,t)|^2 &= \sum_{n\in\mathbb{Z}} \langle n\rangle^{2s} |e^{-in^3t}\widehat{f}^{(x)}(n,t)|^2 \\ &\leq \sum_{n\in\mathbb{Z}} \langle n\rangle^{2s} \|e^{-in^3t}\widehat{f}^{(x)}(n,\cdot)\|^2_{H^b_t} \\ &= \sum_{n\in\mathbb{Z}} \langle n\rangle^{2s} \int_{-\infty}^{\infty} \langle \tau\rangle^{2b} |\widehat{f}(n,\tau+n^3)|^2 d\tau = \|f\|^2_{X_{s,b}}.\end{aligned}$$

The continuity property follows easily.

(6) $Y_s \hookrightarrow C(\mathbb{R}_t; H^s_{\rm per})$. In fact, for $u \in Y_s$ and from $[\widehat{u(t)}^{(x)}(k)]^{\wedge(t)}(\tau) = \hat{u}(k,\tau)$, we have

$$\sup_{t\in\mathbb{R}} \|u(t)\|_{H^s_{\rm per}} \leq \Big[\sum_{n\in\mathbb{Z}} (1+|n|)^{2s} \Big(\int_{-\infty}^{\infty} |\widehat{u}(n,\tau)|\, d\tau\Big)^2\Big]^{1/2}.$$

Therefore we obtain the persistence property. Next, since for any $n \in \mathbb{Z}$, $\int_{-\infty}^{\infty} |\widehat{u}(n,\tau)|\, d\tau < \infty$, we can be assured of the continuity of the time flow of the solutions. Indeed, define $g_k(\tau) = \hat{u}(k,\tau)$. Then we have that the inverse Fourier transform of g_k in time exists. Since,

$$\begin{aligned}\|u(t+h) - u(t)\|^2_{H^s_{\rm per}} &= \sum_{k\in\mathbb{Z}} \langle k\rangle^{2s} |\widehat{u(t+h)}^{(x)}(k) - \widehat{u(t)}^{(x)}(k)|^2 \\ &\leq \sum_{k\in\mathbb{Z}} \langle k\rangle^{2s} \Big[\int_{-\infty}^{\infty} |g_k(\tau)||e^{ih\tau} - 1| d\tau\Big]^2\end{aligned}$$

and

$$\begin{aligned}&\langle k\rangle^{2s} \Big[\int_{-\infty}^{\infty} |g_k(\tau)||e^{ih\tau} - 1| d\tau\Big]^2 \\ &\quad\leq \langle 2k\rangle^{2s} \Big[\int_{-\infty}^{\infty} |g_k(\tau)| d\tau\Big]^2 \equiv a_k\end{aligned}$$

with $\{a_k\} \in \ell^1(\mathbb{Z})$, it follows from the dominated convergence theorem that the flow $t \to u(t) \in H^s_{\rm per}$ is continuous.

REMARK 4.9. The study of periodic initial value problems for nonlinear dispersive evolution equation is usually based on the basic Bourgain space $X_{s,\frac{1}{2}}$, but unfortunately this space does not give us the properties of persistence and continuity of the solutions in the classical Sobolev spaces $H^s_{\rm per}$, which is required in the definition of well-posedness and in our study of stability of periodic travelling wave solutions. The additional norm in the right-hand side of (4.3) appears in fact as a sufficient condition for obtaining these two basic properties of our solutions when they are found in Y_s.

- **Strategy of the proof of l.w.p. for the KdV in $H^s_{\rm per}$, $s \geqq 0$:** The idea is to find, via a fixed-point argument in Y_s, a solution of the integral equation

$$v(t) = \varphi(t)W(t)u_0 - \varphi(t)\int_0^t W(t-\tau)\frac{1}{2}\varphi(\tau)\partial_x(v^2(\tau))d\tau, \tag{4.4}$$

where $W(t) = e^{-\partial_x^3 t}$ and $\varphi \in C_0^\infty(\mathbb{R})$ is the cutoff function such that $\varphi(t) \equiv 1$ for $t \in [-1,1]$ and $\operatorname{supp} \varphi \subseteq [-2,2]$. So, if $v \in Y_s$ satisfies (4.4), then for every $T \in (0,1]$ we find a solution

$$u \equiv v|_{[-T,T]}$$

of the integral equation associated with the KdV equation for $t \in [-T,T]$.

Next we establish the basic steps to obtain a solution to (4.4) in Y_s.

- **Linear estimates:** We establish the basic estimates necessary in performing the contraction principle.

LEMMA 4.10. *For any $s \in \mathbb{R}$, there is $C_s > 0$ such that*

$$\|\varphi(t)W(t)\zeta\|_{Y_s} \leq C_s \|\zeta\|_{H^s}. \tag{4.5}$$

PROOF. Since $[\varphi W(t)\zeta]^\wedge(n,\tau) = \widehat{\zeta}^{(x)}(n)\widehat{\varphi}^{(t)}(\tau - n^3)$, we have

$$\begin{aligned}\|\varphi(t)W(t)\zeta\|_{X_{s,\frac{1}{2}}} &= \|\langle n\rangle^s \langle \tau - n^3\rangle^{1/2}\widehat{\zeta}(n)\widehat{\varphi}(\tau - n^3)\|_{\ell^2(L^2)} \\ &\leqq \|\varphi\|_{H_t^{1/2}}\|\zeta\|_{H^s}\end{aligned}$$

and

$$\|\langle n\rangle^s [\varphi(t)W(t)\zeta]^\wedge(n,\tau)\|_{\ell^2(L^1_\tau)} = \|\widehat{\varphi}\|_{L^1_\tau}\|\zeta\|_{H^s}.$$

This proves the lemma. □

LEMMA 4.11. *For any $s \in \mathbb{R}$,*

$$\|\varphi(t)\int_0^t W(t-\tau)F(\tau)d\tau\|_{Y_s} \leq C\|F\|_{Z_s}.$$

PROOF. The estimate for the norm in $X_{s,\frac{1}{2}}$ has already been made in Bourgain [**66**], so we have for $C_i > 0$,

$$\begin{aligned}\|\varphi(t)\int_0^t W(t-\tau)F(\tau)d\tau\|_{X_{s,\frac{1}{2}}} &\leq C_1\|\langle n\rangle^s\langle\tau - n^3\rangle^{-1}\widehat{F}(\tau)\|_{\ell^2(L^1_\tau)} \\ &\quad + C_2\|\langle n\rangle^s\langle\tau - n^3\rangle^{-1/2}\widehat{F}(\tau)\|_{\ell^2(L^2_\tau)} \\ &\leq C\|F\|_{Z_s}.\end{aligned}$$

Next, we obtain the contribution of the second norm in the right-hand side of (4.3). In fact, for η a smooth cutoff function in the time variable supported in $[-1,1]$ we have

$$\begin{aligned}&\varphi(t)\int_0^t W(t-\tau)F(\tau)d\tau \\ &= \sum_{k\geqq 1}\frac{i^k}{k!}t^k\varphi(t)\Big[\sum_{n\in\mathbb{Z}} e^{i(nx+n^3t)}\int_{-\infty}^{\infty}(\lambda - n^3)^{k-1}\eta(\lambda - n^3)\widehat{F}(n,\lambda)d\lambda\Big] \\ &+ C\varphi(t)\sum_{n\in\mathbb{Z}} e^{inx}\int_{-\infty}^{\infty}(\lambda - n^3)^{-1}(1-\eta)(\lambda - n^3)\widehat{F}(n,\lambda)e^{i\lambda t}d\lambda \\ &+ C\varphi(t)\sum_{n\in\mathbb{Z}} e^{i(nx+n^3t)}\int_{-\infty}^{\infty} -(\lambda - n^3)^{-1}(1-\eta)(\lambda - n^3)\widehat{F}(n,\lambda)d\lambda \\ &\equiv I + II + III.\end{aligned}$$

Next, for

$$h_k(n) = \int_{-\infty}^{\infty} i^{k-1}(\lambda - n^3)^{k-1}\eta(\lambda - n^3)\widehat{F}(n,\lambda)d\lambda$$

it follows immediately that $|h_k(n)| \leq \|\chi_{[-1,1]}(\lambda - n^3)\widehat{F}(n,\lambda)\|_{L^1_\tau}$. So, for $\psi_k(t) = t^k\psi(t)$ we have that

$$\widehat{I}(n,\lambda) = \sum_{k\geqq 1} \frac{1}{k!} h_k(n)\widehat{\psi_k}(\lambda - n^3)$$

and

$$\begin{aligned}
&\|\langle n\rangle^s \widehat{I}(n,\lambda)\|^2_{\ell^2(L^1_\lambda)} \\
&\leqq \sum_{n\in\mathbb{Z}} \langle n\rangle^{2s}\Big(\sum_{k\geqq 1} \frac{1}{k!}\|\widehat{\psi_k}\|_{L^1_\lambda}\|\chi_{[-1,1]}(\lambda - n^3)\widehat{F}(n,\lambda)\|_{L^1_\lambda}\Big)^2 \\
&\leq C\|\langle n\rangle^s\langle\lambda - n^3\rangle^{-1}\widehat{F}(n,\lambda)\|^2_{\ell^2(L^1_\lambda)}.
\end{aligned}$$

Now, by denoting

$$\widehat{\zeta}(n) = -\int_{-\infty}^{\infty} (\lambda - n^3)^{-1}(1-\eta)(\lambda - n^3)\widehat{F}(n,\lambda)d\lambda,$$

we have that $III = \varphi(t)W(t)\zeta$ and so from the proof of Lemma 4.10 we have that

$$\begin{aligned}
\|\langle n\rangle^s \widehat{III}(n,\lambda)\|^2_{\ell^2(L^1_\lambda)} &\leq C\|\zeta\|^2_{H^s_x} \\
&= C\|\langle n\rangle^s\chi_A(\tau - n^3)|\tau - n^3|^{-1}\widehat{F}(n,\tau)\|^2_{\ell^2(L^1_\tau)} \\
&\leq C\|\langle n\rangle^s\langle\tau - n^3\rangle^{-1}\widehat{F}(n,\tau)\|^2_{\ell^2(L^1_\tau)},
\end{aligned}$$

where χ_A is the characteristic function of the set $A = \{\tau : |\tau| \geqq 1\}$.

Now we estimate II. Let $f(n,\lambda) = (\lambda - n^3)^{-1}(1-\eta)(\lambda - n^3)\widehat{F}(n,\lambda)$. Then

$$\widehat{II}(k,\tau) = [\varphi(t)f^{\vee}(x,t)]^{\wedge}(k,\tau) = \widehat{\varphi} \star f(k,\tau).$$

Hence, by Young's inequality for convolutions we obtain that

$$\begin{aligned}
\|\langle n\rangle^s\widehat{II}(n,\tau)\|^2_{\ell^2(L^1_\lambda)} &= \sum_{n\in\mathbb{Z}} \langle n\rangle^{2s}\|\widehat{\varphi}\star f(n,\tau)\|^2_{L^1_\tau} \\
&\leq \sum_{n\in\mathbb{Z}} \langle n\rangle^{2s}\|\widehat{\varphi}\|^2_{L^1_\tau}\|f(n,\tau)\|^2_{L^1_\tau} \\
&= C\|\langle n\rangle^s\chi_A(\tau - n^3)|\tau - n^3|^{-1}\widehat{F}(n,\tau)\|^2_{\ell^2(L^1_\tau)} \\
&\leq C\|\langle n\rangle^s\langle\tau - n^3\rangle^{-1}\widehat{F}(n,\tau)\|^2_{\ell^2(L^1_\tau)}.
\end{aligned}$$

Finally we have for $Q = \varphi(t)\int_0^t W(t-\tau)F(\tau)d\tau$

$$\begin{aligned}
\|\langle n\rangle^s\widehat{Q}(n,\tau)\|_{\ell^2(L^1_\tau)} &\leq C\|\langle n\rangle^s\langle\tau - n^3\rangle^{-1}\widehat{F}(n,\tau)\|_{\ell^2(L^1_\tau)} \\
&\leq c\|F\|_{Z_s},
\end{aligned}$$

which finishes the proof of the lemma. □

Lemma 4.11 above leads us to the problem of obtaining an estimate for $\|F(u)\|_{Z_s}$ in the form

$$\|F(u)\|_{Z_s} \leq C\|u\|^{\beta}_{Y_s}$$

with $\beta \geqq 2$, such that we can perform an iteration argument in Y_s. That is the objective of the following nonlinear estimates.

- **Nonlinear estimates:** Next we establish the classical Bourgain L^4-estimative in the bi-torus [**65**].

LEMMA 4.12. *For a function f on $\mathbb{T}\times\mathbb{T}$ we have*

$$\|f\|_{L^4(\mathbb{T}^2)} \leq C\Big(\sum_{m,n\in\mathbb{Z}} (1+|n-m^3|)^{2/3}|\widehat{f}(m,n)|^2\Big)^{1/2}.$$

The following bilinear estimates can be found in [**65**] and [**162**] (see also [**88**]).

THEOREM 4.13. *We have the following:*

(1) *For $s \geqq 0$, with $u = u(x,t)$ and $v = v(x,t)$ periodic functions of x and having zero x-mean for all t, we have*
$$\|\varphi(t)\partial_x(uv)\|_{X_{s,-\frac{1}{2}}} \leq C\|u\|_{X_{s,\frac{1}{2}}}\|v\|_{X_{s,\frac{1}{2}}}.$$
(2) *For $s \in [-\frac{1}{2}, 0]$, $u = u(x,t)$ and $v = v(x,t)$ periodic functions of x and having zero x-mean for all t, we have*
$$\|\partial_x(uv)\|_{X_{s,-\frac{1}{2}}} \leq C\|u\|_{X_{s,\frac{1}{2}}}\|v\|_{X_{s,\frac{1}{2}}}.$$
(3) *For $s \geqq 0$, $u = u(x,t)$ and $v = v(x,t)$ periodic functions of x and having zero x-mean for all t, we have*
$$\Big\|\langle k\rangle^s\frac{\widehat{\varphi(uv)_x}(k,\tau)}{\langle\tau-k^3\rangle}\Big\|_{\ell^2(L^1_\tau)} \leq C\|u\|_{X_{s,\frac{1}{2}}}\|v\|_{X_{s,\frac{1}{2}}}.$$

Next we give an idea of the proof of Theorem 4.7.

SKETCH OF PROOF OF THEOREM 4.7. Initially we show that equation (4.4) is well-posed with small initial data in the $H^s_{\text{per}}([0,2\pi])$-norm and mean zero. We denote
$$\mathcal{V} = \{h \in Y_s : \int h(x,t)dx = 0, \text{ for all } t\}.$$
For $F(\tau) = \frac{1}{2}\varphi(\tau)\partial_x(u^2)$ and $u_0 \in H^s_{\text{per}}([0,2\pi])$ we define the operator

$$\Phi_{u_0}[u(t)] = \varphi(t)W(t)u_0 + \varphi(t)\int_0^t W(t-\tau)F(\tau)d\tau \tag{4.6}$$

and the ball
$$\mathcal{B} = \{u \in Y_s : \|u\|_{Y_s} \leq K\|u_0\|_{H^s_{\text{per}}}\} \cap \mathcal{V}.$$
Next we show that Φ_{u_0} is a contraction on $\mathcal{B}$ provided $\|u_0\|_{H^s_{\text{per}}}$ is sufficiently small.

We first prove that Φ_{u_0} sends bounded subsets of Y_s into bounded subsets of Y_s. We estimate each part of (4.6). From Lemma 4.10, Lemma 4.11, and Theorem 4.13, we obtain immediately
$$\begin{aligned}\|\Phi_{u_0}[u]\|_{Y_s} &\leq C_1\|u_0\|_{H^s_{\text{per}}} + C_2\|\varphi(t)\partial_x(u^2)\|_{Z_s} \leq C_1\|u_0\|_{H^s_{per}} + C_2\|u\|^2_{X_{s,\frac{1}{2}}}\\ &\leq C_1\|u_0\|_{H^s_{\text{per}}} + C_2\|u\|^2_{Y_s}.\end{aligned}$$
Moreover, for $u \in \mathcal{V}$ we have $\Phi_{u_0}[u] \in \mathcal{V}$.

Now we see that $\Phi_{u_0}(\mathcal{B}) \subseteq \mathcal{B}$ and Φ_{u_0} is a contraction on $\mathcal{B}$. So we will prove that there is a $\beta \in (0,1)$ such that
$$\|\Phi_{u_0}[w] - \Phi_{u_0}[h]\|_{Y_s} \leq \beta\|w-h\|_{Y_s}$$

for all $w, h \in \mathcal{B}$ provided that $\|u_0\|_{H^s_{\text{per}}}$ is sufficiently small. Since $w^2 - h^2 = (w-h)(w+h)$, we have from Lemma 4.11 and Theorem 4.13 that

$$\begin{aligned}\|\Phi_{u_0}[w] - \Phi_{u_0}[h]\|_{Y_s} &\leq C_3(\|w\|_{Y_s} + \|h\|_{Y_s})\|w-h\|_{Y_s} \\ &\leq 2C_3K\|u_0\|_{H^s_{\text{per}}}\|w-h\|_{Y_s}.\end{aligned} \tag{4.7}$$

Now, define $C = \max\{C_1, C_2, C_3\}$ and consider $K > 2C$. So, by choosing $\|u_0\|_{H^s_{\text{per}}}$ sufficiently small such that

$$2CK\|u_0\|_{H^s_{\text{per}}} < 1, \tag{4.8}$$

it follows from (4.7) and (4.8) that $\Phi_{u_0}(\mathcal{B}) \subseteq \mathcal{B}$ and Φ_{u_0} is a contraction on $\mathcal{B}$. Therefore there is a unique $u \in Y_s$ such that $\Phi_{u_0}[v] = v$. So, it follows that for $T \leq 1$ there exists a curve

$$t \in [-T, T] \to u(t)$$

satisfying (4.4).

Therefore we have obtained $u \in C([0,T]; H^s_{\text{per}}([0,2\pi]))$ and satisfied the integral equation associated with the KdV equation with initial data having mean zero and being small. The proof that this solution is unique and without mean zero condition in the initial data requires a little more work. Since this proof follows the arguments from Staffilani [**256**], it will be omitted. The proof of the continuous dependence of the solution upon the data follows standard arguments (see [**256**]). Therefore, we obtain Theorem 4.7 with the small condition on the initial data.

To remove the small condition on the initial data, we consider the following argument: We consider $\phi \in H^s_{\text{per}}([0,2\pi])$. Then the initial value problem associated with the KdV equation is well-posed on a small time interval $[0,\delta]$ if, and only if, the λ-rescaled problem

$$\begin{cases} \partial_t u_\lambda + \partial_x^3 u_\lambda + u_\lambda \partial_x u_\lambda = 0, & x \in [0, 2\lambda\pi], \\ u_\lambda(x,0) = \lambda^2\phi(x/\lambda) \equiv \phi_\lambda(x) \end{cases} \tag{4.9}$$

is well-posed on $[0, \lambda^3\delta]$. Next, for $I(\theta 2\pi) \equiv H^s_{\text{per}}([0,\theta 2\pi])$ we have for $\lambda \geqq 1$ that

$$\|\phi_\lambda\|^2_{I(\lambda 2\pi)} \leq \lambda^{-3}\|\phi\|^2_{I(2\pi)}.$$

Therefore by choosing $\lambda = \lambda(2\pi, \|\phi\|^2_{I(2\pi)})$ sufficiently large, we have

$$2CK\|\phi_\lambda\|_{I(\lambda 2\pi)} \leq 2CK\lambda^{-3/2}\|\phi\|_{I(2\pi)} < 1.$$

This verifies (4.8) for problem (4.9) and so we have the well-posedness of (4.9) on the time interval, for example $[0,1]$. Therefore the KdV equation is locally well-posed for $t \in [0, \lambda^{-3}]$. This finishes the proof. □

4.3. Some Results about Global Well-Posedness

We note that the nonlinear equation appearing in (4.1) has three basic associated functionals which are essential in the study of extending local solutions that are defined in an interval $[-T,T]$ to the interval $(-\infty,\infty)$. Let $\mathcal{I}$ denote $[-L,L]$ or $\mathbb{R}$. For smooth functions u, which suitably vanish at $\pm\infty$ in the case $\mathcal{I} = \mathbb{R}$, we define

$$E_M(u) = \int_{\mathcal{I}} \frac{1}{2}uMu - G(u)\,dx, \quad F(u) = \frac{1}{2}\int_{\mathcal{I}} u^2\,dx, \quad V(u) = \int_{\mathcal{I}} u\,dx, \tag{4.10}$$

where $G'(x) = f(x)$ and $G(0) = 0$. The quantities defined in (4.10) have the property that if $u = u(x,t)$ is a suitable smooth solution of the nonlinear equation in (4.1), then $E_M(u)$, $F(u)$, and $V(u)$ are independent of the temporal variable t; that is, E_M, F, and V are *invariants of the motion* generated by equation (4.1). E_M, F, and V are also called *conservation laws* of equation (4.1). So, if s and u_0 are given as in Theorem 4.2, then the unique solution of (4.1) satisfies $E_M(u(t)) = E_M(u_0)$ and $F(u(t)) = F(u_0)$ for every time t where the solution $u(t)$ is defined. If $\int_{\mathbb{J}} u_0(x)\,dx < \infty$, then it is also true that $V(u(t)) = V(u_0)$.

The property that $E_M(u(t)) = E_M(u_0)$ follows from the following formal computation. Suppose $Mu(t) \in H^s(\mathbb{R})$ and $s > 3/2$. Then $E_M(u_0) < \infty$ and

$$\partial_t E_M(u(t)) = \int_{\mathbb{J}} [Mu - f(u)]u_t\,dx = \int_{\mathbb{J}} [Mu - f(u)][Mu - f(u)]_x\,dx = 0. \tag{4.11}$$

In the last equality we have used the appropriate boundary conditions associated to $Mu(x)$. Therefore $E_M(u(t)) = E_M(u_0)$.

Now, in the case when $u_0 \in H^s(\mathbb{R})$ with $s > 3/2$, Theorem 4.2 shows that $u \in C([-T,T]; H^s(\mathbb{R}))$, so we need to justify the computations made in (4.11), since $Mu(t) \notin H^s(\mathbb{R})$. In this case, we use the continuous dependence property given in Theorem 4.2 and so it is sufficient to approximate the initial data u_0 by a sequence of functions $\{\varphi_n\} \subset \mathcal{S}(\mathbb{R})$ (the Schwartz space; see Appendix A) such that $\varphi_n \to u_0$ in $H^s(\mathbb{R})$ and $\{\varphi_n\} \subset H^r(\mathbb{R})$ for all $r \in \mathbb{R}$. Therefore, from (4.11) we obtain that $E_M(u_n(t)) = E_M(\varphi_n)$ and we get the desired property at the limit. Since E_M is a quantity invariant in time, this basic property is crucial to obtain *a priori* estimates which give global well-posedness results.

In the case of the GKdV equations we have the following conservation law:

$$E_{\text{GKdV}}(u) = \int_{\mathbb{J}} \frac{1}{2}(u_x)^2 - \frac{1}{(p+1)(p+2)} u^{p+2}\,dx, \tag{4.12}$$

which, from the Sobolev embedding $H^\gamma \hookrightarrow L^q$ for every $q \in [2,\infty)$ and $\gamma \geqq 1/2$, will be well-defined. Moreover, the map $E_{\text{GKdV}} : H^1 \to \mathbb{R}$ is continuous. Similarly, we have the following conservation law for the GBO equations:

$$E_{\text{GBO}}(u) = \int_{\mathbb{J}} \frac{1}{2}(D^{\frac{1}{2}}u)^2 - \frac{1}{(p+1)(p+2)} u^{p+2}\,dx, \tag{4.13}$$

which is well-defined and continuous on $H^{\frac{1}{2}}$.

We note that this approach for obtaining global well-posedness results usually works very well when we have Sobolev's integer exponents H^n, $n = 0, 1, 2, \ldots$. What happens, for example, to H^s with $0 < s < 1$ or $s < 0$? In these cases the situation is very delicate, and powerful tools have been developed by Bourgain [**69**] and the I-method by Colliander, Keel, Staffilani, Takaoka, and Tao [**88**] for dealing with these situations.

4.4. Comments

The theories of local and global well-posedness of nonlinear dispersive and wave equations have had a formidable development in the last decade. New techniques have been obtained and so new open problems have been raised for low regularity data. The definition of an equation to be locally well-posed has been strengthened and so today it is natural to ask the following: What is well-posedness?

Following the DispersiveWiki project, we have the following: by well-posedness in H^s we generally mean that there exists a unique solution u for some time T for each set of initial data in H^s, which stays in H^s and depends continuously on the initial data as a map from H^s to H^s. However, there are a couple subtleties involved here.

- Existence: For classical (smooth) solutions it is clear what it means for a solution to exist; for rough solutions one usually asks (as a bare minimum) for a solution to exist in the sense of distributions. (One may sometimes have to write the equation in conservation form before one can make sense of a distribution). It is possible for negative regularity solutions to exist if there is a sufficient amount of local smoothing available.
- Uniqueness: There are many different notions of uniqueness. One common one is uniqueness in the class of limits of smooth solutions. Another is uniqueness assuming certain space-time regularity assumptions on the solution. A stronger form of uniqueness is in the class of all H^s functions. Stronger still is uniqueness in the class of all distributions for which the equation makes sense.
- Time of existence: In subcritical situations the time of existence typically depends only on the H^s norm of the initial data, or at a bare minimum one should get a fixed nonzero time of existence for data of sufficiently small norm. When combined with a conservation law, this can often be extended to global existence. In critical situations one typically obtains global existence for data of small norm and local existence for data of large norm but with a time of existence depending on the profile of the data (in particular, the frequencies where the norm is largest) and not just on the norm itself.
- Continuity: There are many different ways in which the solution map can be continuous from H^s to H^s. One of the strongest is real analyticity (which is what is commonly obtained by iteration methods). Weaker than this are various types of C^k continuity, $k \in \mathbb{N}$. If the solution map is C^k, then this implies that the kth derivative at the origin is in H^s, which roughly corresponds to some iterate (often the kth iterate) lying in H^s. Weaker than this is Lipschitz continuity, and weaker than that is uniform continuity. Finally, there is just plain old continuity. Interestingly, several examples have emerged recently in which one form of continuity holds but not another; in particular we now have several examples (critical wave maps, low-regularity periodic KdV and mKdV, Benjamin-Ono, quasilinear wave equations, ...) where the solution map is continuous but not uniformly continuous.

CHAPTER 5

Definition of Stability

5.1. Introduction

Solitary waves in water were first observed scientifically by Russell [**246**], and in his study of these wave forms he found that individually they appear to be stable states of motion. Propagating along a uniform canal, a solitary wave displays a remarkable property of permanence, such as to give an observer immediate confidence in its stability. The first rigorous mathematical demonstration of this fact was prove by Benjamin in 1972 [**42**]. His original theory, which was initially applied to the KdV equation (2.4) and the Benjamin-Bona-Mahoney equation, has been refined and generalized in various ways by many authors, for example, Cazenave and Lions [**82**], Shatah [**251**], Shatah and Strauss [**253**], Weinstein [**272**], [**274**], Pego and Weinstein [**236**], Grillakis, Shatah, and Strauss [**123**], [**124**], and Martel and Merle [**211**].

Regarding application to the solitary wave solutions, the definition of stability needs care and attention. An immediate consideration is that in order to establish the stability of the solitary wave solutions (for example, those associated to the GKdV equations, ϕ_c, given in (3.4)), it will be necessary to deal with all of the solutions u of (2.5) that are initially close to ϕ_c. Accordingly, to give a rigorous foundation to the theory, we would need to appeal to an *existence and regularity theory* of solutions for the nonlinear evolution equation in question, guaranteeing that solutions exist for the considered class of initial wave forms that start near to the solitary wave solutions and that have sufficient smoothness properties in the stability analysis. We will assume such a theory (well-posedness), in a general form here; details are beyond the scope of this book. We note that these theories of *local or global well-posedness* (see previous chapter) are available with plausible assumptions about the regularity of solutions (for example in Sobolev spaces $H^s(\mathbb{R})$ with an exponent s sufficiently large), and they can be obtained by using the classical theories developed by Kato, in [**150**], for abstract evolution equations, or the parabolic regularization method by Iorio in [**136**]. Finally, more refined techniques of analysis developed by Kenig, Ponce, and Vega [**160**], Bourgain [**65**], and Colliander, Keel, Staffilani, Takaoka, and Tao [**88**] can also be used with the goal of obtaining *low-regularity behavior* for solutions, such as those established in Theorems 4.4 and 4.6 with regard to the GKdV and GBO equations.

5.2. Orbital Stability

Our concept of stability will rely on that introduced by Lyapunov (see Benjamin [**42**]). If we consider ϕ as a travelling wave solution associated to a specific evolution equation, for example (4.1), then the particular motion $\phi(x - ct)$ whose stability is in question has to be compared with a general class of motion $u = u(x, t)$ evolving from initial conditions that are close in some sense to those for ϕ. The stability of ϕ may be considered to mean that if u is made to be close to ϕ at the initial instant $u(0) = u_0$, then u will remain close to ϕ for all subsequent times. So, we need to specify some measure of distance between ϕ and u. If we consider two Banach spaces X, Y of functions defined on $\mathbb{R}$, such that Y is continuously embedding in X (see Section 3.2), and d_1, d_2 are pseudo-metrics on X, then we have the following general concept of stability related to equation (4.1).

DEFINITION 5.1 (Lyapunov Stability). Let $\phi \in X$ be a travelling wave solution of equation (4.1). We say that ϕ is X-stable with regard to the flow of (4.1) and with respect to d_1, d_2 if the following condition is satisfied: for each $\epsilon > 0$ there is a $\delta = \delta(\epsilon) > 0$ such that if $u_0 \in Y$ and

$$d_1(u_0, \phi) < \delta,$$

then the solution $u(t)$ of (4.1) with $u(0) = u_0$ exists for all t, and one has that

$$d_2(u(t), \phi) < \epsilon$$

for all $t \in \mathbb{R}$.

Otherwise, we say that ϕ is X-unstable.

We note that in Definition 5.1 we suppose *a priori* the existence of a local well-posedness theory of the initial value problem (4.1) in Y. As is natural, the choices of d_1 and d_2 are always crucial to the practical purpose of a stability analysis. Here we will consider $d_1 = d_2$, and we will choose d_2 to be the simplest pseudo-metric which gives us a practical criterion of stability for any one of the physical systems to which equation (4.1) is applicable. If we require in Definition 5.1 the stronger condition

$$d_2(u(t), \phi) \to 0 \qquad \text{as } t \to \infty, \tag{5.1}$$

the stability in question is called *asymptotic stability.* This kind of stability is more difficult to prove for equations of the type (4.1) (see Pego and Weinstein [**236**], Martel and Merle [**211**], [**216**], [**217**], and Merle and Vega [**221**]).

Via a simple analysis, we cannot expect to prove the stability of travelling wave solutions with respect to the metric in $H^s(\mathbb{R})$, $s \geqq 0$. More precisely, suppose ϕ_c is a fixed solitary wave solution of the GKdV equations (2.5) given by (3.4) and consider $d_1(u, \phi_c) = \|u - \phi_c\|_1$. If we let $u_0(x) = \phi_\eta$ where $\eta \neq c$, then it follows that $\phi_\eta \to \phi_c$ in the $H^s(\mathbb{R})$-norm, for any $s \geqq 0$, as $\eta \to c$. On the other hand, in this case, the solution u of (2.5) is known explicitly to be

$$u_\eta(x,t) = \phi_\eta(x - \eta t),$$

and hence $\|u_\eta(t) - \phi_c\|$ converges to a positive constant as $t \to \infty$ no matter how close η is to c. Indeed, by using the invariance by translation of the Lebesgue measure and the decay to zero of every solitary wave, we have

$$\|u_\eta(t) - \phi_c\|^2 \to \|\phi_\eta\|^2 + \|\phi_c\|^2 \qquad \text{as } t \to \infty.$$

We note that in this case we can also prove a stronger result, namely,

$$\|\phi_\eta(x - \eta t) - \phi_c(x - ct)\|^2 \to \|\phi_\eta\|^2 + \|\phi_c\|^2$$

as $t \to \infty$. Thus, the strong form of stability put forward here is not expected to happen in the present context. A result of this type of stability is in general not true because the speeds of the propagation of ϕ and that of the solution u may be different. So, small differences in velocity can eventually move two neighboring states very far apart.

One way of avoiding this difficulty is to give up knowing where the solution is in exchange for knowing its *shape* very well. This discrimination leads to a new measure of distance and a new notion of stability called *stability in shape.* Let τ stand for the group of translations along the x-axis. Thus $\tau_r f(x) = f(x+r)$ is defined for every $r \in \mathbb{R}$. Define, for $f, g \in X$,

$$d_2(f,g) = \inf_{r\in\mathbb{R}} \|f - \tau_r g\|_X. \tag{5.2}$$

This "distance" is the closest approach of f and g under the translation group in $\mathbb{R}$. From X we can form a quotient space, X/τ, by identifying the translations τf of each f in X. If we consider functions f, g representing elements of X/τ, we obtain that d_2 represents a *metric* in this set. Note that in X/τ the difference $u - \phi$ between ϕ and the perturbed solution u represents the most vital difference between two wave forms, namely, its *shape.*

In terms of the pseudo-metric (5.2) in X, we can rephrase the definition of stability, Definition 5.1, and obtain the following notion of stability of travelling wave solutions, which will be the object of our future study (see Chapters 6, 7, and 11). Before establishing this notion, we note that if $\phi \in X$ is a travelling wave solution of (4.1), then we have that the map $t \to \tau_{ct}\phi$ represents a trajectory in X, which we will call the *ϕ-orbit generated by ϕ* and will denote by Ω_ϕ, namely,

$$\Omega_\phi = \{\phi(\cdot + r) \ : \ r \in \mathbb{R}\}. \tag{5.3}$$

So, we have the following definition:

DEFINITION 5.2 (Orbital Stability). Let $\phi \in X$ be a travelling wave solution to equation (4.1). We say that ϕ is *orbitally stable* in X by the flow of (4.1) if for each $\epsilon > 0$, there is a $\delta = \delta(\epsilon) > 0$ such that if $u_0 \in Y$ and $\inf_{r\in\mathbb{R}} \|u_0 - \tau_r\phi\|_X < \delta$, then the solution $u(x,t)$ of (4.1) with $u(x,0) = u_0$ exists globally and satisfies

$$\sup_{t\in\mathbb{R}} \inf_{r\in\mathbb{R}} \|u(t) - \tau_r\phi\|_X < \epsilon.$$

Otherwise, we say that ϕ is X-unstable.

In other words, the ϕ-orbit is stable by the flow generated by equation (4.1) if whenever the initial data u_0 is sufficiently near to the ϕ-orbit in the X-norm, then for each instant t_0 there is a translation, $\gamma(t_0)$, such that the shape of the function $x \to u(x + \gamma(t_0), t_0)$ will begin to resemble and remain close to ϕ in the X-norm. Note that if $d_2(f,g)$ in (5.2) is small, then some translation of f lies close to g in the X norm, and the converse is true.

We note that the existence of the other Banach space, Y, in Definition 5.1 is shown by the problem of assuring a theory of well-posedness of the equation in question. Sometimes a theory is not known in the space X where we want to test the stability, but it is possible to obtain information about the initial problem in some subspace Y.

Finally, a more general definition of stability will also be used here, one which will be useful when we work with nonlocal methods for proving stability. This definition arises naturally when we use an analysis of a variational type in our study of the existence and stability of solitary wave solutions (see Chapters 8, 9, and 11). Initially, we establish some necessary notation. For any $\mathcal{O} \subseteq X$ and $\delta > 0$ we define the set $U_\delta(\mathcal{O})$ in X by

$$U_\delta(\mathcal{O}) = \bigcup_{v \in \mathcal{O}} B_\delta(v) = \{z \in X : \inf_{g \in \mathcal{O}} \|z - g\|_X < \delta\},$$

where $B_\delta(v) = \{z \in X : \|z - v\|_X < \delta\}$. The set $U_\delta(\mathcal{O})$ is called a δ-neighborhood of $\mathcal{O}$ in X. So, we can make precise our definition of stability of a set by the flow of an evolution equation as follows:

DEFINITION 5.3. We say that $\mathcal{O} \subset X$ is X-stable by the flow of (4.1) if, and only if, for any $\epsilon > 0$ there exists $\delta > 0$ such that for all $u_0 \in Y \cap U_\delta(\mathcal{O})$ we have that the solution $u(t)$ of (4.1) with $u(0) = u_0$ exists globally and satisfies $u(t) \in U_\epsilon(\mathcal{O})$ for all $t \geqq 0$.

Otherwise, we say that $\mathcal{O}$ is X-unstable.

By using Definition 5.3, we can rephrase Definition 5.2 in the following way: the orbit generated by a specific travelling wave solution ϕ, Ω_ϕ, is stable with regard to equation (4.1) if for every positive number ϵ there is another positive number δ such that if $u_0 \in Y$, then

$$\inf_{g \in \Omega_\phi} \|u_0 - g\|_X < \delta \implies \sup_{t \in \mathbb{R}} \inf_{g \in \Omega_\phi} \|u(t) - g\|_X < \epsilon. \tag{5.4}$$

We note that Definition 5.3 may imply Definition 5.2, as it is established by Theorem 8.20 and Theorem 8.24 in the cases of the stability of solitary wave solutions of the BO equation.

Finally, we note that in the case of the standing wave solutions (3.12) or (3.40) associated with the NLS equation (2.12), the orbit to be studied will be

$$\mathcal{O}_{\phi_\omega} = \{e^{i\theta}\phi_\omega(\cdot + x_0) : (x_0, \theta) \in \mathbb{R}^n \times [0, 2\pi)\},$$

so we have the following definition.

DEFINITION 5.4. We say that $\mathcal{O}_{\phi_\omega}$ is stable by the periodic flow of the NLS equation (2.12) if for all $\varepsilon > 0$, there is a $\delta(\varepsilon) > 0$ such that if

$$\inf_{(y,\theta) \in \mathbb{R} \times [0,2\pi)} ||u_0 - e^{i\theta}\phi_\omega(\cdot + y)||_{H^1_{\text{per}}} < \delta, \tag{5.5}$$

then the solution $u(x, t)$ associated to the NLS equation with initial data u_0 satisfies

$$\inf_{(y,\theta) \in \mathbb{R} \times [0,2\pi)} ||u(\cdot, t) - e^{i\theta}\phi_\omega(\cdot + y)||_{H^1_{\text{per}}} < \varepsilon, \tag{5.6}$$

for all $t \in \mathbb{R}$.

5.3. Comments

Definitions 5.3 or 5.4 coincide sometimes with Definition 5.2. As we will see later, a stability result in the form established in Definition 5.3 appears when the travelling wave solutions are obtained via a variational problem, and the set $\mathcal{O}$ is formed by many orbits generated by the critical points. So, when this orbit is unique, we obtain Definition 5.2. At this point, a uniqueness theory of travelling wave solutions is sometimes required.

There are other notions of stability which are very close to orbital stability and they can give some information about instability mechanisms. For instance, we consider the abstract evolution equation

$$U_t = F(U(t)), \tag{5.7}$$

with $F : Y \subset X \to X$, X, Y Banach spaces, and $U(t) \in X$. We say that $\Phi \in X$ is an equilibrium point for (5.7) provided $F(\Phi) = 0$, and we refer to the linearization of equation (5.7) around Φ, to the linearized problem

$$V_t = F'(\Phi)V, \tag{5.8}$$

$V(t) \in X$ and $F'(\Phi)$ being a (closed) linear operator in the space X. So, we say the following:

(1) The nonlinear wave Φ is *spectrally stable* if

$$\sigma(F'(\Phi)) \subset \{\lambda \in \mathbb{C} : \mathrm{Re}\lambda \leqq 0\}$$

where $\sigma(F'(\Phi))$ represents the spectrum of $F'(\Phi)$ (see Appendix B).

(2) The equilibrium $V^* = 0$ of (5.8) is linearly stable if for every $\epsilon > 0$ there exists $\delta > 0$ such that

$$\text{if } \|V(0)\| < \delta, \text{ then } \|V(t)\| < \epsilon, \ t \geqq 0.$$

Otherwise, we say that V^* is linearly unstable.

In general, linear stability does not prove stability of Ω_Φ in the sense of Definition 5.2. In fact the question is very difficult to answer directly. On the other hand, linear instability can imply nonlinear instability (see Chapter 10).

Part 3

Stability Theory

Laplace, Pierre-Simon (1749–1827)

All the effects of Nature are only the mathematical consequence of a small number of immutable laws.

Hilbert, David (1862–1943)

It is difficult and often impossible to judge the value of a problem correctly in advance; for the final award depends upon the gain which science obtains from the problem. Nevertheless, we can ask whether there are general criteria which mark a good mathematical problem. An old French mathematician said: 'A mathematical theory is not to be considered complete until you have made it so clear that you can explain it to the first man whom you meet on the street'. This clarity and ease of comprehension, here insisted on for a mathematical theory, I should still more demand for a mathematical problem if it is to be perfect; for what is clear and easily comprehended attracts, the complicated repels us.

In "Mathematical Problems", delivered by Hilbert before the Second International Congress of Mathematicians at Paris in 1900. Bulletin of the American Mathematical Society, vol. 8, 1902, pp. 437–445, 478–479 (see also in *Hilbert–Courant*, Constance Reid, Springer-Verlag, 1986).

CHAPTER 6

Orbital Stability—the Classical Method

6.1. Introduction

In this chapter we present the basic ideas of Boussinesq and Benjamin, [**70**] and [**42**], about the theory of stability for solitary wave solutions of nonlinear evolution equations. We remember that the first exact mathematical theory for the stability of solitary waves was given in Benjamin's article [**42**] (we note that Bona, in [**58**], makes some basic corrections to the theory in [**42**]), which was devoted to the Korteweg-de Vries equation

$$u_t + uu_x + u_{xxx} = 0$$

and to the regularized long-wave equation, or BBM equation (Benjamin, Bona, and Mahoney equation),

$$u_t + uu_x - u_{xxt} = 0.$$

In the setting of the KdV equation, Benjamin's idea, for obtaining orbital stability of the explicit solitary wave solutions

$$\phi_c(\xi) = 3c\,\mathrm{sech}^2\Big(\frac{1}{2}\sqrt{c}\,\xi\Big)$$

by the flow of the KdV equation, is to see the persistence of the solutions ϕ_c under small perturbations of initial data. Namely, one imagines a given wave form $u(x,0) = u_0(x)$ specified for all values of x and near to one of the solitary wave profiles $\phi_c(\cdot + y)$, and one tries to show that the solution u of the KdV equation emanating from the initial data u_0 will always resemble ϕ_c in shape (see Definition 5.2). Thus, by making a suitable translation of the solitary wave profile ϕ_c, one can nearly match it to the solution u. Hence, the assertion that the solitary wave ϕ_c is orbitally stable means that this solution will be stable in shape. The crucial idea in the proof of stability given by Benjamin in the case of the KdV equation is essentially to show that ϕ_c is a *local minimizer* of the functional

$$E_{\mathrm{KdV}}(u) = \frac{1}{2}\int_{-\infty}^{\infty} \Big[(u_x)^2 - \frac{1}{3}u^3\Big]\,dx \tag{6.1}$$

over the set of all admissible functions u satisfying

$$F(u) \equiv \frac{1}{2}\int_{-\infty}^{\infty} u^2\,dx = \frac{1}{2}\int_{-\infty}^{\infty} \phi_c^2\,dx. \tag{6.2}$$

More precisely, he showed that if $u \in H^1(\mathbb{R})$ is sufficiently close to ϕ_c in $H^1(\mathbb{R})$-norm and $F(u) = F(\phi_c)$, then

$$E_{\mathrm{KdV}}(u) - E_{\mathrm{KdV}}(\phi_c) \geq A \inf_{y\in\mathbb{R}} \|u - \phi_c(\cdot + y)\|_1^2 \tag{6.3}$$

where A denotes a positive constant which is independent of u. From this estimate and from an elaboration of Boussinesq's original argument, Benjamin deduced the following stability result:

(Benjamin [**42**]) *For every* $\epsilon > 0$, *there exists* $\delta = \delta(\epsilon) > 0$ *such that if* $u_0 \in H^1(\mathbb{R})$ *and* $\|u_0 - \phi_c\|_1 < \delta$, *then the solution* $u(x,t)$ *of the KdV equation* (2.4) *with initial data* u_0 *satisfies*

$$\inf_{y\in\mathbb{R}} \|u(t) - \phi_c(\cdot + y)\|_1 < \epsilon \quad \textit{for all} \quad t \in \mathbb{R}.$$

Benjamin points out in [**42**] that his analysis could also be applied to solitary wave solutions of other equations of Hamiltonian form. Others soon took up this suggestion and extended the theory to a more general setting by simplifying substantially what is necessary in the theory and establishing more accurate conditions that imply stability or instability of travelling waves for general nonlinear evolution equations. We refer the reader to the works of Weinstein [**274**], [**276**], Shatah and Strauss [**253**], Grillakis, Shatah, and Strauss [**123**], [**124**], Pego and Weinstein [**236**]. Examples of these advances are shown in the next chapters. As we will see in this chapter and in the following ones that the heart of the analysis to obtain a stability theory of travelling wave solutions is to get the main inequality (6.3).

The main interest in this chapter is to illustrate that the form as the local analysis method, established by Benjamin, Bona, and Weinstein in the study of orbital stability, works in the case of obtaining the nonlinear stability for the solitary wave solutions in (3.4) when $p < 4$ by the flow of the GKdV equation

$$u_t + u^p u_x + u_{xxx} = 0. \tag{6.4}$$

We also present here how this Lyapunov method can be used to get information about large-time asymptotic behavior of solutions for equations of the form

$$u_t + u^p u_x - D^\beta u_x = 0, \tag{6.5}$$

where $\widehat{D^\beta u}(\xi) = |\xi|^\beta \widehat{u}(\xi)$ and $p = 2\beta$, $\beta \geqq 1$. More precisely, we will establish a result of "*stability of the blow-up*" near to unstable solitary wave solutions for equation (6.5). We remember that it is conjectured that the solitary wave solutions of equation (6.5) are unstable in the case $p = 2\beta$. For $p > 2\beta$ it is known that we have instability (see [**64**] or Chapter 10 of this book where a different approach is used to obtained this result). For $\beta = 2$ and $p = 4$, equation (6.5) is called *the critical KdV equation*, and as has been shown recently by Martel and Merle in [**212**] and [**214**], the corresponding solitary wave solutions are in fact unstable.

6.2. Stability of Solitary Wave Solutions for the GKdV

In this section we present a theory of stability of the solitary wave solutions, ϕ_c, for the Generalized Korteweg-de Vries equations (6.4) given in (3.4). Note that the GKdV equations have translation symmetry, i.e., if $u(x,t)$ solves the GKdV equations, then $u(x+\gamma,t)$ solves the GKdV equations for any $\gamma \in \mathbb{R}$. So, our notion of stability to be discussed here will be stability modulo these symmetries. Therefore, we are interested in the study of the stability of the *orbit* generated by a solitary wave solution ϕ_c, namely the set

$$\Omega_{\phi_c} = \{\phi_c(\cdot + y) : y \in \mathbb{R}\}.$$

Our goal here will be to present a self-contained exposition of the classical Lyapunov method which will give us stability properties of Ω_{ϕ_c}. In Appendix B we establish the required spectral theory necessary in our exposition.

THEOREM 6.1. *Let $c > 0$ and $p < 4$. Then the orbit Ω_{ϕ_c} generated by the solitary wave solution ϕ_c in (3.4) is orbitally stable in $H^1(\mathbb{R})$ with respect to the flow of the GKdV equations (6.4). Namely, for every $\epsilon > 0$ there is $\delta = \delta(\epsilon) > 0$ such that if $\|u_0 - \phi_c\|_1 < \delta$, then the solution $u(x,t)$ of the GKdV equation with initial data u_0 satisfies*

$$\|u(t) - \phi_c(\cdot - \gamma)\|_1 < \epsilon$$

for all $t \in \mathbb{R}$ and $\gamma = \gamma(t)$.

PROOF. The proof is based on the following continuous functional defined on $H^1(\mathbb{R})$:

$$\mathcal{B}(u) \equiv E_{\text{GKdV}}(u) + cF(u)$$

where E_{GKdV} and F are the conserved laws defined in (4.12) and (6.2). To simplify notation, we will write E_{GKdV} as E. We note that $\mathcal{B}$ is also a conserved functional by the flow of the GKdV equations. Hence, if $u_0 \in H^1(\mathbb{R})$ and $t \to u(t) \in H^1(\mathbb{R})$ is the solution of (6.4) with $u(0) = u_0$ (see Theorem 4.4), then $\mathcal{B}(u(t)) = \mathcal{B}(u_0)$ for all $t \in \mathbb{R}$.

We consider the perturbation of the solitary wave solution $\phi = \phi_c$

$$u(x+\gamma, t) = \phi(x) + v(x,t),$$

where v is a real function and $\gamma = \gamma(t)$ will be chosen later (see Lemma 6.3 below) so as to minimize the functional

$$\Omega_t(\gamma) = ||u_x(\cdot + \gamma, t) - \phi'||^2 + c||u(\cdot + \gamma, t) - \phi||^2.$$

Now, from (6.4), the representation of u above, and the invariance of Lebesgue's measure by translations, one calculates that

$$\begin{aligned}
\Delta\mathcal{B}(t) &\equiv \mathcal{B}(u(\cdot,t)) - \mathcal{B}(\phi) = \mathcal{B}(\phi + v(\cdot,t)) - \mathcal{B}(\phi) \\
&= \frac{1}{2}\int_{-\infty}^{\infty} [(v_x)^2 + cv^2 - \phi^p v^2]\, dx + \int_{-\infty}^{\infty} v[c\phi - \phi'' - \frac{1}{p+1}\phi^{p+1}]\, dx \\
&\quad - \frac{1}{(p+1)(p+2)} \sum_{k=3}^{p+2} C_k(p) \int_{-\infty}^{\infty} \phi^{p+2-k} v^k\, dx \qquad (6.6) \\
&\geqq \frac{1}{2}\langle \mathcal{L}v, v\rangle - \sum_{k=3}^{p+2} C_k(p,c)\|v\|_1^k.
\end{aligned}$$

The inequality in (6.6) is obtained by using the classical embedding $H^1(\mathbb{R}) \hookrightarrow L^r(\mathbb{R})$ for every $r \geqq 2$, which gives us the estimate

$$\left|\int_{-\infty}^{\infty} \phi^{p+2-k} v^k\, dx\right| \leqq \|\phi\|_\infty^{p+2-k} \int_{-\infty}^{\infty} |v|^k\, dx \leqq C_0(c,p)\|v\|_1^k.$$

The closed, unbounded, self-adjoint operator on $L^2(\mathbb{R})$, $\mathcal{L}$, appearing in (6.6), is defined by

$$\mathcal{L}v = -\frac{d^2}{dx^2}v + cv - \phi^p v.$$

We note that $\mathcal{L}$ is related to the differential equation (3.3) satisfied by ϕ. In fact, if we take a derivative in (3.3) with regard to the spacial variable, we obtain that

$\mathcal{L}\phi' = 0$. It follows that $\mathcal{L}$ has zero as an eigenvalue with eigenfunction ϕ'. The property that zero is a simple eigenvalue will be important in our analysis. We note that the functional $\mathcal{B}$ has been chosen so that *ϕ is a critical point of it*, namely, $\mathcal{B}'(\phi) = 0$.

The main ingredient to obtain our result of stability will be to obtain a suitable lower bound on the quadratic form $\langle \mathcal{L}v, v\rangle$ of the form

$$\langle \mathcal{L}v, v\rangle \geqq \beta_0\|v\|_1^2 - \beta_1\|v\|_1^{2+\theta_1} - \beta_2\|v\|_1^{2+\theta_2} \tag{6.7}$$

with $\beta_i > 0$ and $\theta_i > 0$ (see inequality (6.3)). At this point a careful study of the spectral structure of the linear operator $\mathcal{L}$ will be necessary, and use of the Sturm-Liouville theory (see Appendix B) will arise as a crucial ingredient in the proof of stability. Inequality (6.7) is an effective lower bound from which one may infer that $\|v(t)\|_1$ is small for all t provided it is small enough at $t = 0$. In other words, if the perturbation of the solitary wave is not too large, then it is stable in shape. Therefore, if $\gamma(t)$ is determined to be a minimum of $\Omega_t(\gamma)$, then we obtain that

$$\|u(t) - \phi(\cdot - \gamma(t))\|_1 < \epsilon$$

for all $t \in \mathbb{R}$, if $\|u_0 - \phi\|_1$ is small enough.

With the idea of showing inequality (6.7), we will establish some basic lemmas.

LEMMA 6.2. *Suppose that for some $t \in [0,\infty)$ and some $\gamma_0 \in \mathbb{R}$,*

$$\Omega_t(\gamma_0) < ||(-\partial_x^2 + c)^{1/2}\phi||^2. \tag{6.8}$$

Then it follows that

$$\inf\{\Omega_t(\gamma)|\gamma \in \mathbb{R}\} \tag{6.9}$$

is attained at least once in $\mathbb{R}$.

PROOF. It is immediate that $\Omega_t(\gamma)$ is a continuous function of $\gamma \in \mathbb{R}$. Moreover, using the Lebesgue Dominated Convergence Theorem, we obtain

$$\begin{aligned} \lim_{|\gamma|\to\infty} \Omega_t(\gamma) &= ||u_x(t)||^2 + c||u||^2 + ||\phi'||^2 + c||\phi||^2 \\ &= ||u_x(t)||^2 + c||u||^2 + ||(-\partial_x^2 + c)^{1/2}\phi||^2. \end{aligned} \tag{6.10}$$

So, from hypothesis (6.8), the continuity of Ω_t, and (6.10), it follows that there is an interval $I = [-\gamma_1, \gamma_1]$ with $\gamma_1 > \gamma_0$, such that $\Omega_t\Big|_I$ is continuous and therefore $\operatorname{Inf}_{\gamma\in\mathbb{R}} \Omega_t(\gamma) = \operatorname{Inf}_{\gamma\in I} \Omega_t(\gamma)$, which is obviously attained. This completes the proof of the lemma. □

Next, it is established that the minimum in (6.9) is achieved at finite values of γ at least for all t in some interval of time $[0, T]$.

LEMMA 6.3. *There exits an interval of time $\mathcal{J} = [0, T]$, such the $\inf \Omega_t$ is attained in $\gamma(t)$ for every $t \in \mathcal{J}$.*

PROOF. From Lemma 6.2 we have that it is sufficient to obtain condition (6.8) in some interval of time. Let $\epsilon > 0$ be such that

$$\max\{1, c\}\epsilon^2 < \|(-\partial_x^2 + c)^{1/2}\phi\|^2.$$

The solitary wave solution $u(x,t) = \phi(x - ct)$ is globally defined, and hence by the continuous dependence theory for the GKdV equations in $H^1(\mathbb{R})$, we have for the value of ϵ just specified and for $T > 0$ that there exists a $\delta > 0$ such that if

$||u_0 - \phi||_1 < \delta$, then the solution u of the GKdV equations with initial data u_0 exists at least for $0 \leqq t \leqq T$ and in addition,

$$\|u(\cdot,t) - \phi(\cdot - ct)\|_1^2 < \epsilon^2$$

for all $t \in [0,T]$. Therefore, from the definition of Ω_t it follows that

$$\Omega_t(ct) \leqq \max\{1,c\}\ \|u(\cdot,t) - \phi(\cdot - ct)\|_1^2 < \max\{1,c\}\epsilon^2 < \|(-\partial_x^2 + c)^{1/2}\phi\|^2$$

for all $t \in [0,T]$. Thus, the infimum (6.9) is obtained at values $\gamma(t)$ throughout the time interval $[0,T]$. This finishes the proof. □

Now, we take such values of γ obtained in Lemma 6.3 as providing a meaning for the definition of v given in the beginning of the proof of Theorem 6.1, at least for $t \in [0,T]$.

The result of Lemma 6.3 immediately gives us that for $t \in [0,T]$ and $\gamma = \gamma(t)$ a minimum of Ω_t, $\frac{d}{dr}\Omega_t(r)\Big|_{r=\gamma} = 0$. Hence, using the fact that ϕ satisfies equation (3.3), we obtain that

$$\frac{d}{dr}\Omega_t(r)\Big|_{r=\gamma} = 2\int_{-\infty}^{\infty} [c\phi'(x) - \phi'''(x)]v(x,t)dx = 2\int_{-\infty}^{\infty} \phi^p(x)\phi'(x)v(x,t)dx,$$

which provides us with a compatibility relation on v, namely

$$\int_{-\infty}^{\infty} \phi^p(x)\phi'(x)v(x,t)dx = 0 \tag{6.11}$$

for all $t \in [0,T]$. As we will see later, the orthogonality condition (6.11) appears as a crucial ingredient in the proof of stability.

The issue of obtaining the lower bound (6.7) is addressed in the next few lemmas. The first one is an abstract result from spectral theory (Weinstein [**274**]). In the next $\mathcal{N}(A)$ will represent the kernel of a linear operator A (see Appendix B).

LEMMA 6.4. *Let A be a self-adjoint operator on $L^2(\mathbb{R})$ having exactly one negative simple eigenvalue λ with corresponding ground-state eigenfunction $\varphi_\lambda \geqq 0$ and zero being an isolated eigenvalue. Let $\psi \in \mathcal{N}(A)^\perp$ such that $\langle\psi, \varphi_\lambda\rangle \neq 0$ and*

$$-\infty < \alpha \equiv \min_{\langle f,\psi\rangle=0, \|f\|=1} \langle Af, f\rangle.$$

If $\langle A^{-1}\psi, \psi\rangle \leqq 0$, then it must be the case that $\alpha \geqq 0$.

REMARK 6.5. The number $\langle A^{-1}\psi, \psi\rangle$ is independent of $h \in A^{-1}\psi$. Indeed, the proof is very easy: let $Ah_1 = Ah_2 = \psi$; then $A(h_1 - h_2) = 0$ and so $\langle h_1 - h_2, \psi\rangle = 0$.

PROOF. We consider the functionals

$$G_1(f) = \langle Af, f\rangle, \quad G_2(f) = \langle f, f\rangle, \quad G_3(f) = \langle f, \psi\rangle.$$

If α is attained in f^*, it follows that $G_1(f^*) = \alpha$, $G_2(f^*) = 1$, and $G_3(f^*) = 0$. Moreover, from Lagrange's multiplier theory (see [**203**]) it follows that

$$\delta G_1(f^*) = \mu\delta G_2(f^*) + \beta\delta G_3(f^*)$$

where $\delta H(g)$ represents the Fréchet derivative of H at g and μ, β are real constants. So, we get the main relation

$$Af^* = \mu f^* + \beta\psi. \tag{6.12}$$

Hence, from (6.12) we obtain that $\langle Af^*, f^*\rangle = \mu$. We note that from the property satisfied by λ, it follows that $\langle Af, f\rangle \geqq \lambda$ for every $f \in D(A)$ with $\|f\| = 1$. So,

$\mu \geqq \lambda$. Therefore, to obtain the relation $\alpha \geqq 0$, it is sufficient to see that $\mu < 0$ is not possible. We will consider the following cases:

(1) Suppose $\mu = \lambda$. Then by (6.12) and the hypothesis, it follows that $\lambda\langle f^*, \varphi_\lambda\rangle = \lambda\langle f^*, \varphi_\lambda\rangle + \beta\langle \psi, \varphi_\lambda\rangle$. So $\beta\langle\psi, \varphi_\lambda\rangle = 0$. Since $\langle\psi, \varphi_\lambda\rangle \neq 0$, it follows that $\beta = 0$. Then $Af^* = \lambda f^*$, and since λ is a simple eigenvalue, it follows that $f^* = r\varphi_\lambda$, where $r \neq 0$. Then $0 = \langle f^*, \psi\rangle = r\langle\varphi_\lambda, \psi\rangle$, which is a contradiction.

(2) Suppose that $\lambda < \mu < 0$. If $\beta = 0$ in (6.12), then $Af^* = \mu f^*$ and therefore μ is an eigenvalue, which is not true. Hence, $\beta \neq 0$ and we have that $f^* = \beta(A - \mu)^{-1}\psi = \beta R_\mu(A)\psi$, where $R_\mu(A)$ represents the resolvent of A. Now we consider the mapping

$$H(\theta) = \begin{cases} \langle R_\theta(A)\psi, \psi\rangle, & \theta \in (\lambda, 0), \\ \langle A^{-1}\psi, \psi\rangle, & \theta = 0. \end{cases}$$

Then $H(\mu) = \langle R_\mu(A)\psi, \psi\rangle = \frac{1}{\beta}\langle f^*, \psi\rangle = 0$. Now, we will see that this relation is not possible if we have that $\langle A^{-1}\psi, \psi\rangle \leqq 0$. In fact, initially we have the following properties associated with the function H.

(a) H is continuous in $(\lambda, 0]$. In fact, it follows from Theorem B.17 in Appendix B that the mapping $\theta \to R_\theta(A)$ is an analytic function on each component of $\rho(A)$ (resolvent of A) and therefore H is continuous on $(\lambda, 0)$. Now, we see that

$$\lim_{\theta \to 0^-} H(\theta) = \langle A^{-1}\psi, \psi\rangle.$$

In fact, define $D_1 \equiv D(A) \cap \mathcal{N}(A)^\perp$. Then $A(D_1) \subset \mathcal{N}(A)^\perp$, since $\mathcal{N}(A)^\perp = \overline{\mathcal{R}(A)}$. Moreover, if we denote by A_1 the restriction of A to D_1, then $A_1 : D_1 \to \mathcal{N}(A)^\perp$ is self-adjoint and satisfies $\mathcal{N}(A_1) = \{0\}$. Now, if $0 \in \sigma(A_1)$ (spectrum of A_1), then zero is a isolated point of $\sigma(A_1)$ and therefore from Theorem B.53 in Appendix B, it follows that zero will be an eigenvalue of A_1, which is false. Then we need to have that $0 \in \rho(A_1)$ and so $A_1^{-1} : \mathcal{N}(A)^\perp \to D_1$ is bounded. Hence, applying Theorem B.17 to the operator A_1, we have that the mapping defined on $\rho(A_1)$, $\theta \to R_\theta(A_1)$, is continuous. So, since $\rho(A_1)$ is an open subset of $\mathbb{C}$, it follows that for each $\psi \in \mathcal{N}(A)^\perp$ the following limit makes sense:

$$\lim_{\theta \to 0^-} \langle (A - \theta)^{-1}\psi, \psi\rangle = \lim_{\theta \to 0^-} \langle (A_1 - \theta)^{-1}\psi, \psi\rangle = \langle A_1^{-1}\psi, \psi\rangle = \langle A^{-1}\psi, \psi\rangle.$$

Therefore, we have that H is a continuous function in $(\lambda, 0]$.

(b) H is strictly increasing in $(\lambda, 0)$. In fact, we will show that

$$H'(\theta) = \|(A - \theta)^{-1}\psi\|^2 > 0$$

for all $\theta \in (\lambda, 0)$. In fact, by using the first resolvent formula on the resolvent (Theorem B.17) with θ, $\theta + h$ in $(\lambda, 0)$, we have

$$R_{\theta+h}(A) - R_\theta(A) = hR_{\theta+h}(A)R_\theta(A).$$

Hence,

$$\frac{H(\theta + h) - H(\theta)}{h} = \langle R_{\theta+h}(A)R_\theta(A)\psi, \psi\rangle$$

and so $H'(\theta) = \langle R_\theta(A)\psi, R_\theta(A)\psi\rangle = \|(A - \theta)^{-1}\psi\|^2 > 0$ for each $\theta \in (\lambda, 0)$.

Hence, from properties (a) and (b) satisfied by H and $\langle A^{-1}\psi, \psi\rangle \leqq 0$, we must have that $H(\mu) < 0$. This finishes the proof of the lemma. □

Now, we have a main result about the spectral structure of the linear operator $\mathcal{L}$ in (6.6). The proof of the following lemma is based in the Sturm-Liouville theory and it can be seen in Theorem B.61 of Appendix B.

LEMMA 6.6. *Suppose that $\phi \in L^2(\mathbb{R})$ satisfies the differential equation* (3.3) *with $c > 0$. Then the differential operator*

$$\mathcal{L}\psi = -\psi'' + [c - \phi^p]\psi$$

defined in $L^2(\mathbb{R})$ has exactly one negative simple eigenvalue λ_0 with strictly positive eigenfunction, the eigenvalue 0 is simple with associated ϕ' eigenfunction, and there exists $\delta > 0$ such that for all $\lambda \in \sigma(\mathcal{L}) - \{\lambda_0, 0\}$ it satisfies $\lambda > \delta$.

LEMMA 6.7. *Let $\mathcal{L} = -\frac{d^2}{dx^2} + c - \phi^p$ with ϕ a solution of* (3.3). *Then*

$$\alpha \equiv \min_{\langle f,\phi\rangle=0,\|f\|=1} \langle \mathcal{L}f, f\rangle = 0, \qquad \textit{for } p \leqq 4.$$

PROOF. Because ϕ is bounded, we can infer that α is finite. Moreover, since $\langle \phi', \phi\rangle = 0$ and $\mathcal{L}\phi' = 0$, it follows that $\alpha \leqq 0$. Now, we shall show that there is an $f^* \in H^1(\mathbb{R})$ such that $\alpha = \langle \mathcal{L}f^*, f^*\rangle$, $\|f^*\| = 1$, and $\langle f^*, \phi\rangle = 0$. In fact, since α is finite, let $\{f_j\}$ be a sequence of $H^1(\mathbb{R})$ functions with $||f_j|| = 1$, $\langle f_j, \phi\rangle = 0$, and

$$\lim_{j\to\infty} \langle \mathcal{L}f_j, f_j\rangle = \alpha.$$

Then for any $\epsilon > 0$ there is a J such that for $j > J$,

$$0 < c \leqq \|f_j'\|^2 + c\|f_j\|^2 \leqq \int_{-\infty}^{\infty} \phi^p f_j^2 dx + \epsilon + \alpha. \tag{6.13}$$

Since $\|\phi\|_\infty < \infty$, (6.13) implies that $||f_j||_1$ is uniformly bounded as j varies. Therefore, there exists a subsequence of $\{f_j\}$, which we denote again by $\{f_j\}$, and $f^* \in H^1(\mathbb{R})$ such that $f_j \rightharpoonup f^*$ weakly in $H^1(\mathbb{R})$, which by compact embedding and a Cantor diagonalization argument will converge strongly in $L^2_{\text{loc}}(\mathbb{R})$. The function f^* satisfies the condition $\langle f^*, \phi\rangle = 0$. A consequence of the recently mentioned properties of the sequence f_j and the decay of ϕ to 0 as $|x| \to \infty$ is that

$$\int_{-\infty}^{\infty} \phi^p (f_j)^2\, dx \to \int_{-\infty}^{\infty} \phi^p (f^*)^2\, dx$$

as $j \to \infty$. Taking the limit in (6.13) as $j \to \infty$ yields

$$0 < c \leqq p \int_{-\infty}^{\infty} \phi^p (f^*)^2 dx + \epsilon + \alpha.$$

As ϵ is arbitrary, it must be the case that $f^* \neq 0$.

It is now shown that the infimum is achieved. Indeed, weak convergence is lower continuous, so

$$||f^{*\prime}|| \leqq \liminf_{j\to\infty} ||f_j'||.$$

Since $\langle \phi^p f_j, f_j\rangle \to \langle \phi^p f^*, f^*\rangle$ as $j \to \infty$, as noted above, it follows that

$$\langle \mathcal{L}f^*, f^*\rangle = \int_{-\infty}^{\infty} (f^{*\prime})^2 + [c - \phi^p](f^*)^2\, dx \leqq \liminf_{j\to\infty} \langle \mathcal{L}f_j, f_j\rangle = \alpha. \tag{6.14}$$

Also, since $f_j \rightharpoonup f^*$ weakly in $L^2(\mathbb{R})$, it follows that $\|f^*\| \leqq 1$. Suppose now that $\alpha = 0$. Since $f^* \neq 0$, define $g^* = \frac{f^*}{||f^*||}$. Then, we have $||g^*|| = 1$, $\langle g^*, \phi\rangle = 0$, and from (6.14) it follows that $\langle \mathcal{L}g^*, g^*\rangle = 0$. Therefore, g^* is a minimum. Suppose $\alpha < 0$ and $\|f^*\| < 1$. Then

$$\alpha \leqq \langle \mathcal{L}g^*, g^*\rangle \leqq \frac{\alpha}{||f^*||^2},$$

and therefore $\alpha \leqq \frac{\alpha}{||f^*||^2}$, which is a contradiction. Hence, $\|f^*\| = 1$ and so f^* is a minimum.

Now, we want to apply Lemma 6.4 with $A = \mathcal{L}$ and $\psi \equiv \phi$, and so we obtain that $\alpha \geqq 0$. Indeed, from Lemma 6.6 it follows that the linear operator $\mathcal{L}$ has the necessary spectral properties required by Lemma 6.4. Now, we need to find a χ such that $\mathcal{L}\chi = \phi$ and $\langle \chi, \phi\rangle \leqq 0$. We will show here two ways to find such a function χ.

(1) *First Choice*: From (3.3) and (3.4) we have that $\phi = \phi_c$ is a solitary wave solution of the GKdV equations for every $c > 0$, and the mapping $c \in (0, \infty) \to \phi_c \in H^1(\mathbb{R})$ is of class C^1. So, by differentiation with respect to c in (3.3) we obtain that $\chi = -\frac{d}{dc}\phi_c$ satisfies

$$\mathcal{L}\Big(-\frac{d}{dc}\phi_c\Big) = \phi_c. \tag{6.15}$$

Therefore, for every $c > 0$ fixed, we have that

$$\begin{aligned} \langle \mathcal{L}^{-1}\phi_c, \phi_c\rangle &= -\langle \frac{d}{dc}\phi_c, \phi_c\rangle = -\frac{1}{2}\frac{d}{dc}\int_{-\infty}^{\infty} \phi_c^2(\xi)\, d\xi \\ &= -\Big[\frac{(p+1)(p+2)}{2}\Big]^{2/p}\int_{-\infty}^{\infty} \operatorname{sech}^{4/p}(\xi)d\xi \cdot \frac{d}{dc}\, c^{\frac{2}{p}-\frac{1}{2}} \\ &= K(p)\Big(\frac{1}{2} - \frac{2}{p}\Big)c^{\frac{2}{p}-\frac{3}{2}}, \end{aligned} \tag{6.16}$$

where $K(p)$ is a positive constant. Therefore, $\langle \mathcal{L}^{-1}\phi_c, \phi_c\rangle \leqq 0$ if and only if $p \leqq 4$.

(2) *Second Choice*: We consider the function

$$\chi(\xi) = -\frac{1}{cp}\phi_c(\xi) - \frac{1}{2c}\xi\phi_c'(\xi).$$

Since $\mathcal{L}\phi_c' = 0$ and ϕ_c satisfies equation (3.4), we have after some calculations that $\mathcal{L}\chi = \phi_c$. So, by using integration by parts, we obtain

$$\begin{aligned} \langle \mathcal{L}^{-1}\phi_c, \phi_c\rangle = \langle \chi, \phi_c\rangle &= -\frac{1}{cp}\int_{-\infty}^{\infty} \phi_c^2(\xi)d\xi - \frac{1}{2c}\int_{-\infty}^{\infty} \xi\phi_c'(\xi)\phi_c(\xi)d\xi \\ &= \Big(-\frac{1}{cp} + \frac{1}{4c}\Big)\int_{-\infty}^{\infty} \phi_c^2(\xi)d\xi. \end{aligned} \tag{6.17}$$

Therefore, $\langle \mathcal{L}^{-1}\phi_c, \phi_c\rangle \leqq 0$ if and only if $p \leqq 4$.

Then, from Lemma 6.4 we obtain that $\alpha \geqq 0$. This finishes the proof of the lemma. □

REMARK 6.8. For $p = 4$, we note that χ defined in the second choice of Lemma 6.4 satisfies $\langle \chi, \phi\rangle = 0$ and $\langle \mathcal{L}\chi, \chi\rangle = \langle \phi, \chi\rangle = 0$. Moreover,

$$\int_{-\infty}^{\infty} \phi^p(\xi)\phi'(\xi)\chi(\xi)d\xi = 0 \qquad \text{for each } p \in \mathbb{N}.$$

Next we show a basic inequality in the proof of (6.7).

LEMMA 6.9. *Consider the same notation as in Lemma 6.7. Then, for every* $p < 4$ *we have that*

$$\inf\{\langle \mathcal{L}f, f\rangle \;:\; \|f\| = 1, \langle f, \phi\rangle = 0, \; and \; \langle f, \phi^p\phi'\rangle = 0\,\} \equiv \beta > 0. \tag{6.18}$$

PROOF. Because of Lemma 6.7, we can infer that $\beta \geqq 0$. Suppose that $\beta = 0$. Then, using the same analysis as in the proof of Lemma 6.7, we can guarantee the existence of a function f^* such that $\|f^*\| = 1$, $\langle f^*, \phi\rangle = 0$, $\langle f^*, \phi^p\phi'\rangle = 0$, and $\langle \mathcal{L}f^*, f^*\rangle = 0$. A consequence of the reasoning just put forward is that there exists at least one nontrivial critical point $(f^*, \gamma, \theta, \nu)$ for the Lagrange multiplier problem

$$\begin{cases} \mathcal{L}f = \gamma f + \theta\phi + \nu\phi^p\phi', \\ \text{subject to } \|f\| = 1, \; \langle f, \phi^p\phi'\rangle = 0, \; \text{ and } \; \langle f, \phi\rangle = 0. \end{cases} \tag{6.19}$$

Using the fact $\langle \mathcal{L}f^*, f^*\rangle = 0$, it is easily seen that (6.19) implies $\gamma = 0$. Moreover, since $\mathcal{L}\phi' = 0$, we have that $\langle \mathcal{L}f^*, \phi'\rangle = \langle f^*, \mathcal{L}\phi'\rangle = \nu\int(\phi')^2\phi^p dx = 0$, which implies $\nu = 0$. It is thereby concluded that

$$\mathcal{L}f = \theta\phi$$

has nontrivial solutions (f^*, θ) satisfying the constraints. But if χ is the auxiliary function arising in the proof of Lemma 6.7, we have that $\mathcal{L}\chi = \phi$ and so $\mathcal{L}(f^* - \theta\chi) = 0$. Then $f^* - \theta\chi \in \mathcal{N}(\mathcal{L})$. Since $p < 4$, it follows from (6.16) or (6.17) that $\langle \chi, \phi\rangle \neq 0$, and so $\theta = 0$. Therefore, for some nonzero $\lambda \in \mathbb{R}$, it is true that $f^* = \lambda\phi'$, which is a contradiction since such a function cannot be orthogonal to $\phi^p\phi'$. Therefore, the minimum in (6.18) is positive and the proof of the lemma is completed. ☐

We note that from (6.18) and from the specific form of $\mathcal{L}$, we have that if $f \in H^1(\mathbb{R})$ satisfies $\langle f, \phi\rangle = 0$ and $\langle f, \phi^p\phi'\rangle = 0$, then

$$\langle \mathcal{L}f, f\rangle \geqq \beta_0\|f\|_1^2.$$

Continuation of the proof of Theorem 6.1. We now turn our attention to a way of estimating the term $\langle \mathcal{L}v, v\rangle$ in (6.6), where v satisfies the compatibility relation (6.11). Initially we consider the normalization

$$\|u_0\| = \|\phi\|.$$

Then from (4.10), $\int[u(x,t)]^2 dx = \int \phi^2 dx$ for every t in the interval of time $[0, T]$, where T is determined by Lemma 6.3, and so from relation $u(x + \gamma(t), t) = \phi(x) + v(x)$, it follows that $\int v^2(t) + 2v(t)\phi dx = 0$. Then

$$\langle v(t), v(t)\rangle = -2\langle v(t), \phi\rangle. \tag{6.20}$$

Next, we suppose without loss of generality that $\|\phi\| = 1$. Define $v_{||}$ and $v_\perp$ to be

$$v_{||} = \langle v, \phi\rangle\phi \quad \text{and} \quad v_\perp = v - v_{||}.$$

It follows then from the properties of v in (6.11) and (6.20) that for $t \in [0, T]$,

$$v_{||} = -\frac{1}{2}\|v\|^2\phi, \quad v_\perp = v + \frac{1}{2}\|v\|^2\phi,$$

$$\langle v_\perp, \phi\rangle = 0, \quad \langle v_\perp, \phi^p\phi'\rangle = \int_{-\infty}^{\infty}(v + \frac{1}{2}\|v\|^2\phi)\phi^p\phi' dx = 0.$$

Thus, from Lemma 6.9, the Cauchy-Schwarz inequality, and the properties of $v, v_{||}$, and $v_\perp$ that

$$\text{(6.21)} \quad \begin{cases} \langle \mathcal{L}v_\perp, v_\perp\rangle \geqq \beta\|v_\perp\|^2 = \beta\|v\|^2 - \beta\frac{1}{4}\|v\|^4\|\phi\|^2 \geqq \beta\|v\|^2 - \widetilde{\beta}_3\|v\|_1^4, \\ \langle \mathcal{L}v_{||}, v_{||}\rangle = \frac{1}{4}\|v\|^4\langle \mathcal{L}\phi, \phi\rangle \geqq -\widetilde{\beta}_4\|v\|_1^4, \\ \langle \mathcal{L}v_{||}, v_\perp\rangle = -\frac{1}{2}\|v\|^2\ \langle \mathcal{L}\phi, v_\perp\rangle \geqq -\widetilde{\beta}_5\|v\|^3 \geqq -\widetilde{\beta}_2\|v\|_1^3, \end{cases}$$

where $\widetilde{\beta}_i > 0$, $i = 2, 3, 4, 5$. We note that the second inequality in (6.21) follows from the property that $\langle \mathcal{L}\phi, \phi\rangle < 0$. Hence, with (6.21) in mind, it follows easily from the specific form of the operator $\mathcal{L}$ that for each $t \in [0, T]$,

$$\text{(6.22)} \qquad \langle \mathcal{L}v(t), v(t)\rangle \geqq \beta_0\|v(t)\|_1^2 - \beta_1\|v(t)\|_1^3 - \beta_2\|v(t)\|^4,$$

which is the desired estimate (6.7) with $\beta_i > 0$.

Next, using the fact that the norm of $H^1(\mathbb{R})$ and the norm $\|f\|_{1,c}^2 = \|f\|^2 + c\|f'\|^2$ are equivalent, it follows from (6.6) and (6.22) that

$$\text{(6.23)} \qquad \Delta\mathcal{B}(t) \geqq \eta\|v(t)\|_{1,c}^2 - \sum_{k=3}^{p+2} C_k(p,c)\|v\|_{1,c}^k \equiv g(\|v\|_{1,c})$$

where

$$g(s) = \eta s^2 - \sum_{k=3}^{p+2} C_k(p,c)s^k,$$

with η and $C_k(p,c) > 0$. Note that $\inf_{r\in\mathbb{R}} \Omega_t(r) = \|v(t)\|_{1,c}^2$ for $t \in [0,T]$, $g(0) = 0$, and $g(s) > 0$ for s small. Now, from the equality

$$\Delta\mathcal{B}(t) = \Delta\mathcal{B}(0) = \mathcal{B}(u_0) - \mathcal{B}(\phi)$$

and from the continuity of $\mathcal{B}$ on the set $S = \{f \in H^1(\mathbb{R}) : \ \|f\| = \|\phi\|\}$, it follows that given $\epsilon > 0$, there exists $\delta_2 > 0$ such that if $u_0 \in S$ and $\|u_0 - \phi\|_{1,c} < \delta_2 < \epsilon$, then

$$g(\|v(t)\|_{1,c}) \leqq \Delta\mathcal{B}(0) < \ g(\epsilon), \qquad t \in [0,T].$$

Since $t \to \|v(t)\|_{1,c}$ is a continuous function of $t \in [0,T]$ (see Lemma 2 in [**58**]) and $\|v(0)\|_{1,c} = \|u_0 - \phi\|_{1,c} < \epsilon$, it follows from the specific form of g that $\|v(t)\|_{1,c} < \epsilon$, provided that δ_2 is chosen small enough at the outset. So, by definition of v we obtain the orbital stability of the solitary wave solution ϕ_c with regard to the flow of the GKdV equations with $p < 4$, at least on the interval of time $[0,T]$, namely,

$$\text{(6.24)} \quad \inf_{r\in\mathbb{R}} \|u(t) - \phi_c(\cdot + r)\|_{1,c}^2 = \|u(\cdot + \gamma(t), t) - \phi_c\|_{1,c}^2 = \|v(t)\|_{1,c}^2 < \epsilon^2$$

with $t \in [0,T]$.

Now we see that the inequality in (6.24) is still true for all $t > 0$. Following Bona [**58**], let

$$\mathcal{A} = \{t|\ \text{the infimum in (6.9) is attained at finite values of } \gamma\ \}.$$

As has been shown above, $[0,T] \subset \mathcal{A}$. Let T_1 be the largest value such that $[0,T_1) \subset \mathcal{A}$ and suppose that $T_1 < \infty$. Then from (6.24) and choosing

$$\epsilon^2 < \|(-\partial_x^2 + c)^{1/2}\phi\|^2/2,$$

we obtain that

$$\inf_{r\in\mathbb{R}} \Omega_t = \|v(t)\|_{1,c}^2 < \frac{1}{2}\,\|(-\partial_x^2+c)^{1/2}\phi_c\|^2.$$

Since $t\in(0,\infty)\to\inf_{r\in\mathbb{R}}\ \Omega_t(r)$ is continuous, there is a $T_0>0$ such that for $t\in[T_1,T_1+T_0]$

$$\inf_{r\in\mathbb{R}} \Omega_t < \frac{3}{4}\,\|(-\partial_x^2+c)^{1/2}\phi_c\|^2.$$

So, Lemma 6.2 implies that the infimum in (6.9) is taken at finite values of γ_0 and this contradicts the choice of T_1. Therefore, $T_1=\infty$ and so Theorem 6.1 is established if $\|u_0\|=\|\phi_c\|$.

Now, attention is diverted to the general case wherein the initial data is not necessarily such that $\|u_0\|=\|\phi_c\|$. First, we remark that if ϕ_c is a solution of (3.3) for $c>0$, then $R(\xi)=c^{-\frac{1}{p}}\phi_c(c^{-\frac{1}{2}}\xi)$ satisfies

$$-R''+R-\frac{1}{p+1}R^{p+1}=0.$$

Moreover,

$$\|R\|^2 = c^{\frac{p-4}{2p}}\,\|\phi_c\|^2. \tag{6.25}$$

Note that for $p=4$ in (6.25), the L^2-norm of ϕ_c is always conserved for each $c>0$. Next, let $u_0\in H^1(\mathbb{R})$ obey the restriction

$$\|u_0-\phi_c\|_{1,c}<\delta_0$$

where further restrictions on δ_0 will be determined presently. Initially, we choose $\delta_0<\frac{1}{3}\delta_2$ and c_1 such that

$$\|\phi_{c_1}\|^2 = c_1^{\frac{4-p}{2p}}\,\|R\|^2 = \|u_0\|^2.$$

The idea now is to apply the preceding theory to the case $c=c_1$, and then by using the triangle inequality, we can conclude the desired result for the given values of c and u_0. Initially we will see that $\|u_0-\phi_{c_1}\|_{1,c_1}<\delta_2$. In fact, the continuity of the mapping $c\in(0,\infty)\to\phi_c\in(H^1(\mathbb{R}),\|\cdot\|_{1,c_1})$ assures the existence of a positive number $0<\eta_1<c$ such that if $|c_0-c_1|<\eta_1$, then $\|\phi_{c_0}-\phi_{c_1}\|_{1,c_1}<\delta_2/3$. We have now determined that we can choose δ_0 small enough such that $|c_1-c|<\eta_1$. In fact, from the choice of c_1 it follows that

$$\begin{aligned}\Big|\frac{c_1^{2/p}}{c_1^{1/2}}-\frac{c^{2/p}}{c^{1/2}}\Big| &= \frac{1}{\|R\|^2}\Big|\|u_0\|^2-\|\phi_c\|^2\Big| \leqq \frac{1}{\|R\|^2}\|u_0-\phi_c\|(\|u_0\|+\|\phi_c\|)\\ &\leqq \frac{1}{\|R\|^2}\frac{\delta_0}{\sqrt{c}}\Big(\frac{\delta_0}{\sqrt{c}}+2\|\phi_c\|\Big) \leqq \frac{1}{\|R\|^2}\Big(\frac{1}{c}+\frac{2\|\phi_c\|}{\sqrt{c}}\Big)\delta_0 \equiv K(R,c)\delta_0.\end{aligned} \tag{6.26}$$

So, by choosing δ_0 small in (6.26), we have that $|c_1-c|<\eta_1$ and therefore $|\frac{c_1}{c}|\leqq 2$ and $|\frac{c}{c_1}-1|\leqq 1$. Then, from the triangle inequality we have that

$$\begin{aligned}\|u_0-\phi_{c_1}\|_{1,c_1} &\leqq \|u_0-\phi_c\|_{1,c_1}+\|\phi_c-\phi_{c_1}\|_{1,c_1}\\ &\leqq \max\Big\{1,\frac{c_1}{c}\Big\}\|u_0-\phi_c\|_{1,c}+\|\phi_c-\phi_{c_1}\|_{1,c_1} < 2\delta_0+\frac{\delta_2}{3}<\delta_2.\end{aligned}$$

Therefore, applying the preceding theory to the case $c = c_1$, it follows that if $u(x,t)$ is the solution of the GKdV equations with initial data $u(x,0) = u_0(x)$, then

$$\begin{aligned}\inf_{\gamma\in\mathbb{R}} \|u(\cdot+\gamma,t)-\phi_c\|_{1,c} &\leqq \max\left\{1,\frac{c}{c_1}\right\}\inf_{\gamma\in\mathbb{R}} \|u(\cdot+\gamma,t)-\phi_c\|_{1,c_1}\\ &\leqq \|\phi_c-\phi_{c_1}\|_{1,c_1} + 2\inf_{\gamma\in\mathbb{R}} \|u(\cdot+\gamma,t)-\phi_{c_1}\|_{1,c_1} < \frac{\delta_2}{3} + 2\epsilon < 3\epsilon.\end{aligned}$$

We have now proven Theorem 6.1. □

REMARK 6.10. We note that using the strictly increasing property of the mapping $c \to \|\phi_c\|^2$ and the triangle inequality, we can also obtain the case when the initial data is not necessarily such that $\|u_0\| = \|\phi_c\|$ (see Lemma 7.9).

The method used to prove Theorem 6.1 can be applied to a large variety of nonlinear evolution equations or systems. For instance, we can apply it to the Benjamin-Ono equation (2.6) ([**49**]), the interaction equation between a short and long wave given by (2.15) ([**35**]), and the nonlinear Schrödiger equation (Cazenave [**79**], Weinstein [**273**])

$$iu_t + \Delta u + |u|^p u = 0.$$

We remember that in this last example the travelling wave solutions in question are of the form $u(x,t) = e^{i\omega t}\phi_\omega$, $\omega > 0$, where ϕ_ω is a positive, radial, smooth solution. The orbit to be studied is the set

$$\Omega_{\phi_\omega} = \{e^{iy}\phi_\omega(\cdot+\gamma) : y \in [0,2\pi),\ \ \gamma \in \mathbb{R}^n\}.$$

In this case it is known that Ω_{ϕ_ω} is stable if $0 < p < 4/n$.

If we consider the general class of nonlinear evolution equations of Korteweg-de Vries-type given in (2.1), namely,

$$u_t + (f(u))_x - Mu_x = 0$$

where M is defined as a multiplier operator with symbol $\alpha(\xi)$ (see (2.2)–(2.3)), we can infer from the proof of Theorem 6.1 that three basic pieces of information are necessary to get a stability theory of solitary wave solutions associated with this general model, namely,

(1) a local or global well-posedness theory of the initial value problem associated with (2.1),

(2) existence of solutions of the pseudo-differential equation

$$M\phi(\xi) + c\phi(\xi) - f(\phi(\xi)) = 0,$$

with c in a determined domain, and

(3) the fact that the following linear operator, defined on a dense subset of $L^2(\mathbb{R})$,

$$\mathcal{L} \equiv M + c - f'(\phi),$$

satisfies the following properties: it has a single negative eigenvalue which is simple, the zero eigenvalue is simple with eigenfunction ϕ', and the remainder of the spectrum of $\mathcal{L}$ is positive and bounded away from zero (see Theorem B.43 in Appendix B, which contains a spectral study of this operator).

As we will see in the next chapter, these three basic assumptions are the heart of a general theory of stability established by Grillakis, Shatah, and Strauss in [**123**] and [**124**]. Moreover, from relation (6.16) it is inferred that if we have a curve

$c \to \phi_c$ of class C^1 of solitary wave solutions for (2.1), then a sufficient condition for stability will be

$$\frac{d}{dc}\|\phi_c\|^2 > 0.$$

(see Theorem 7.1 in Chapter 7).

6.3. "Stability of the Blow-up" for a Class of KdV Equations

In this section we illustrate that the local method established in the previous section can be used, by obtaining information of the large-time asymptotic behavior of solutions, for the following class of Korteweg-de Vries equations :

$$u_t + u^p u_x - D^\beta u_x = 0 \tag{6.27}$$

on $\mathbb{R}$ and $\widehat{Du}(\xi) = |\xi|^\beta \widehat{u}(\xi)$ with $\beta \geqq 1$. Initially, via the *Concentration-Compactness Method* (see Lions [**193**], [**194**], Albert [**8**], Weinstein [**276**] or Chapter 9 of this book), we can prove the existence of solitary wave solutions of the form $u(x,t) = S(x - \eta t)$ of (6.27) for every wave speed $\eta > 0$, where $S = S_\eta$ satisfies the nonlinear equation

$$D^\beta S + \eta S - \frac{1}{p+1} S^{p+1} = 0. \tag{6.28}$$

As we will see in the next chapter (Theorem 7.11), the orbit generated by these solitary wave solutions will be stable when $p < p_{\text{crit}} = 2\beta$; meanwhile, they are unstable if $p > p_{\text{crit}}$ ([**64**] or Chapter 10, Theorem 10.19, of this book).

For the class of equations (6.27), it is well known that the initial value problem is globally well-posed provided the exponent p is less than the same critical power $p_{\text{crit}} = 2\beta$ that arises in the question of orbital stability (see Chapter 5). For $p \geqq p_{\text{crit}}$, it is suspected for equation (6.27) that large data need not lead to globally defined solutions. An affirmative result in this issue has been shown recently by Merle in [**220**], where the existence of blow-up solutions for the critical generalized KdV equation was established. More precisely, there exists $\alpha > 0$ such that the following property is true: let $u_0 \in H^1(\mathbb{R})$ and $u(t)$ be a solution of (6.27) with $\beta = 2$ and $p = 4$. Assume that

$$E_{\text{GKdV}}(u_0) = \int_{-\infty}^{\infty} [(u_0')^2 - \frac{1}{15} u_0^6]\, dx < 0 \quad \text{and} \quad \|u_0\|^2 < \|S\|^2 + \alpha$$

with S satisfying (6.28). Then the solution $u(t)$ blows up in $H^1(\mathbb{R})$ in finite or infinite time. Namely, there is $t^* > 0$ such that

$$\lim_{t\to t^*} \|u(t)\|_1 = +\infty.$$

There are some results providing information about the large-time asymptotic behavior of solutions of equation (6.27) in the critical case $p_{\text{crit}} = 2\beta$ ([**30**], [**176**], [**215**], [**273**], [**275**]). It is our purpose here to show that the method established to obtain Theorem 6.1 can be used to get a result of "*stability of the blow-up*" near to unstable solitary wave solutions for equations (6.27) in the case $p = p_{\text{crit}}$. We will give the basic ingredients to obtain the main result, Theorem 6.12 below, and we refer the reader to the recent work of Angulo, Bona, Linares, and Scialom [**30**] to see a more accurate proof of the results.

Initially, let $\mathcal{L}$ be the linear, self-adjoint, closed, unbounded operator defined on the dense subset $H^{\beta}(\mathbb{R})$ of $L^2(\mathbb{R})$, given by

$$\mathcal{L} = D^{\beta} + \eta - S^{p}. \tag{6.29}$$

Suppose $\mathcal{L}$ and S satisfy the following properties.

(H_0) The positive solitary wave solution $S = S_\eta$ of (6.28) in $H^{\frac{\beta}{2}}(\mathbb{R})$ is unique up to sign and translations.

(H_1) The operator $\mathcal{L}$ has a single negative eigenvalue which is simple, with eigenfunction $g_\eta > 0$, the zero eigenvalue is simple with eigenfunction S'_η, and the remainder of the spectrum of $\mathcal{L}$ is positive and bounded away from zero.

(H_2) The curve $\eta \to S_\eta$ is C^1 with values in $H^{1+\frac{\beta}{2}}(\mathbb{R})$.

Initially, we have a global well-posedness theory associated with equations (6.27) when $p = 2\beta$.

THEOREM 6.11. *Let $p > 0$ be an integer and suppose that $p = 2\beta$. If $\beta \geqq 2$, then the initial value problem associated with* (6.27) *in $H^{\frac{\beta}{2}}(\mathbb{R})$ is globally well-posed in the open ball $B_r(0)$ in $H^{\frac{\beta}{2}}(\mathbb{R})$ for $r = \|S\|$, where S is the solitary wave profile that solves* (6.28). *Moreover, if $u_0 \in \overline{B_r(0)}$, then E defined by*

$$E(u) = \frac{1}{2}\int_{-\infty}^{\infty}\left[uD^{\beta}u - \frac{2}{(p+1)(p+2)}u^{p+2}\right]dx \tag{6.30}$$

satisfies $E(u_0) \geqq 0$.

PROOF. The proof is based on the following sharp inequality of Gagliardo–Nirenberg type (see Theorem 2.2 in [**30**]). Let $\beta \geqq 1$. If $f \in H^{\frac{\beta}{2}}(\mathbb{R})$, then for any $r \geq 2$, $f \in L^r(\mathbb{R})$, and there is a constant $C_{r,\beta}$ such that

$$|f|_r^r \leqq C_{r,\beta}\|D^{\frac{\beta}{2}}f\|^{\frac{r-2}{\beta}}\|f\|^{\frac{2+r(\beta-1)}{\beta}}. \tag{6.31}$$

The smallest constant $C_{r,\beta}$ for which this inequality is valid is

$$C_{r,\beta} = \frac{r\beta}{2+r(\beta-1)}\left[\left(\frac{2+r(\beta-1)}{r-2}\right)^{\frac{1}{\beta}}\frac{1}{\|\Psi\|^2}\right]^{\frac{r-2}{2}} \tag{6.32}$$

where Ψ is a function in $H^{\frac{\beta}{2}}(\mathbb{R})$ satisfying

$$D^{\beta}\Psi + \Psi - |\Psi|^{r-2}\Psi = 0. \tag{6.33}$$

Next, since the $L^2(\mathbb{R})$-norm of a solution u of the IVP associated to (6.27) is conserved, it is only necessary to establish an appropiate *a priori* bound for $\|D^{\beta/2}u(t)\|$. Observe that the conserved functional E in (6.30) implies

$$\frac{1}{2}\|D^{\beta/2}u(t)\|^2 \leqq E(u_0) + \frac{1}{(p+1)(p+2)}\int_{-\infty}^{\infty}|u(t)|^{p+2}dx.$$

Using the inequality above with $r = p+2$, a calculation shows that

$$\|D^{\beta/2}u(t)\|^2\left[1 - \left(\frac{\|u_0\|}{\|S\|}\right)^{2\beta}\right] \leqq 2E(u_0),$$

and the result follows. □

Now, we introduce the auxiliary functions

$$\psi(x,t) = \mu(t)^{-\frac{1}{2}} u(\mu(t)^{-1}x, t) \tag{6.34}$$

where

$$\mu(t) = \frac{||D^{\beta/2}u(t)||^{2/\beta}}{||D^{\beta/2}S||^{2/\beta}}, \tag{6.35}$$

$\mu(0) = 1$, and $0 \leqq t < t^*$ with t^* the maximal time of existence of the solution of (6.27) under consideration (if the solution is global, $t^* = +\infty$). Note that unless u is the zero-solution, $\mu(t) \in (0, \infty)$ for $0 < t < t^*$. The normalization $\mu(0) = 1$ is a temporary one made to simplify the presentation of the argument. It will be dispensed with later. By using E defined in (6.30), it is easy to check that the function ψ verifies the identities

$$(1) \quad \|\psi(\cdot,t)\| = \|u(\cdot,t)\| = \|u_0\|, \tag{6.36}$$

$$(2) \quad \langle \psi(\cdot,t), D^\beta \psi(\cdot,t)\rangle = \langle S, D^\beta S\rangle, \tag{6.37}$$

$$(3) \quad E(\psi(\cdot,t)) = \frac{1}{\mu(t)^\beta} E(u(\cdot,t)). \tag{6.38}$$

Since the stability considered here is with respect to form, i.e., up to translation in space, we introduce the pseudo-metric

(6.39)

$$\rho_\eta(\psi(\cdot,t),S)^2 = \inf_{r\in\mathbb{R}}\{||D^{\beta/2}\psi(\cdot+r,t) - D^{\beta/2}S(\cdot)||^2 + \eta||\psi(\cdot+r,t) - S(\cdot)||^2\}$$

on $H^{\beta/2}(\mathbb{R})$. Define the set $\mathcal{K}$ to be either

$$\mathcal{K} = \{u_0 : u_0 \in H^{\beta/2}(\mathbb{R}) \text{ and } E(u_0) \leqq 0\}$$

if $\beta \geqq 2$ (see Theorem 4.3) or, if $1 \leq \beta < 2$,

$$\mathcal{K} = \{u_0 : u_0 \in H^s(\mathbb{R}) \text{ and } E(u_0) \leqq 0\}$$

where $s > 3/2$ is fixed (see Theorem 4.2), but otherwise arbitrary.

The next theorem is a stability result which belongs to the spatial structure of the solutions of (6.27) in the critical case $p = 2\beta$.

THEOREM 6.12. *Let $p \geq 1$ be an integer and let $p = 2\beta$. For $\eta > 0$, let $S = S_\eta$ be a solitary-wave solution of* (6.28). *For any $\epsilon > 0$ there is a $\delta = \delta(\epsilon) > 0$ such that if $u_0 \in \mathcal{K}$ with $\rho_\eta(u_0, S) < \delta$ and if u is the solution of* (6.27) *corresponding to the initial value u_0, then $u \in C([0,t^*); H^{\beta/2}(\mathbb{R}))$ and*

$$\begin{aligned}&\inf_{r\in\mathbb{R}}\Big\{\eta\|u(\cdot,t) - \mu(t)^{\frac{1}{2}} S(\mu(t)(\cdot - r))\|^2 \\ &\qquad + \frac{1}{\mu(t)^\beta}\|D^{\beta/2}u(\cdot,t) - \mu(t)^{\frac{1}{2}} D^{\beta/2}S(\mu(t)(\cdot - r))\|^2\Big\} < \epsilon\end{aligned} \tag{6.40}$$

for all $t \in [0,t^*)$, where t^* is the maximal existence time for the solution u and μ is as in (6.35).

REMARK 6.13. We observe about Theorem 6.12 that for $\beta = 2$ (the critical generalized KdV equation) a result similar to this one was established by Martel and Merle [**215**] by making use of a concentration-compactness argument as in Weinstein [**273**]. The difference here, other than the broader range of β, is two-fold. In [**215**], the parameter $\lambda(t)$ which corresponds to our $\mu(t)$ is defined implicitly via a minimization problem or by using the Implicit Function Theorem. Here, the function $\mu(t)$ is defined directly in term of the solution u. Thus our proof may be made by direct calculation and thereby includes the case $E(u_0) = 0$ as well as the case $E(u_0) < 0$.

PROOF. Suppose at the outset that $\mu(0) = 1$. The proof is based on the time-dependent functional

$$B_t[u] = \frac{1}{\mu(t)^\beta} E(u(\cdot,t)) + \frac{\eta}{2}\left(\frac{||u(\cdot,t)||}{||S||}\right)^{2k} (||u(\cdot,t)||^2 - ||S||^2)$$

where $k \in \mathbb{N}$ will be chosen later. From the definition of B_t, it is clear that if u is a solution of (6.27), then $B_t[u] = B_t[u_0]$. Using (6.34), (6.35), and (6.36), (6.38), we may write $B_t[u]$ in terms of ψ as:

$$\tilde{B}_t[\psi] = E(\psi(\cdot,t)) + \frac{\eta}{2}\left(\frac{||\psi(\cdot,t)||}{||S||}\right)^{2k} (||\psi(\cdot,t)||^2 - ||S||^2) \tag{6.41}$$

where the explicit dependence on μ disappears. As will be argued presently, if it is established that, modulo translations, the inequalities

$$\Delta\widetilde{B}_t \leqq c_0\,||u_0 - S|| \tag{6.42}$$

and

$$\begin{aligned} \Delta\widetilde{B}_t \geqq\ & c_1\,||\psi(\cdot,t) - S||^2_{\beta/2} - c_2 \sum_{j=1}^{p-1} ||\psi(\cdot,t) - S||^{j+2}_{\beta/2} \\ & - \sum_{j=1}^{2k} c_{k,j}\,||\psi(\cdot,t) - S||^{j+2}_{\beta/2} \end{aligned} \tag{6.43}$$

hold for $\Delta\widetilde{B}_t = \widetilde{B}_t[\phi] - \widetilde{B}_t[S]$, where $c_i, c_{k,j}$ are fixed constants, then the result in Theorem 6.12 follows in the same form as in Theorem 6.1. Hence, attention is turned to establishing these bounds. The upper bound (6.42) is a straightforward consequence of $E(u_0) \leqq 0$ and $E(S) \geqq 0$, where the constant c_0 depends on $\|S\|$ (and on an upper bound for the choice of δ). To prove (6.43), consider the perturbation of the solitary wave solution S

$$\psi(x+\gamma,t) = S(x) + a(x,t), \tag{6.44}$$

where a is a real function and $\gamma = \gamma(t)$ minimizes the functional

$$\Pi_t(\gamma) = ||D^{\beta/2}\psi(\cdot+\gamma,t) - D^{\beta/2}S(\cdot)||^2 + \eta||\psi(\cdot+\gamma,t) - S(\cdot)||^2.$$

Using the representation (6.44), one calculates that

$$\begin{aligned} \Delta\widetilde{B}_t &= \widetilde{B}_t[S+a] - \widetilde{B}_t[S] \\ &= E(S+a) - E(S) + \frac{\eta}{2}\left(\frac{||S+a||}{||S||}\right)^{2k} (||S+a||^2 - ||S||^2) \\ &\geqq \frac{1}{2}\langle \mathcal{L}a, a\rangle + \frac{2k\eta}{||S||^2}\langle a, S\rangle^2 - c_2(\eta)\sum_{j=1}^{p-1} ||a||^{j+2}_{\beta/2} - \sum_{j=1}^{2k} c_{k,j}(\eta)||a||^{j+2}_{\beta/2}. \end{aligned} \tag{6.45}$$

The inequality in (6.45) is obtained using the definition (6.29) of $\mathcal{L}$, the Cauchy-Schwarz inequality, and interpolation.

A suitable lower bound on the quadratic form $\mathcal{L}$ is the next order of business. Initially, since $\gamma = \gamma(t)$ minimizes the functional $\Pi_t(\gamma)$ (see Lemma 6.2), we have the compatibility relation on a, namely,

$$\int_{-\infty}^{\infty} S^p(x)S'(x)a(x,t)\,dx = 0 \tag{6.46}$$

for all t in an interval of time $[0, T]$.

The issue of obtaining the lower bound (6.43) for the right-hand side of inequality (6.45) is addressed in the next few lemmas.

LEMMA 6.14. *Let $\beta \geq 1$ and let $\mathcal{L} = D^\beta + \eta - S^p$ satisfy hypothesis (H_1). Then there exists $\sigma < 0$ such that if $\widetilde{h} = S - \sigma D^\beta S$, then*

$$\min_{\langle f,\widetilde{h}\rangle=0, ||f||=1} \langle \mathcal{L}f, f\rangle = 0.$$

PROOF. For any given value σ, define the function f_0 by

$$f_0(x) = -\frac{1}{2\beta\eta}S(x) - \frac{1+\eta\sigma}{\beta\eta}xS'(x).$$

Then using the relations $\mathcal{H}(xS'') = x\mathcal{H}S''$, $D^{\beta-1}(x\mathcal{H}S'') = (\beta-1)D^\beta S + xD^\beta S'$ and (6.29), we obtain that $\mathcal{L}f_0 = S - \sigma D^\beta S = \widetilde{h}$ and, consequently, that

$$\langle f_0, \mathcal{L}f_0\rangle = \langle f_0, \widetilde{h}\rangle = \left(\frac{1}{2\beta}||S||^2 + \frac{1}{2\eta}||D^{\beta/2}S||^2\right)\sigma + \left(\frac{\beta-1}{2\beta}||D^{\beta/2}S||^2\right)\sigma^2.$$

It is thus obvious that for small negative values of σ, it is possible to have both

$$\langle \widetilde{h}, g_\eta\rangle = \int g_\eta S\,dx - \sigma\int g_\eta D^\beta S\,dx \neq 0$$

and

$$\langle \mathcal{L}^{-1}\widetilde{h}, \widetilde{h}\rangle = \langle f_0, \widetilde{h}\rangle < 0.$$

Since $\mathcal{N}(\mathcal{L}) = \text{Span}\{S'\}$ and $\widetilde{h} \in \mathcal{N}(\mathcal{L})^\perp$, it follows immediately from Lemma 6.4 that

$$\theta = \min\Big\{\langle \mathcal{L}f, f\rangle : \ ||f|| = 1 \text{ and } \langle f, \widetilde{h}\rangle = 0\Big\} = 0.$$

This completes the proof of the lemma. □

LEMMA 6.15. *If $\widetilde{h} \equiv S - \sigma D^\beta S$ with $\sigma < 0$ chosen as in the last lemma, then*

$$\inf\Big\{\langle \mathcal{L}f, f\rangle : ||f|| = 1, \langle f, \widetilde{h}\rangle = 0, f \perp S^pS'\Big\} \equiv \nu > 0. \tag{6.47}$$

PROOF. See Lemma 3.6 in [**30**] or Lemma 6.9 above. □

Continuation of the proof of Theorem 6.12. Attention is now turned to estimating the term

$$\frac{1}{2}\langle \mathcal{L}a, a\rangle + \frac{2k\eta}{||S||^2}\langle a, S\rangle^2$$

in (6.45), where a satisfies the compatibility relation (6.46). We continue to carry over the notation from Lemma 6.14 and Lemma 6.15. In particular, σ is chosen so that the conclusions of Lemma 6.14 are valid. Define $a_{||}$ and $a_\perp$ to be

$$a_{||} = \frac{\langle a, \widetilde{h}\rangle}{||\widetilde{h}||^2}\widetilde{h} \quad \text{and} \quad a_\perp = a - a_{||}.$$

It follows from the properties of a and $\widetilde{h} = S - \sigma D^\beta S$ that $\langle a_\perp, \widetilde{h}\rangle = 0$ and

$$\int S^p S' a_\perp \, dx = 0.$$

Without loss of generality, take $\langle a, \widetilde{h}\rangle < 0$. Thus, from Lemma 6.15, the Cauchy-Schwarz inequality, and the properties of a, $a_\perp$, $a_{||}$, and $\widetilde{h}$, it follows that

$$\tag{6.48} \begin{cases} \langle \mathcal{L}a_\perp, a_\perp\rangle \geqq D_1 ||a_\perp||^2, \quad \langle \mathcal{L}a_{||}, a_{||}\rangle = \dfrac{||a_{||}||^2}{||\widetilde{h}||^2}\langle \widetilde{h}, \mathcal{L}\widetilde{h}\rangle, \\ \langle \mathcal{L}a_{||}, a_\perp\rangle = \dfrac{\langle a, \widetilde{h}\rangle}{||\widetilde{h}||^2}\langle \mathcal{L}\widetilde{h}, a_\perp\rangle \geqq -D_2 ||a_\perp|| \; ||a_{||}|| \end{cases}$$

for some positive constants D_1 and D_2. Identity (6.37) and elementary properties of Hilbert spaces imply that

$$-2\langle a, D^\beta S\rangle = ||D^{\beta/2} a||^2.$$

Thus, from the Cauchy-Schwarz inequality we obtain (remember, σ and $\langle a, \widetilde{h}\rangle$ are both negative)

$$\tag{6.49} \begin{aligned} \frac{2k\eta}{||S||^2}\langle a, S\rangle^2 &\geqq \frac{2k\eta}{||S||^2}\left(\langle a, \widetilde{h}\rangle^2 - \sigma\langle a, \widetilde{h}\rangle ||D^{\beta/2} a||^2\right) \\ &\geqq \frac{2k\eta}{||S||^2}||\widetilde{h}||^2 ||a_{||}||^2 + 2k\eta\sigma D_3 ||a||^3_{\beta/2}, \end{aligned}$$

with $D_3 > 0$. We choose $\theta > 0$ so that $D_1 - \theta D_2 \equiv D_4 > 0$. By Young's inequality,

$$||a_\perp|| ||a_{||}|| \leqq \theta ||a_\perp||^2 + \frac{1}{\theta}||a_{||}||^2.$$

Finally, fix k in such a way that

$$\frac{2k\eta}{||S||^2}||\widetilde{h}||^2 + \frac{\langle \widetilde{h}, \mathcal{L}\widetilde{h}\rangle}{||\widetilde{h}||^2} - \frac{D_2}{\theta} \equiv D_5 > 0.$$

With these choices, it follows from (6.48) and (6.49) that

$$\tag{6.50} \begin{aligned} \frac{1}{2}\langle \mathcal{L}a, a\rangle + \frac{2k\eta}{||S||^2}\langle a, S\rangle^2 &\geqq D_5 ||a_{||}||^2 + D_4 ||a_\perp||^2 + 2k\eta\sigma D_3 ||a||^3_{\beta/2} \\ &\geqq D' ||a||^2 - D'' ||a||^3_{\beta/2} \end{aligned}$$

for some positive constants D' and D''. With (6.50) in hand, it follows easily from the specific form of the operator $\mathcal{L}$ that

$$\tag{6.51} \frac{1}{2}\langle \mathcal{L}a, a\rangle + \frac{2k\eta}{||S||^2}\langle a, S\rangle^2 \geqq \widetilde{D}_1 ||a||^2_{\beta/2} - \widetilde{D}_2 ||a||^3_{\beta/2},$$

with $\widetilde{D}_1, \widetilde{D}_2 > 0$. Finally, using (6.51) in conjunction with (6.45), we obtain

$$\begin{aligned} \Delta \widetilde{B}_t &\geqq \widetilde{D}_1 ||a||^2_{\frac{\beta}{2}} - \widetilde{D}_2 ||a||^3_{\beta/2} - c_2(\eta)\sum_{j=1}^{p} ||a||^{j+2}_{\beta/2} - \sum_{j=1}^{2k} c_{k,j}(\eta) ||a||^{j+2}_{\beta/2} \\ &\geqq c_1 ||a||^2_{\beta/2} - c_2 \sum_{j=1}^{p} ||a||^{j+2}_{\beta/2} - \sum_{j=1}^{2k} c_{k,j} ||a||^{j+2}_{\beta/2} \end{aligned}$$

where $c_1, c_2, c_{k,j}$ are positive constants which depend only on η.

Next, suppose first that u_0 lies in the set $\mathcal{K}$ of "negative-energy" initial values and suppose $||u_0 - S||_{\beta/2} = \delta$. Then at least for $t \in [0,T]$, it follows from (6.42) and (6.43) that

$$q(\rho_\eta(\psi(\cdot,t),S)) \leqq \Delta\widetilde{B}_t \leqq c_0\delta \tag{6.52}$$

where

$$q(x) = c_1x^2 - c_2\sum_{j=1}^{p} x^{j+2} - \sum_{j=1}^{2k} c_{k,j}x^{j+2}.$$

Since

$$||a(\cdot,t)||^2_{\beta/2} = \rho_\eta(\psi(\cdot,t),S)^2$$

is a continuous function of $t \in [0,t^*)$ (see Lemma 2 in [**58**]), it follows from the inequality

$$q(\rho_\eta(\psi(\cdot,0),S)) \leqq c_0\delta \tag{6.53}$$

and (6.52) that given $\epsilon > 0$, then for all $t \in [0,T]$,

$$\rho_\eta(\psi(\cdot,t),S) \leqq \epsilon, \tag{6.54}$$

provided that δ is chosen small enough at the outset. To finish the proof, we need to show that inequality (6.54) is still true for $t \in [0,t^*)$. This part is shown using a method similar to that of the proof of Theorem 6.1. Therefore, the stability in Theorem 6.12 is established if $\mu(0) = 1$. The general case, wherein the initial data is not necessarily such that $\mu(0) = 1$, requires a little more work as in the case of Theorem 6.1, and therefore we refer the reader to the reference [**30**]. This completes the proof of Theorem 6.12. □

REMARK 6.16. We note that for the focusing nonlinear Schrödinger equations

$$iu_t + \Delta u + |u|^p u = 0$$

posed on $\mathbb{R}^n$, a similar result to Theorem 6.12 can be obtained in the critical case $p = \frac{4}{n}$. We refer the reader to the works of Angulo, Bona, Linares, and Scialom in [**29**] and [**30**].

6.4. Comments

In this section we establish a review of some improvements made in the stability theory for solitary wave solutions of Korteweg-de Vries-type equations.

(1) In the proof of Theorem 6.1, it was actually shown that there is a concrete choice of $\gamma = \gamma(t)$ (Lemma 6.3) for which

$$||u_x(\cdot+\gamma,t) - \phi'(\cdot)||^2 + c||u(\cdot+\gamma,t) - \phi(\cdot)||^2 = \inf_{r\in\mathbb{R}} \Omega_t(r) \leqq \epsilon \tag{6.55}$$

for all $t \in \mathbb{R}$. In fact, a choice of γ for which (6.55) holds may be determined via the orthogonality condition

$$\int_{-\infty}^{\infty} u(x+\gamma,t)\phi^p(x)\phi'(x)\,dx = 0. \tag{6.56}$$

By an application of the Implicit Function Theorem, Bona and Soyeur, in [**63**], have shown in the case of the KdV equation that as long as u satisfies (6.55), there is a unique continuously differentiable choice of the value $\gamma(t)$ that achieves (6.56). The principal result with regard to the behavior of the parameter γ in the case of the KdV equation is stated in the following result:

- (Bona and Soyeur in [**63**]) Let ϕ_c be a solitary wave solution of the KdV equation (2.4). For any $\epsilon > 0$, there exists $\delta = \delta(\epsilon) > 0$ such that if $u_0 \in H^1(\mathbb{R})$ and $\|u_0 - \phi_c\|_1 < \delta$, then there exists a C^1 map $\gamma : \mathbb{R} \to \mathbb{R}$ such that

$$\begin{array}{ll} (1) & \|u(\cdot + \gamma(t), t) - \phi_c\|_1 \leqq \epsilon \textit{ and} \\ (2) & |\gamma(t) - ct| \leqq C\epsilon, \end{array}$$

  for $t \in \mathbb{R}$, where C depends only on ϕ_c.

(2) Numerical simulations for solutions of many nonlinear dispersive wave propagations have provided the conjecture that a small perturbation of a stable solitary wave solution will resolve itself into a solitary wave whose speed is very nearly that of the unperturbed wave plus a small residual that separates from the bulk of the wave and is left behind. A first result in favor of this scenario, called *asymptotic stability* (see (4.6)), has been shown initially by Pego and Weinstein [**236**] in the case of the KdV equation and the modified KdV equation, for the initial data with exponential decay as $x \to +\infty$. To complete the exposition, we establish the following results of asymptotic stability:

- (Pego and Weinstein in [**236**]) Let $\phi_c(x - ct + \gamma)$, $c > 0, \gamma \in \mathbb{R}$, be a solitary wave solution of the KdV equation (2.4). Suppose $a \in (0, \sqrt{c/3})$ and $0 < b < a(c - a^2)$. Then there exists $A > 0$ such that if $\epsilon > 0$ is sufficiently small and we consider the initial value problem for the KdV equation with data $u_0(x) = \phi_c(x + \gamma) + v_0(x)$, such that $v_0 \in H^2(\mathbb{R})$ and $e^{ax}v_0 \in H^1(\mathbb{R})$ with

$$\|v_0\|_1 + \|e^{ax}v_0\|_1 < \epsilon,$$

  then there exist $c_+ > 0$, $\gamma_+ \in \mathbb{R}$ satisfying $|c - c_+| < A\epsilon$, $|\gamma - \gamma_+| < A\epsilon$ and for all $t \geqq 0$

$$\begin{aligned} &\|u(t) - \phi_{c_+}(\cdot - c_+t + \gamma_+)\|_1 \leqq A\epsilon \textit{ and} \\ &\|e^{ax}[u(\cdot + c_+t - \gamma_+, t) - \phi_{c_+}]\|_1 \leqq A\epsilon\, e^{-bt}. \end{aligned}$$

  Exactly the same result is true for the modified KdV equation.
- (Martel and Merle in [**211**]) In (6.4), let $p = 1, 2, 3$. There exists a $\delta_{0,c}$ such that for $\delta_0 \leqq \delta_{0,c}$, if $\|u_0 - \phi_c\|_1 \leqq \delta_0$, then there exists $\gamma(t)$ such that for $u(t)$, a solution of (6.4) with initial data u_0,

$$u(x + \gamma(t), t) \rightharpoonup \phi_{c_0} \quad \text{in } H^1(\mathbb{R}) \text{ as } t \to +\infty$$

  where $|c_0 - c| \leqq \epsilon(\delta_0)$ and $\epsilon(\delta_0) \to 0$ as $\delta_0 \to 0$.
- (Martel and Merle in [**216**]) In (6.4), let $p = 1, 2, 3$. There exists $\alpha_0 > 0$ such that if $u(t)$ is a global ($t \geqq 0$) $H^1(\mathbb{R})$ solution of (6.4) satisfying

$$\text{for all } t \geqq 0, \ \inf_{r \in \mathbb{R}} \|u(t) - \phi_c(\cdot - r)\|_1 < \alpha_0,$$

  then the following holds: There exist $t \to c(t) \in (0, +\infty)$, $t \to \rho(t) \in \mathbb{R}$ such that

$$u(t) - \phi_{c(t)}(\cdot - \rho(t)) \to 0 \quad \text{in } H^1(x > \frac{ct}{10}) \text{ as } t \to +\infty.$$

- (Merle and Vega in [**221**]) In (6.4), let $p = 1$. Assume that u is a solution of the KdV equation with initial data u_0. Then, we have
(1) *Orbital stability in* $L^2(\mathbb{R})$: Given $\epsilon > 0$, there exists $\delta > 0$ such that if $\|u_0 - \phi_c\| < \delta$, then for all $t \in \mathbb{R}$, there exists $\gamma(t)$ such that
$$\|u(\cdot + \gamma(t), t) - \phi_c\| < \epsilon.$$
(2) *Asymptotic stability in* $L^2(\mathbb{R})$: There is a $\delta_{0,c}$ such that for $\|u_0 - \phi_c\| < \delta_{0,c}$, there are c_0 and $\gamma(t)$ such that
$$u(x + \gamma(t), t) \to \phi_{c_0} \quad \text{in } L^2_{\text{loc}}(\mathbb{R}) \text{ as } t \to +\infty.$$

(3) We recall that the problem of the collision of two or more solitons (see [**2**] and [**3**]) is a classical problem in nonlinear wave propagation. In the case of the KdV and mKdV equations, it is well known that there exists an explicit multi-soliton solution describing the elastic collision of several solitons (see [**132**], [**179**], [**223**], [**204**]). Indeed, in the case of the KdV equation we have the explicit form of a 2-soliton with speeds c_1, c_2 in the following way:
$$u(x,t) = 12\frac{d^2}{dx^2}\log(1 + e^{2\eta_1(x,t)} + e^{2\eta_2(x,t)} + Ae^{2\eta_1(x,t)+2\eta_2(x,t)}),$$
where $\eta_i(x,t) = -b_i x + 4b_i^3 t$ for $i = 1, 2$, $A = (b_2 - b_1)^2/(b_1 + b_2)^2$, and $c_i = 4b_i^2$.

Because the second logarithmic derivative annihilates pure exponential functions, we find that as $t \to +\infty$, the double soliton splits into two separate 1-soliton's that are far apart in the following sense: if we consider
$$w(x,t) = u(x,t) - 2b_1^2\text{sech}^2 b_1(x - b_1^2 t) - 2b_2^2\text{sech}^2 b_2(x - b_2^2 t + \rho),$$
where $\rho = \log A$ is the phase shift, then
$$\lim_{t\to+\infty} \|w(t)\|_{k,p} = 0, \ \ 1 \leqq p \leqq \infty, \ k \in \mathbb{N},$$
where $\|\cdot\|_{k,p}$ denotes the norm in the Sobolev space $W^{k,p}(\mathbb{R})$.

In the case of the BO equation (Matsuno [**218**]), the double soliton with speeds $c_1 > 0$ and $c_2 > 0$, $c_1 < c_2$, is
$$u(x,t) = 4\frac{c_2\tau_1^2 + c_1\tau_2^2 + (c_1 + c_2)c_{12}}{(\tau_1\tau_2 - c_{12})^2 + (\tau_1 + \tau_2)^2},$$
where $\tau_i = c_i(x - c_i t)$, $i = 1, 2$, and $c_{12} = (c_1 + c_2)^2/(c_1 - c_2)^2$.

For a rigorous study on the large-time behavior and stability in the multi-soliton case for KdV-type equations, we refer to Maddocks and Sachs [**204**], Martel [**208**], Martel and Merle [**217**], Lopes and Neves [**202**], and reference therein.

CHAPTER 7

Grillakis-Shatah-Strauss's Stability Approach

7.1. Introduction

In this chapter we will give the basic ideas of the theory of stability of travelling wave solutions to equations of evolution in the Hamiltonian form

$$u_t = JE'(u(t)), \tag{7.1}$$

developed by Grillakis, Shatah, and Strauss in [**123**]. Our interest here will be to illustrate through specific examples how this theory works in the case of solitary wave solutions and periodic travelling waves. A wide variety of applications of this theory has been obtained in the last decade. Thus the theory can be applied to different equations or systems which appear in the physical descriptions of phenomena, for example, in the dynamic of fluid, internal waves; in nonlinear interactions in shallow-water ocean surface waves; in optical, hydrodynamical systems; and in plasma physics.

Our principal application here will be to establish a general theory of stability of solitary wave solutions associated with the following class of equations (see 2.1)

$$u_t + u^p u_x - Mu_x = 0 \tag{7.2}$$

where $u = u(x,t)$ is real valued, $x, t \in \mathbb{R}$, $p \in \mathbb{N}$, $p \geqq 1$, and M is a linear operator defined as a Fourier multiplier operator by $\widehat{Mu}(\xi) = \alpha(\xi)\widehat{u}(\xi)$. Here the symbol $\alpha(\xi)$ is a measurable, even function on $\mathbb{R}$ and satisfies

$$a_1|\xi|^{\beta_1} \leqq \alpha(\xi) \leqq a_2(1+|\xi|)^{\beta_2}, \quad \text{for } \xi \in \mathbb{R} \tag{7.3}$$

where $a_1, a_2 > 0$ and $\beta_2 \geqq \beta_1 \geqq 1$. We note that the general equation (7.2) contains the basic models of GKdV equations (2.5), GBO equations (2.7), and the intermediate long-wave ILW equation (2.8). By considering

$$c > \inf_{\xi\in\mathbb{R}} \alpha(\xi),$$

we obtain that $M + c$ represents a positive operator. The shape of the solitary wave solutions $u(x,t) = \phi_c(x - ct)$ of (7.2) must satisfy the equation

$$M\phi_c + c\phi_c - \frac{1}{p+1}\phi_c^{p+1} = 0. \tag{7.4}$$

7.2. Geometric Overview of the Theory

Suppose that E in (7.1) is a real-valued smooth functional defined on a real Hilbert space X embedded in a space of type L^2. We assume that the solutions of (7.1) are invariant under the symmetry of translations, namely, if $T(r)$ represents the translation operator defined for $\psi \in X$ by

$$T(r)\psi(x) = \psi(x + r),$$

we have that $T(r)u(x,t) = u(x+r,t)$ is a solution of (7.1) provided $u(x,t)$ is a solution.

We assume that (7.1) has a smooth curve $c \to \phi_c$ of solutions of travelling wave type

$$u(x,t) = \phi_c(x - ct) \tag{7.5}$$

for a real profile $\phi_c : \mathbb{R} \to \mathbb{R}$ and $c \in I \subseteq \mathbb{R}$. Moreover, we also suppose that there is another smooth functional $F : X \to \mathbb{R}$ such that E and F are invariants by translations and are conserved quantities by the flow of (7.1). We also assume that ϕ_c is a critical point for the functional $H = E + cF$, $H'(\phi_c) = 0$, for every c admissible. So, from the formal relation $H'(T(r)\phi_c) = T(r)H'(\phi_c)$, we obtain that

$$H'(T(r)\phi_c) = 0, \quad \text{for every } r \in \mathbb{R}. \tag{7.6}$$

Next, since $\{T(r)\}_{r\in\mathbb{R}}$ represents a one-parameter group of unitary operators on X with infinitesimal generator $T'(0) = \frac{d}{dx}$, it follows from (7.6) that for $\mathcal{L} = H''(\phi_c)$,

$$\mathcal{L}\Big(\frac{d}{dx}\phi_c\Big) = 0. \tag{7.7}$$

Therefore, we have that $\frac{d}{dx}\phi_c$ belongs to the kernel of the linear operator $\mathcal{L}$. This operator is closed, self-adjoint, unbounded, and defined on a specific dense subspace of L^2.

In Chapter 6 we saw that the main point to obtain the orbital stability of a specific travelling wave ϕ_c is to determine the following property:

$$\begin{cases} \text{there are } \eta > 0 \text{ and } D > 0 \text{ such that} \\ \qquad E(u) - E(\phi_c) \geqq D[d(u;\Omega_{\phi_c})]^2 \\ \text{for } d(u;\Omega_{\phi_c}) < \eta \text{ and } F(u) = F(\phi_c). \end{cases} \tag{7.8}$$

In other words, ϕ_c is a constraint local minimum of E. So, from (7.8) and from the continuity of the functionals E, F and of the flow $t \to u(t)$, we immediately obtain the stability of Ω_{ϕ_c} by initial perturbation in the manifold

$$\mathcal{M} = \{u : F(u) = F(\phi_c)\}.$$

For general perturbations of Ω_{ϕ_c} we use the continuity of the curve $c \to \phi_c$, the increasing property of $c \to F(\phi_c)$, and the triangular inequality (see Lemma 7.9).

Now, for obtaining (7.8), we need to get sufficient conditions for the quadratic form $\langle \mathcal{L}f, f\rangle$ to be positive defined. From the calculus of variations one possible condition is that

$$\langle \mathcal{L}f, f\rangle \geqq \beta\|f\|_X^2 \quad \text{for every } f \in T_{\phi_c}\mathcal{M}, \tag{7.9}$$

where $T_{\phi_c}\mathcal{M}$ is the tangent space to $\mathcal{M}$ in ϕ_c. Since the curve $t \to T(t)\phi_c$ belongs to $\mathcal{M}$ and

$$\frac{d}{dt}T(t)\phi_c|_{t=0} = \frac{d}{dx}\phi_c \in T_{\phi_c}\mathcal{M},$$

we have

$$\mathrm{Ker}(\mathcal{L}) \cap T_{\phi_c}\mathcal{M} \neq \{0\}.$$

Therefore condition (7.9) is not sufficient in our case. However, there are two further pieces of informations that in general emerge in our stability analysis. First, the range $R(\mathcal{L})$ is closed and so $\mathrm{Ker}(\mathcal{L})^\perp = R(\mathcal{L})$. Second, the null space $\mathrm{Ker}(\mathcal{L})$ is the tangent space to the family of critical point Ω_{ϕ_c} in ϕ_c. So we will see below that the condition

$$\langle \mathcal{L}f, f\rangle \geqq \beta\|f\|_X^2 \quad \text{for every } f \in T_{\phi_c}\mathcal{M} \cap \mathrm{Ker}(\mathcal{L})^{\perp} \tag{7.10}$$

is sufficient to imply (7.8). The direct check of condition (7.10), however, is in general extremely inconvenient, in part because no requirement is directly related to the number of negative eigenvalues of $\mathcal{L}$ and because in general this operator has a nontrivial negative eigenspace. In fact, consider

$$\mathcal{L} = M + c - \phi_c^p$$

and ϕ_c satisfying equation (7.4). Suppose that $\phi \quad _c > 0$; then

$$\langle \mathcal{L}\phi_c, \phi_c\rangle = -\frac{p}{p+1}\int \phi_c^{p+2}(x)dx < 0.$$

So the heart of this chapter will describe how the approach developed by Weinstein, Grillakis, Shatah, and Strauss [**276**], [**123**] finesses this difficulty and provides a computable test that guarantees when (7.10) is satisfied.

7.3. Stability of Solitary Wave Solutions

In this section we establish a theory of stability of the solutions of equation (7.4). In discussing the properties of equation (7.2), the linear space $\mathcal{W}$ defined by

$$\mathcal{W} = \{f \in L^2(\mathbb{R}) : \|f\|_{\mathcal{W}} = \Big(\int_{-\infty}^{\infty}[1+\alpha(\xi)]|\hat{f}(\xi)|^2 d\xi\Big)^{1/2} < +\infty\}$$

will be necessary. Its dual, $\mathcal{W}^*$, may be understood as the space of all tempered distributions Φ whose Fourier transform $\widehat{\Phi}$ is given by a measurable function for which

$$\|\Phi\|_{\mathcal{W}^*} = \Big(\int_{-\infty}^{\infty}\frac{|\widehat{\Phi}(\xi)|^2}{1+\alpha(\xi)}d\xi\Big)^{1/2} < +\infty.$$

Note that this is a natural extension of the classical Sobolev space $H^s(\mathbb{R})$. The pairing between $\mathcal{W}$ and $\mathcal{W}^*$, which will be used below, is denoted for $f \in \mathcal{W}$ and $\Phi \in \mathcal{W}^*$ as $\Phi(f)$, and it will be writen as (Φ, f). If Φ is given by an $L^2(\mathbb{R})$ function φ, then via the Plancherel theorem, $(\varphi, f) = \langle \varphi, f\rangle$ is the usual $L^2(\mathbb{R})$ inner product. We note that because of condition (7.3) on the symbol α of the operator M, the space $\mathcal{W}$ is continuously embedded in $H^{\frac{1}{2}}(\mathbb{R})$.

We note that by using the conservation law $E_M : \mathcal{W} \to \mathbb{R}$ defined in (4.10), with

$$G(u) = \frac{1}{(p+2)(p+1)}u^{p+2},$$

we can write equation (7.2) formally in the Hamiltonian form

$$u_t(t) = -\partial_x E'_M(u(t)). \tag{7.11}$$

Next we look at the basic assumptions of the theory of Grillakis *et al.* [**123**] in the case of equation (7.2).

Assumption 1 (Existence of solutions). There is a Banach space $(Y, \|\cdot\|_Y)$ which is continuously embedding in $\mathcal{W}$, such that for each $u_0 \in Y$ there exist $T = T(\|u_0\|_Y)$ and a unique solution $u \in C([-T,T]; Y)$ of (7.2) satisfying

(1) $u(0) = u_0$ and
(2) $E_M(u(t)) = E_M(u_0)$ and $F(u(t)) = F(u_0)$ for $t \in [0,T]$.

Assumption 2 (Existence of solitary wave solutions). There exist real ω_1, ω_2 such that

(1) the mapping $c \to \phi_c$ from the open interval (ω_1, ω_2) into $\mathcal{W} \subseteq H^{\frac{\beta_1}{2}}(\mathbb{R})$ is C^1 and
(2) $E'_M(\phi_c) + cF'(\phi_c) = 0$, *i.e.*, ϕ_c is a critical point of the functional $E_M(u) + cF(u)$.

Assumption 3 (Spectral structure). For each $c \in (\omega_1, \omega_2)$, the self-adjoint, closed, unbounded linear operator $\mathcal{L}_c$, defined on a dense subspace of $L^2(\mathbb{R})$ as

$$\mathcal{L}_c \equiv M + c - \phi_c^p, \tag{7.12}$$

enjoys the following spectral properties: it has a single negative eigenvalue which is simple, the zero eigenvalue is simple with eigenfunction ϕ'_c, and the remainder of the spectrum of $\mathcal{L}_c$ is positive and bounded away from zero.

For any $\eta > 0$ we define the following η-neighborhood, U_η, of the orbit Ω_{ϕ_c} generated by ϕ_c as

$$U_\eta = \{u : u \in \mathcal{W} \text{ and } \inf_{r \in \mathbb{R}} \|u - \tau_r \phi_c\|_{\mathcal{W}} < \eta\}.$$

Then, if we define the scalar function $d : (\omega_1, \omega_2) \to \mathbb{R}$ by

$$d(c) = E_M(\phi_c) + cF(\phi_c), \tag{7.13}$$

we have the following main theorem of orbital stability for (7.2).

THEOREM 7.1. *Assume Assumptions 1, 2, 3. Then the ϕ_c-orbit is $\mathcal{W}$-stable by the flow of equation* (7.2) *if the function $d(\cdot)$ is strictly convex, namely, $d''(c) > 0$.*

REMARK 7.2. We note that the interval (ω_1, ω_2) can be of the form $(-\infty, a)$ or $(b, +\infty)$ or $(\alpha, \beta) \cup (\gamma, \theta)$.

REMARK 7.3. If ϕ_c satisfies (7.4), then a simple bootstrap argument shows that $\phi_c \in H^s(\mathbb{R})$ for each $s \in \mathbb{R}$. Moreover, differentiating equation (7.4) with regard to the spatial variable, we obtain that

$$\mathcal{L}_c(\phi'_c) = 0. \tag{7.14}$$

Thus, ϕ'_c is an eigenvector of $\mathcal{L}_c$ with eigenvalue 0.

REMARK 7.4. From Assumption 2 it follows that $d'(c) = F(\phi_c)$, so the sufficient condition for obtaining the orbital stability given by Theorem 7.1 is reduced to

$$d''(c) = \frac{1}{2}\frac{d}{dc}\|\phi_c\|^2 > 0. \tag{7.15}$$

REMARK 7.5. By differentiating (7.4) with regard to the variable c, we obtain that

$$\mathcal{L}_c\Big(-\frac{d}{dc}\phi_c\Big) = \phi_c. \tag{7.16}$$

Note that (7.16) implies that $\frac{d}{dc}\phi_c \in H^s(\mathbb{R})$ for every $s \in \mathbb{R}$. Now, if for some $\psi \in L^2(\mathbb{R})$ we have that $\mathcal{L}_c\psi = \phi_c$, then from (7.16) it follows that

$$\mathcal{L}_c\Big(\frac{d}{dc}\phi_c + \psi\Big) = 0.$$

Hence $\frac{d}{dc}\phi_c + \psi = \theta\phi'_c$ and therefore

$$\langle \psi, \phi_c \rangle = -\frac{1}{2}\frac{d}{dc}\|\phi_c\|^2.$$

So, we have that the conclusion of Theorem 7.1 remains valid if the condition $d''(c) > 0$ is replaced by the condition

$$\text{if } \mathcal{L}_c\psi = \phi_c, \text{ then } \langle \psi, \phi_c \rangle = \langle \mathcal{L}_c^{-1}\phi_c, \phi_c \rangle < 0. \tag{7.17}$$

We note that the condition in (7.17) is the same as the condition for obtaining the main Lemma 6.9 in the proof of Theorem 6.1 (see also Lemma 6.4 and Lemma 6.7). Condition (7.17) is useful in situations where the family of solitary waves ϕ_c does not depend smoothly on c (see Section 7.6 below where an example of this situation is given).

REMARK 7.6. If we suppose that the symbol α, defining operator M, is continuous, that $\alpha'(\xi) \neq 0$ a.e. $\mathbb{R}$, and that $\inf_{\xi\in\mathbb{R}} \alpha(\xi) = 0$, then by Theorem B.43 in Appendix B, we have that $\mathcal{L}_c$ defined in (7.12) is a closed, unbounded, self-adjoint operator on $L^2(\mathbb{R})$ whose spectrum consists of the essential part $[c, \infty)$ together with a finite number of discrete eigenvalues (with finite-dimensional eigenspaces) in the interval $(-\infty, c)$.

Next we start the proof of Theorem 7.1. We will split the proof into a series of lemmas.

LEMMA 7.7. *There exist $\epsilon > 0$ and a C^1 map $\alpha : U_\epsilon \to \mathbb{R}$ such that, for all $u \in U_\epsilon$ and $r \in \mathbb{R}$,*

(1) $\langle u(\cdot + \alpha(u)), \phi_c' \rangle = 0$,
(2) $\alpha(u(\cdot + r)) = \alpha(u) - r$,
(3) $\alpha(\phi_c) = 0$.

PROOF. Consider the functional

$$G(u, \alpha) = \int_{-\infty}^{\infty} u(x+\alpha)\phi_c'(x)\, dx,$$

for $(u, \alpha) \in \mathcal{W} \times \mathbb{R}$. Since $G(\phi_c, 0) = 0$ and

$$\partial_\alpha G(\phi_c, 0) = \|\phi_c'\|^2 \neq 0,$$

by the Implicit Function Theorem there is a unique C^1 functional $\alpha(u)$ satisfying $G(u, \alpha(u)) = 0$ for u in a ball $B_\epsilon(\phi_c)$ of ϕ_c. Now we show that (2) above is also true within $B_\epsilon(\phi_c)$. Let $u \in B_\epsilon(\phi_c)$ and $\eta \in \mathbb{R}$ such that $u(\cdot + \eta) \in B_\epsilon(\phi_c)$. Then since $0 = G(u, \alpha(u)) = G(\tau_\eta u, \alpha(u) - \eta)$ by the uniqueness of $\alpha(u)$, $\alpha(\tau_\eta u) = \alpha(u) - \eta$.

Finally, we extend the definition of $\alpha(u)$ to $u \in U_\epsilon$. If $\|u - \tau_{s_0}\phi_c\|_{\frac{\beta}{2}} \leqq \epsilon$ for some $s_0 \in \mathbb{R}$, we define $\alpha(u) \equiv \alpha(\tau_{-s_0}u) - s_0$. This definition does not depend on the choice of s_0. In fact, let s_1 be such that $\|u - \tau_{s_1}\phi_c\|_{\frac{\beta}{2}} \leqq \epsilon$. Then $\tau_{-s_0}u$ and $\tau_{-s_1}u$ belong to $B_\epsilon(\phi_c)$. Since (2) is true within $B_\epsilon(\phi_c)$, we have $\alpha(\tau_{s_0-s_1}\tau_{-s_0}u) = \alpha(\tau_{-s_0}u) - (s_0 - s_1)$. Therefore, $\alpha(\tau_{-s_1}u) - s_1 = \alpha(\tau_{-s_0}u) - s_0$. So, $\alpha(u)$ is defined for all $u \in U_\epsilon$ and satisfies properties (1) and (2). □

In the next lemma we will use the special structure of the operator $\mathcal{L}_c$ given by Assumption 3.

LEMMA 7.8. *Let $d''(c) > 0$. Define*

$$\mathcal{A} = \{\psi \in \mathcal{W} : \langle \psi, \phi_c \rangle = \langle \psi, \phi_c' \rangle = 0 \text{ and } \langle \psi, \psi \rangle = 1\}.$$

Then $(\mathcal{L}_c\psi, \psi) > 0$. Moreover,

$$\zeta = \inf\{(\mathcal{L}_c\psi, \psi) \mid \psi \in \mathcal{A}\} > 0.$$

Therefore, there is a positive constant D such that for all $\psi \in \mathcal{A}$

$$(\mathcal{L}_c\psi, \psi) \geqq D\|\psi\|_{\mathcal{W}}^2.$$

PROOF. We first show that $\zeta \geqq 0$. In fact, from (7.15) and (7.16) we have that

$$0 < d''(c) = -(\mathcal{L}_c \frac{d}{dc}\phi_c, \frac{d}{dc}\phi_c) = -\langle \mathcal{L}_c \frac{d}{dc}\phi_c, \frac{d}{dc}\phi_c \rangle. \tag{7.18}$$

Now from Assumption 3, we obtain the following decomposition of $L^2(\mathbb{R})$:

$$L^2(\mathbb{R}) = [\chi_c] \oplus [\phi_c'] \oplus P \tag{7.19}$$

(see Kato's book [**153**]) where $\chi_c \in D(\mathcal{L}_c)$ satisfies $\|\chi_c\| = 1$ and $\mathcal{L}_c\chi_c = -\lambda^2\chi_c$ with $\lambda > 0$. We remember that $P_c = D(\mathcal{L}_c) \cap P$ is called the positive subspace of $\mathcal{L}_c$ and that there exists $\eta > 0$ such that

$$\langle \mathcal{L}_c p, p \rangle \geqq \eta \|p\|^2 \qquad \text{for } p \in P_c.$$

Now, making the spectral decomposition

$$\frac{d}{dc}\phi_c = a_0\chi_c + b_0\phi_c' + p_0,$$

where $p_0 \in P_c$ and $a_0, b_0 \in \mathbb{R}$, it follows from the relation $\mathcal{L}_c\phi_c' = 0$ that

$$\langle \mathcal{L}_c p_0, p_0 \rangle = \langle \mathcal{L}_c \frac{d}{dc}\phi_c, p_0 \rangle = \langle \frac{d}{dc}\phi_c, \mathcal{L}_c \frac{d}{dc}\phi_c \rangle + a_0^2\lambda^2.$$

Hence from (7.18), $\langle \mathcal{L}_c p_0, p_0 \rangle < a_0^2\lambda^2$. Now, let $\psi \in \mathcal{A}$ and write $\psi = a\chi_c + p$ with $p \in P$. Therefore, $p \in \mathcal{W}$. Then, from (7.16) we get that

$$0 = -\langle \phi_c, \psi \rangle = \langle \mathcal{L}_c \frac{d}{dc}\phi_c, \psi \rangle = -a_0 a\lambda^2 + \langle \mathcal{L}_c p_0, p \rangle.$$

So, using the fact that $(\mathcal{L}_c p, \chi_c) = \langle p, \mathcal{L}_c\chi_c \rangle = 0$, it is inferred that

$$\begin{aligned}(\mathcal{L}_c\psi, \psi) &= -a^2\lambda^2 + (\mathcal{L}_c p, p) \geqq -a^2\lambda^2 + \frac{|(\mathcal{L}_c p_0, p)|^2}{(\mathcal{L}_c p_0, p_0)} \\ &> -a^2\lambda^2 + \frac{(a_0 a\lambda^2)^2}{a_0^2\lambda^2} = 0,\end{aligned} \tag{7.20}$$

where we have used the fact that the function $\Phi(f, g) = (\mathcal{L}_c f, g)$, defined for $f, g \in \mathcal{W} \cap P$, is a nonnegative sesquilinear form on P. Therefore we have the Schwarz inequality

$$|\Phi(f, g)|^2 \leqq (\mathcal{L}_c f, f)(\mathcal{L}_c g, g).$$

So, we conclude that $\zeta \geqq 0$.

To see that $\zeta > 0$, we suppose that $\zeta = 0$. Then there is a sequence $\{\psi_n\} \subseteq \mathcal{A}$ such that $(\mathcal{L}_c\psi_n, \psi_n) \to 0$. Then for $\psi_n = a_n\chi_c + p_n$ with $p_n \in P$ and $(\mathcal{L}_c p_n, p_0) = a_0 a_n\lambda^2$ it follows from (7.20) that

$$(\mathcal{L}_c\psi_n, \psi_n) \geqq a_n^2\Big(\frac{a_0^2\lambda^4}{(\mathcal{L}_c p_0, p_0)} - \lambda^2\Big) > 0.$$

Hence, we conclude that $a_n^2 \to 0$. Therefore $(\mathcal{L}_c p_n, p_n) \to 0$ and it implies that $\|p_n\| \to 0$. So

$$1 = \|\psi_n\|^2 = a_n^2\|\chi_c\|^2 + \|p_n\|^2 \to 0,$$

which is a contradiction. Therefore, for $\psi \in \mathcal{W}$ such that $\langle \psi, \phi_c \rangle = \langle \psi, \phi_c' \rangle = 0$ we have

$$(\mathcal{L}_c\psi, \psi) \geqq \zeta\|\psi\|^2. \tag{7.21}$$

Finally, it follows easily from the specific form of the operator $\mathcal{L}_c$ that we can change the $L^2(\mathbb{R})$-norm in the right-hand side of (7.21) by the $\mathcal{W}$-norm. This finishes the proof of the lemma. □

Now we obtain the key inequality to establish the stability result.

LEMMA 7.9. *Let $d''(c) > 0$. There exist constants $D > 0$ and $\epsilon > 0$ such that*

$$E_M(u) - E_M(\phi_c) \geqq D\|u(\cdot + \alpha(u)) - \phi_c\|^2_{\mathcal{W}} \tag{7.22}$$

for all $u \in U_\epsilon$ which satisfy $F(u) = F(\phi_c)$.

PROOF. Write u in the form $u(\cdot + \alpha(u)) = (1+a)\phi_c + \psi$ where $\langle \psi, \phi_c \rangle = 0$ and a is a scalar. Then, by the translation invariance of F and Taylor's theorem

$$F(\phi_c) = F(u) = F(\phi_c) + \langle \phi_c, u(\cdot + \alpha(u)) - \phi_c \rangle + O(\|u(\cdot + \alpha(u)) - \phi_c\|^2_{\mathcal{W}}).$$

The middle term is precisely $a\|\phi_c\|$, so that $a = O(\|u(\cdot + \alpha(u)) - \phi_c\|^2_{\mathcal{W}})$. Now, define $L : \mathcal{W} \to \mathbb{R}$ by $L(u) = E_M(u) + cF(u)$. Then, another Taylor expansion gives

$$L(u) = L(u(\cdot + \alpha(u))) = L(\phi_c) + (L'(\phi_c), v) + \frac{1}{2}(L''(\phi_c)v, v) + o(\|v\|^2_{\mathcal{W}}), \tag{7.23}$$

where $v \equiv u(\cdot + \alpha(u)) - \phi_c = a\phi_c + \psi$. Since $F(u) = F(\phi_c)$, $L'(\phi_c) = 0$, and $L''(\phi_c) = \mathcal{L}_c$, we have that (7.23) can be written as

$$E_M(u) - E_M(\phi_c) = \frac{1}{2}(\mathcal{L}_c\psi, \psi) + O(a^2) + O(a\|v\|_{\mathcal{W}}) + o(\|v\|^2_{\mathcal{W}}) = \frac{1}{2}(\mathcal{L}_c\psi, \psi) + o(\|v\|^2_{\mathcal{W}}).$$

From Lemma 7.7 we have $\langle \psi, \phi'_c \rangle = 0$. Therefore Lemma 7.8 implies that

$$E_M(u) - E_M(\phi_c) \geqq \frac{1}{2} D_1 \|\psi\|^2_{\frac{\beta}{2}} + o(\|v\|^2_{\mathcal{W}}),$$

for some constant $D > 0$. Since $\|\psi\|_{\mathcal{W}} \geqq \|v\|_{\mathcal{W}} - |a|\|\phi_c\|_{\mathcal{W}} \geqq \|v\|_{\mathcal{W}} - O(\|v\|^2_{\mathcal{W}})$, we have for small $\|v\|_{\mathcal{W}}$ that

$$E_M(u) - E_M(\phi_c) \geqq D\|v\|^2_{\mathcal{W}},$$

which proves (7.22). Thus Lemma 7.9 is established. □

Next we prove Theorem 7.1:

PROOF OF THEOREM 7.1. Assume $d''(c) > 0$ and Ω_{ϕ_c} is $\mathcal{W}$-unstable (see Definition 5.2). Then we can choose initial data $w_k \equiv u_k(0) \in U_{1/k} \cap Y$ and $\epsilon > 0$ such that

$$\inf_{r \in \mathbb{R}} \|w_k - \phi_c(\cdot + r)\|_{\mathcal{W}} \to 0 \quad \text{but} \quad \sup_{t \in \mathbb{R}} \inf_{r \in \mathbb{R}} \|u_k(t) - \phi_c(\cdot + r)\|_{\mathcal{W}} \geqq \epsilon,$$

where $u_k(t)$ is the solution of equation (7.2) with initial datum w_k. Now, by continuity in t, we can pick the first time t_k such that

$$\inf_{r \in \mathbb{R}} \|u_k(t_k) - \phi_c(\cdot + r)\|_{\mathcal{W}} = \frac{\epsilon}{2}. \tag{7.24}$$

Since E_M and F are continuous functionals over $\mathcal{W}$, by conservation laws for (7.2) and translation invariance, we get

$$E_M(u_k(t_k)) = E_M(w_k) \to E_M(\phi_c) \quad \text{and} \quad F(u_k(t_k)) \to F(\phi_c)$$

as $k \to \infty$. Next we can choose $v_k \in U_\epsilon$ so that $F(v_k) = F(\phi_c)$ and $\|v_k - u_k(t_k)\|_{\mathcal{W}} \to 0$ as $k \to \infty$. In fact, since $F(w_k) \neq 0$ for k sufficiently large, define

$\{\alpha_k\} \subset \mathbb{R}$ such that $F(\alpha_k w_k) = F(\phi_c)$. Then, $\alpha_k^2 \to 1$ as $k \to \infty$. Now, define $v_k \equiv \alpha_k u_k(t_k)$. Then, we have the following:

(1) $F(v_k) = \alpha_k^2 F(u_k(t_k)) = \alpha_k^2 F(w_k) = F(\phi_c)$.

(2) Since $\|u_k(t_k)\|_{\mathcal{W}} \leqq C_0$, with $k \in \mathbb{N}$, we have

$$\|v_k - u_k(t_k)\|_{\mathcal{W}} = |\alpha_k - 1| \|u_k(t_k)\|_{\mathcal{W}} \to 0, \qquad k \to \infty.$$

(3) We have that $v_k \in U_\epsilon$. In fact, $\|v_k - \phi_c(\cdot + r)\|_{\mathcal{W}} \leqq \frac{\epsilon}{3} + \|u_k(t_k) - \phi_c(\cdot + r)\|_{\mathcal{W}}$ implies

$$\inf_{r\in\mathbb{R}} \|v_k - \phi_c(\cdot + r)\|_{\mathcal{W}} \leqq \frac{\epsilon}{3} + \inf_{r\in\mathbb{R}} \|u_k(t_k) - \phi_c(\cdot + r)\|_{\mathcal{W}} = \frac{5\epsilon}{6} < \epsilon.$$

Then, by Lemma 7.9 we have that

$$0 \leftarrow E_M(v_k) - E_M(\phi_c) \geqq D\|v_k(\cdot + \alpha(v_k)) - \phi_c\|_{\mathcal{W}}^2 = D\|v_k - \phi_c(\cdot - \alpha(v_k))\|_{\mathcal{W}}^2.$$

Therefore, the inequality

$$\|u_k(t_k) - \phi_c(\cdot - \alpha(v_k))\|_{\mathcal{W}} \leqq \|u_k(t_k) - v_k\|_{\mathcal{W}} + \|v_k - \phi_c(\cdot - \alpha(v_k))\|_{\mathcal{W}}$$

implies that

$$\|u_k(t_k) - \phi_c(\cdot - \alpha(v_k))\|_{\mathcal{W}} \to 0 \qquad \text{as } k \to \infty,$$

which contradicts (7.24). So, we must have that Ω_{ϕ_c} is $\mathcal{W}$-stable. This finishes the proof. □

REMARK 7.10. We note that by using condition (7.17) in place of condition $d''(c) > 0$, Lemma 7.8 and Lemma 7.9 are still true.

7.4. Stability of Solitary Waves for KdV-Type Equations

In this section we apply Theorem 7.1 to equation (7.2) when the pseudo-differential operator M is homogeneous, more precisely,

$$\widehat{Mu}(\xi) = |\xi|^\beta \widehat{u}(\xi) \tag{7.25}$$

with fixed $\beta \geqq 1$. If we assume Assumptions 1, 2, and 3, it follows from Theorem 7.1 that we only need to find an expression for $d''(c)$. In fact, by considering the new variable

$$\varphi(\xi) = c^{-\frac{1}{p}} \phi_c(c^{-\frac{1}{\beta}}\xi),$$

we see that φ is a solution of

$$M\varphi + \varphi - \frac{1}{p+1}\varphi^{p+1} = 0.$$

Note the independence of φ with regard to the wave speed c. Hence, we have the following result.

THEOREM 7.11. *Let M be defined by (7.25). Then the orbit generated by the solitary wave solution ϕ_c of (7.4) is $H^{\frac{\beta}{2}}(\mathbb{R})$-stable if $p < 2\beta$.*

PROOF. Since $\phi_c(\xi) = c^{\frac{1}{p}}\varphi(c^{\frac{1}{\beta}}\xi)$, it follows from (7.15) that

$$\begin{aligned} d''(c) &= \frac{1}{2}\frac{d}{dc}\int_{-\infty}^{\infty} c^{\frac{2}{p}}\varphi^2(c^{\frac{1}{\beta}}\xi)\,d\xi = \frac{1}{2}\|\varphi\|^2 \frac{d}{dc}c^{\frac{2}{p}-\frac{1}{\beta}} \\ &= \frac{1}{2}\Big(\frac{2}{p} - \frac{1}{\beta}\Big)c^{\frac{2}{p}-\frac{1}{\beta}-1}\|\varphi\|^2. \end{aligned} \tag{7.26}$$

Therefore, $d''(c) > 0 \Leftrightarrow p < 2\beta$. □

REMARK 7.12. Assumption 1 in the case of equations (7.2) with M given by (7.25) can be proved by Theorem 4.2 if we choose $Y = H^s(\mathbb{R})$ for some $s > 3/2$. We note that it can also be improved with the space Y being $H^{\frac{\beta}{2}}(\mathbb{R})$ and $\beta \geqq 2$, as has been explained in Chapter 4, Theorem 4.3.

REMARK 7.13. Regarding Assumption 2, for equations (7.2) with M given by (7.25), we only know an explicit smooth curve of solitary wave solutions for (7.4), depending on the wave speed $c \in (0, \infty)$, in the cases of the GKdV equations (see (3.4)) and the BO equations (see (3.6)). In the general case, we can show via the Concentration-Compactness Method (Chapter 9) that for every $c > 0$, equation (7.4) has a solution ϕ_c. Unfortunately, we do not know if this approach can give us information about the smoothness of the map $c \to \phi_c$ (note that in the cases of the GKdV and BO equations, this property is obtained from the uniqueness, modulo translations and signal, of the solitary wave solutions). But, as we will see in Chapter 9, we can still obtain stability results for these solitary wave solutions based on condition (7.17).

REMARK 7.14. Assumption 3 for equations (7.2) with M given by (7.25) is well know to be true in the cases of the GKdV equations (see Theorem B.61 in Appendix B) and in the case of the BO equation (see [**49**]). We note that by using the results in Albert [**6**] and Albert and Bona [**10**] (see Section 7.5 below), we can also obtain the required spectral information in the cases of the KdV and BO equations. In the other cases of the operator M, a proof of this assumption has not been given yet.

7.5. On Albert-Bona's Spectrum Approach

As we have seen in the theory of Grillakis *et al.* explained in Section 7.3, Assumption 3 established for the linear operator $\mathcal{L}_c$ in (7.12) is the main point to be verified. In general, it is not easy to prove this condition and it is an especially delicate business to see that zero is a simple eigenvalue. A more accurate study on this point of the theory has been conducted by Albert in [**6**] and Albert and Bona in [**10**], where they state that the operator $\mathcal{L}_c$ will satisfy Assumption 3 whenever the Fourier transforms of ϕ_c and ϕ_c^p belong to a special class of functions defined by Karlin in [**149**]. For the reader's convenience, we will state here the main results of [**6**] and [**10**]. We say that a function $\mathcal{K} : \mathbb{R} \to \mathbb{R}$ is in the class $PF(2)$ if

(1) $\mathcal{K}(x) > 0$ for $x \in \mathbb{R}$,
(2) $\mathcal{K}(x_1 - y_1)\mathcal{K}(x_2 - y_2) - \mathcal{K}(x_1 - y_2)\mathcal{K}(x_2 - y_1) \geqq 0$ for $x_1 < x_2$ and $y_1 < y_2$,
(3) strict inequality holds in (2) whenever the intervals (x_1, x_2) and (y_1, y_2) intersect.

So, we have the following theorem (see Theorem 3.2 in [**6**]).

THEOREM 7.15. *Let ϕ_c be a positive and even solution of* (7.4). *Suppose that $\widehat{\phi_c} > 0$ on $\mathbb{R}$ and $\widehat{\phi_c^p} = \mathcal{K} \in PF(2)$. Then, Assumption 3 holds for the operator $\mathcal{L}_c$ defined in* (7.12).

The proof of the following result can be found in [**10**].

THEOREM 7.16. *Suppose f is a positive, twice-differentiable function on $\mathbb{R}$ satisfying*

$$\frac{d^2}{dx^2}(\log f(x)) < 0 \qquad \textit{for } x \neq 0.$$

Then $f \in PF(2)$.

We will illustrate Theorem 7.15 and Theorem 7.16 in the cases of the families of solitary wave solutions associated with the KdV, BO, and ILW equations, given in (3.4), (3.6), and (3.8), respectively. First, we obtain the Fourier transforms

$$\widehat{\phi_{c,\mathrm{KdV}}}(\xi) = 12\pi \frac{\xi}{\sinh(\pi\xi/\sqrt{c})}, \quad \widehat{\phi_{c,\mathrm{BO}}}(\xi) = 4\pi e^{-c|\xi|},$$
$$\widehat{\phi_{c,H}}(\xi) = 2\pi \frac{\sinh(\delta\xi/2a)}{\sinh(\pi\xi/2a)} \tag{7.27}$$

where c and H are positive and $\delta \in (-\pi, \pi)$ satisfies

$$\cos\delta = \frac{16a^2 - b^2}{16a^2 + b^2}.$$

From (7.27) we immediately get the first condition given in Theorem 7.15. Using the formula for $\widehat{\phi_{c,\mathrm{BO}}}$, it is easy to verify directly that it belongs to $PF(2)$. The property that $\widehat{\phi_{c,\mathrm{KdV}}}$ and $\widehat{\phi_{c,H}}$ belong to $PF(2)$ follows from Theorem 7.16.

Finally, we note that from Theorem 7.15 applied to $\phi_{c,H}$ and from the relation

$$\frac{d}{dc}\|\phi_{c,H}\|^2 > 0$$

for any $c, H > 0$ (see Theorem 11 in [**10**]), we obtain, by using the theory established in Section 7.3, that the solitary wave solutions $\phi_{c,H}$ for the ILW equation are $\mathcal{W}$-stable for any $c, H > 0$. Here, the Banach space $\mathcal{W}$ is the space induced by the symbol α_H that generates the ILW equation (see Section 7.3).

7.6. Comments

There are a large number of publications devoted to the question of stability of solitary wave solutions for equations with a form different from that given by (7.2) by applying the theory presented above. Assumptions 3 could be difficult to verify when we work with systems or when there is more than one parameter determining the solitary wave solutions. In this last case, the approach given by Grillakis *et al.* in [**124**] is the best-known tool for handling this situation.

We note that the first study about the stability of periodic travelling waves in the spatial variable for the KdV equation is that of Benjamin in [**45**]. But its proposed proof of stability of cnoidal wave solutions is not clear. In Chapter 11 we will give a complete theory about the stability of these specific waves.

Next, we establish some other applications of the Grillakis *et al.* theory in [**123**] just with the idea of explaining how condition (7.17) is basic in some situations.

(1) A very interesting class of solitary wave solutions has been found by Albert in [**6**], one associated with equations of the form (7.2) where $M = M_{n,p}$ is a differential operator of order $2n$. The solitary waves in question are of the form

$$\varphi(\xi) = [\mathrm{sech}(\xi)]^r \tag{7.28}$$

with $r = 2n/p$. The operators $M_{n,p}$ are defined based on the following result: for each $n \in \mathbb{N}$ and $p > 0$, there is a unique vector $A = (a_0, a_1, ..., a_n)$

in $\mathbb{R}^{n+1}$ such that

$$\sum_{i=0}^{n} a_i(\partial^{2i}\varphi) = \frac{1}{p+1}\varphi^{p+1}. \tag{7.29}$$

So, we define the differential operator $M_{n,p}$ by

$$M_{n,p} \equiv \sum_{i=1}^{n} a_i\partial^{2i}.$$

Therefore, defining $c = a_0$, we have from (7.29) that φ is a solution of

$$(M_{n,p} + c)\varphi = \frac{1}{p+1}\varphi^{p+1}. \tag{7.30}$$

The following condition of stability of the solitary waves φ in (7.28) can be found in [**6**].

- Let $n \in \mathbb{N}$ and $0 < p < 4n$. Consider the solitary wave solutions $\varphi(\xi) = [\text{sech}(\xi)]^r$ of equation (7.30) with $r = \frac{2n}{p}$. Suppose that there exists a $\psi \in L^2(\mathbb{R})$ such that $(M_{n,p}+c-\varphi^p)\psi = \varphi$. If $I = \langle\psi,\varphi\rangle < 0$, then φ is an $H^n(\mathbb{R})$-stable solution of equation (7.2).

Note that in this case the condition of stability is not given by (7.15) but by (7.17). The reason for this situation is that we do not have a differentiable family of solitary wave solutions depending on the wave speed c. In Theorem 4.10 in [**6**], an explicit formula for I was found, but unfortunately this relation does not give a simple relation between the parameters n and p. For example, the solitary wave solution

$$u(x,t) = \text{sech}^4(x - \frac{12}{35}t)$$

($n = 2$ and $p = 1$) of the equation

$$u_t + uu_x + \frac{13}{420}u_{xxx} - \frac{1}{1680}u_{xxxxx} = 0$$

is stable.

We note that when $n = 1$, then $I < 0$ if and only if $p < 4$, as we have already seen in the proof of Theorem 7.11 (see (7.26)). For $n = 2$ the value of p at which I changes sign was found to be approximately 4.82, while for $n = 3$ it is approximately 5.26. Finally, we note that when $n = 2$ and $p = 5$ the corresponding solitary wave φ is unstable (see Souganidis and Strauss [**255**] or Chapter 10 of this book). But since $p < 4n$, the equation

$$u_t + u^5u_x - M_{2,5}u_x = 0$$

has a global solution in $H^s(\mathbb{R})$ for any $s > 3/2$ (see Chapter 4, Theorem 4.2). Therefore, *although φ is unstable, solutions with initial data near to φ do not blow up in finite time.* We refer the reader to Chapter 10 of this book where other examples of this situation are presented.

(2) A theory of existence and stability of solitary wave solutions for equations in (7.2), with M given by (7.25), can be given by using condition (7.17) and considering Assumption 3 to be true. In fact, by using the Concentration-Compactness Method (see Chapter 9), we can show that for every $c > 0$

the minimization problem

$$\theta(\lambda) = \inf\Big\{ \int_{-\infty}^{\infty} [uD^{\beta}u + cu^2]\, dx : u \in H^{\frac{\beta}{2}}(\mathbb{R}), \int_{-\infty}^{\infty} u^{p+2}\, dx = \lambda \Big\} \tag{7.31}$$

has a solution ψ for every $\lambda > 0$. So, by the Lagrange multiplier principle ([**203**]), there exists $\theta \in \mathbb{R}$ such that ψ is a solution of the equation

$$D^{\beta}\psi + c\psi = \theta(p+2)\psi^{p+1} \tag{7.32}$$

in the sense of distributions. Moreover, by using a bootstrapping argument, we see that $\psi \in H^{\infty}(\mathbb{R})$ and so (7.32) holds in the pointwise sense. Now, if we define $\phi = \eta\psi$ with $\eta^p = \theta(p+1)(p+2)$, we obtain that ϕ satisfies (7.2) for $c > 0$.

Next, we consider the linear operator $\mathcal{L} = D^{\beta} + c - \phi^p$ and $\varphi \in D(\mathcal{L})$ given by

$$\varphi(x) = -\frac{1}{pc}\phi(x) - \frac{1}{\beta c}x\phi'(x).$$

Then by using the relations $\mathcal{H}(x\phi'') = x\mathcal{H}\phi''$,

$$D^{\beta-1}(x\mathcal{H}\phi'') = (\beta - 1)D^{\beta}\phi + xD^{\beta}\phi',$$

and $\mathcal{L}\phi' = 0$, we obtain that $\mathcal{L}\varphi = \phi$. Now, we estimate the product $\langle \varphi, \phi \rangle$. By using integration by parts, we get

$$\langle \varphi, \phi \rangle = \Big(\frac{1}{2c\beta} - \frac{1}{cp} \Big) \|\phi\|^2 < 0 \Leftrightarrow p < 2\beta.$$

Therefore from the theory established in Section 7.3 it follows that the Ω_{ϕ}-orbit is $H^{\frac{\beta}{2}}(\mathbb{R})$-stable (see Theorem 7.11).

(3) We note that the theory of nonlinear stability of the solitary wave solutions (1.3) for the classical Boussinesq equation,

$$u_{tt} = u_{xx} - (u^2 + u_{xx})_{xx}$$

(see (1.2)), was obtained by Bona and Sachs in [**62**] following the framework established in Section 7.3.

Part 4

The Concentration-Compactness Principle in Stability Theory

Euler, Leonhard (1707–1783)

[upon losing the use of his right eye]

Now I will have less distraction.

In H. Eves, *Mathematical Circles*, Boston: Prindle, Weber & Schimdt, 1969.

LaGrange, Joseph-Louis (1736–1813)

The reader will find no figures in this work. The methods which I set forth do not require either constructions or geometrical or mechanical reasonings: but only algebraic operations, subject to a regular and uniform rule of procedure.

In *Mécanique Analytique*, 4th ed., 2 v., Paris: Gauthier-Villars et fils, 1888–1889.

CHAPTER 8

Existence and Stability of Solitary Waves for the GBO

8.1. Introduction

In Chapters 6 and 7 we saw that the stability of solitary wave solutions to nonlinear dispersive equations, for example those of the form (2.1), is proved by verifying an inequality of the form (6.7) or (7.22). This means that one must show that the solitary wave solution is a local constrained minimizer of a Hamiltonian functional associated with (2.1), and this is done basically by obtaining basic spectral information for a specific operator obtained by linearizing the solitary wave equation. In practice this spectral analysis is particularly difficult to perform (see Chapter 7). To avoid these difficulties, an alternative method to prove the stability of solitary waves, one which does not rely on local analysis, was developed by Cazenave and Lions in [**82**] using Lions's method of *Concentration-Compactness.* In this method, instead of starting with a given travelling wave and trying to prove that it realizes a local minimum of a constrained variational problem (as, for example, in the proofs of Theorems 6.1 and 7.1), one starts with the constrained variational problem and looks for global minimizers. For example, consider the problem

$$
(8.1)\qquad \begin{cases} \text{minimize} \\ E_{\text{GKdV}}(\psi) = \dfrac{1}{2}\displaystyle\int_{-\infty}^{\infty} (\psi_x)^2 - \frac{2}{(p+1)(p+2)}\psi^{p+2}\,dx, \\ \text{subject to} \\ F(\psi) = \dfrac{1}{2}\displaystyle\int_{-\infty}^{\infty} \psi^2 = \lambda > 0. \end{cases}
$$

We can see via the Concentration-Compactness Principle (Lemma 8.1 below) that this problem has a solution for $p < 4$, one which will satisfy equation (3.3) and therefore will be a solitary wave solution of the GKdV equations (2.5), with a wave speed $c > 0$ being the Lagrange multiplier associated with problem (8.1). When the method works, it shows not only the existence of global minimizers, but also that every minimizing sequence associated with (8.1) is relatively compact up to translations. Moreover, based on the main property that the functionals E_{GKdV} and F are conserved quantities for the GKdV equations, we obtain that the set of global minimizers for (8.1) is a stable set for the flow associated with the GKdV equations. It is stable in the sense that a solution which starts near the set of minimizers will remain near it all the time (see Definition 5.3).

We would like to note that while the Concentration-Compactness Method gives us proof of stability of solitary waves without requiring a detailed analysis such as that given in the local methods of Benjamin, Weinstein, and Grillakis *et al.*, it also produces a weaker result since it only proves the stability of a set of minimizing solutions without providing information on the structure of that set, or distinguishing among its different members, called orbits. For example, it is not known if the set of solitary wave solutions obtained by de Bouard and Saut in [**95**] for the KP-I equation (2.13), via the Concentration-Compactness Method, contains the explicit lump solitary wave solution (3.16). In the special case of the set of minimizers obtained for the variational problem (8.1), it is not difficult to see (based on the uniqueness of solutions for equation (3.3) modulo translations and signal [**54**] and [**174**]) that this set contains a single orbit, $\{\psi(\cdot+y) : y \in \mathbb{R}\}$, and that this function, ψ, is indeed the explicit solitary wave solution to (3.4). We refer the reader to the work of Albert [**8**], where it is illustrated that this method works for obtaining the stability results of solitary wave solutions of nonlinear evolution equations of the general form (2.1). We note that *this approach for proving stability works whenever the functionals involved in the variational analysis are conserved quantities for the evolution equation in question.*

In the last couple of decades, a series of applications and variants of the Concentration-Compactness Method has been obtained for a great range of dispersive evolution equations. We cite, for example, the works of Weinstein [**276**], Albert, Bona, and Saut [**14**], de Bouard and Saut [**95**], Kichenassamy [**164**], Lopes [**195**], [**197**], [**199**], Angulo [**21**], and Albert and Angulo [**9**].

For the benefit of the reader, we establish the Concentration-Compactness Principle, which is the key tool in our analysis (see Lemma 1.1 in Lions [**193**]).

LEMMA 8.1 (The Concentration-Compactness Principle). *Let $\{\rho_n\}_{n\geq 1}$ be a sequence of nonnegative functions in $L^1(\mathbb{R})$ satisfying $\int_{-\infty}^{\infty}\rho_n(x)dx = \lambda$ for all n and some $\lambda > 0$. Then there exists a subsequence $\{\rho_{n_k}\}_{k\geq 1}$ satisfying one of the following three conditions:*

(1) **(Compactness)** *There are $y_k \in \mathbb{R}$ for $k = 1, 2, ...,$ such that $\rho_{n_k}(\cdot + y_k)$ is tight, i.e., for any $\epsilon > 0$, there is $R > 0$ large enough such that*

$$\int_{|x-y_k|\leq R} \rho_{n_k}(x)\, dx \geq \lambda - \epsilon.$$

(2) **(Vanishing)** *For any $R > 0$,*

$$\lim_{k\to\infty} \sup_{y\in\mathbb{R}} \int_{|x-y|\leq R} \rho_{n_k}(x)dx = 0.$$

(3) **(Dichotomy)** *There exists $\tilde{\alpha} \in (0, \lambda)$ such that for any $\epsilon > 0$, there exists $k_0 \geq 1$ and $\rho_k^1, \rho_k^2 \in L^1(\mathbb{R})$, with $\rho_k^1, \rho_k^2 \geq 0$, such that for $k \geq k_0$,*

$$|\rho_{n_k} - (\rho_k^1 + \rho_k^2)|_{L^1} \leq \epsilon,$$

$$|\textstyle\int_{-\infty}^{\infty}\rho_k^1\, dx - \tilde{\alpha}| \leq \epsilon, \quad |\int_{-\infty}^{\infty}\rho_k^2\, dx - (\lambda - \tilde{\alpha})| \leq \epsilon,$$

$$\text{supp } \rho_k^1 \cap \text{supp } \rho_k^2 = \emptyset, \;\; \text{dist}(\text{supp } \rho_k^1, \text{ supp } \rho_k^2) \to \infty \text{ as } k \to \infty.$$

REMARK 8.2. In Lemma 8.1 above, the condition $\int_{-\infty}^{\infty}\rho_n(x)\, dx = \lambda$ can be replaced by $\int_{-\infty}^{\infty}\rho_n(x)\, dx = \lambda_n$ where $\lambda_n \to \lambda > 0$ as $n \to \infty$. It is enough to replace ρ_n by ρ_n/λ_n and apply the lemma.

Our interest in this chapter will be to show how the Concentration-Compactness Method can still be used to prove in the stability of solitary wave solutions if *the functionals involved in the variational problem are not conserved quantities.* This approach has been put forward by Levandosky, in [**182**], in which the stability of a fourth-order wave equation is studied (see Section 9.4). We note that Liu and Wang, in [**192**], have applied this method with success to study the nonlinear stability of solitary wave solutions of the GKP-I equations in (2.14). In this book, we will discuss Levandosky's approach in the case of the Generalized Benjamin-Ono equations (GBO equations),

$$u_t + u^p u_x - \mathcal{H} u_{xx} = 0 \tag{8.2}$$

with $p \in \mathbb{N}$ or $p = \frac{m}{n}$, where m is any even integer and n any odd integer. In particular, we give a theorem of existence of solitary wave solutions for the GBO equations and we prove the orbital stability of the solitary waves (3.6) associated with the BO equation ([**49**]).

8.2. Solitary Waves for the GBO

Here we are looking for solutions of (8.2) of the form $u(x,t) = \phi(x - ct)$ such that $\phi^{(n)}(\xi) \to 0$ as $|\xi| \to \infty$. Inserting this form into (8.2) and integrating once, we see that ϕ must satisfy

$$\mathcal{H}\phi'(\xi) + c\phi(\xi) - \frac{1}{p+1}\phi^{p+1}(\xi) = 0. \tag{8.3}$$

When $p = 1$, we have that equation (8.3) has the explicit solution (3.6) for every $c > 0$. For $p \neq 1$, explicit solutions are not known. We note that by using the results in Bona and Li [**61**], we can see that every solution of (8.3) has the following asymptotic behavior:

$$\lim_{|\xi| \to \infty} |\xi|^{2+n} \phi^{(n)}(\xi) = C_n$$

for $n \in \mathbb{N} \cup \{0\}$ and $C_n \in \mathbb{R}$.

We establish the main theorem of this section.

THEOREM 8.3. *For every $c > 0$, equation (8.3) has a solution $\phi \in H^\infty(\mathbb{R})$ which is an even, strictly positive, decreasing function of $|\xi|$.*

The proof of Theorem 8.3 will be a consequence of the Concentration-Compactness Principle and the theory of symmetric decreasing rearrangement of functions (see [**74**], [**157**], or Appendix A).

We start by defining the nonlinear continuous functional J on $H^{\frac{1}{2}}(\mathbb{R})$:

$$J(\psi) = \frac{1}{2}\int_{-\infty}^{\infty} [D^{\frac{1}{2}}\psi(x)]^2 + c\psi^2(x)\, dx = \frac{1}{2}\int_{-\infty}^{\infty} \psi\mathcal{H}\psi' + c\psi^2\, dx,$$

and the following two constrained minimization problems on $H^{\frac{1}{2}}(\mathbb{R})$:

$$(\mathrm{P}_1) \qquad \begin{cases} \text{minimize} \quad J(\psi) \\ \text{subject to } \ H(\psi) \equiv \displaystyle\int_{-\infty}^{\infty} \psi^{p+2}(x)\, dx = \lambda > 0 \end{cases} \tag{8.4}$$

and

$$\text{(8.5)} \qquad (\mathrm{P}_2) \qquad \begin{cases} \text{minimize} \quad J(\psi) \\ \text{subject to} \;\; K(\psi) \equiv \displaystyle\int_{-\infty}^{\infty} |\psi(x)|^{p+2}\, dx = \lambda > 0. \end{cases}$$

In association with these two minimization problems we consider the families of minimization problems:

$$\text{(8.6)} \qquad I_\lambda = \inf\{J(\psi) : \psi \in H^{\frac{1}{2}}(\mathbb{R}) \text{ and } H(\psi) = \lambda\}$$

and

$$\text{(8.7)} \qquad M_\lambda = \inf\{J(\psi) : \psi \in H^{\frac{1}{2}}(\mathbb{R}) \text{ and } K(\psi) = \lambda\}.$$

Also, we consider the set of minimizers G_λ and N_λ, for I_λ and M_λ, respectively, defined by

$$\text{(8.8)} \qquad G_\lambda = \{\psi \in H^{\frac{1}{2}}(\mathbb{R}) : J(\psi) = I_\lambda,\ H(\psi) = \lambda\}$$

and

$$\text{(8.9)} \qquad N_\lambda = \{\psi \in H^{\frac{1}{2}}(\mathbb{R}) : J(\psi) = M_\lambda,\ K(\psi) = \lambda\}.$$

We see below that the existence part in Theorem 8.3 follows immediately once we show that $G_\lambda \neq \emptyset$. For technical reasons, which arise in a natural way from the use of the Concentration-Compactness Principle, it is better to work with problem (P_2) instead of problem (P_1). In Proposition 8.17 we establish a relation between the sets G_λ and N_λ and between the numbers I_λ and M_λ.

We call a sequence $\{\psi_n\} \subset H^{\frac{1}{2}}(\mathbb{R})$ a minimizing sequence to M_λ if

$$\begin{cases} \lim\limits_{n\to\infty} J(\psi_n) = M_\lambda, \\ K(\psi_n) = \lambda, \qquad \text{for all } n. \end{cases}$$

We similarly define a minimizing sequence to I_λ. The following theorem shows that $N_\lambda \neq \emptyset$.

THEOREM 8.4. *Let $c > 0$, and let $\{\psi_n\}$ be a minimizing sequence to M_λ. Then there is a subsequence $\{\psi_{n_k}\}$ and a sequence of real numbers $\{y_{n_k}\}$ such that $\psi_{n_k}(\cdot + y_{n_k})$ converges strongly in $H^{\frac{1}{2}}(\mathbb{R})$ to some $\psi \in H^{\frac{1}{2}}(\mathbb{R})$. The limit ψ is a minimizer for M_λ; i.e, $N_\lambda \neq \emptyset$.*

The proof of Theorem 8.4 will be a consequence of Lemma 8.1. For the benefit of the reader we will give the details of the method. We begin with the following.

LEMMA 8.5. *For all $\lambda > 0$, M_λ is finite and positive. Moreover, each minimizing sequence $\{\psi_n\}$ for M_λ is bounded in $H^{\frac{1}{2}}(\mathbb{R})$.*

PROOF. Let $\psi \in H^{\frac{1}{2}}(\mathbb{R})$ and $K(\psi) = \lambda$. Then from the classical Sobolev embedding $H^{\frac{1}{2}}(\mathbb{R}) \hookrightarrow L^r(\mathbb{R})$ for $r \geqq 2$, we have

$$\lambda^{2/(p+2)} = |\psi|_{p+2}^2 \leqq C_0 \|\psi\|_{\frac{1}{2}}^2 \leqq CJ(\psi).$$

Hence $M_\lambda > 0$ and $M_\lambda < \infty$. Now, if $\{\psi_n\}$ is a minimizing sequence to M_λ, then $J(\psi_n) \to M_\lambda$. Therefore, since $J(\psi)$ represents an equivalent norm to $\|\psi\|_{\frac{1}{2}}^2$, it follows that there exists $B > 0$ such that $\|\psi_n\|_{\frac{1}{2}}^2 \leqq B$. □

Let $\{\psi_n\}$ be a minimizing sequence to problem (P$_2$). Then we have $K(\psi_n) = \lambda$ for all n and $J(\psi_n) \to M_\lambda$ as $n \to \infty$. In general, because the inclusion of $H^{\frac{1}{2}}(\mathbb{R})$ into $L^{p+2}(\mathbb{R})$ is not compact, one cannot extract a subsequence of $\{\psi_n\}$ which converges in $L^{p+2}(\mathbb{R})$. This difficulty will be gotten around in fact by Lions's Concentration-Compactness Principle. Overall, Lions's principle provides a method for proving that a subsequence of $\{\psi_n\}$ can be found such that, after being suitably translated, each function in the subsequence is *concentrated* on a fixed bounded interval. This enables one to apply the compactness of the inclusion of $H^{\frac{1}{2}}(\Omega)$ into $L^{p+2}(\Omega)$ when Ω is a bounded set.

Now, we start by associating to each minimizing sequence $\{\psi_n\}$ for problem (P$_2$) the following sequence of nondecreasing functions: $Q_n : [0, \infty) \to [0, \lambda]$ defined by

$$Q_n(\omega) = \sup_{y \in \mathbb{R}} \int_{y-\omega}^{y+\omega} |\psi_n(x)|^{p+2}\, dx.$$

Since $|\psi_n|_{p+2}$ remains bounded, then $\{Q_n\}$ comprises a uniformly bounded sequence of nondecreasing functions on $[0, \infty)$. A standard argument implies that $\{Q_n\}$ must have a subsequence, which we also denote by $\{Q_n\}$, that converges pointwise and uniformly on compact sets to a nondecreasing limit function on $[0, \infty)$. Let Q be this limit function, and define

$$\alpha = \lim_{\omega \to \infty} Q(\omega). \tag{8.10}$$

From the assumption that $|\psi_n|_{p+2}^{p+2} = \lambda$, it follows that $0 \leqq \alpha \leqq \lambda$. The Concentration-Compactness Method distinguishes three cases: $\alpha = \lambda$, called the case of *compactness*; $\alpha = 0$, called the case of *vanishing*; and $0 < \alpha < \lambda$, called the case of *dichotomy*. Our goal is to show that for minimizing sequences of (8.7), only the case of compactness can occur. It will follow, by a standard argument, that every minimizing sequence is relatively compact, after suitable translations (see Theorem 8.14 below). Later, we will show that this compactness property is also enjoyed by problem (8.6) (Theorem 8.18). Typically, one proves compactness by ruling out the last two possibilities. We begin by ruling out *vanishing* using an idea of Brezis (see Lied [**186**], Albert, Bona, and Saut [**14**]).

LEMMA 8.6. *Let $\zeta \in C^\infty(\mathbb{R})$ be given such that $0 \leqq \zeta \leqq 1$, $\zeta(x) = 0$ for $x \notin [-2, 2]$, and $\sum_{j \in \mathbb{Z}} \zeta(x - j) = 1$ for all $x \in \mathbb{R}$. Then there exists a positive constant A_0 such that for all $f \in H^{\frac{1}{2}}(\mathbb{R})$,*

$$\sum_{j \in \mathbb{Z}} \|\zeta(x - j) f\|_{\frac{1}{2}}^2 \leqq A_0 \|f\|_{\frac{1}{2}}^2. \tag{8.11}$$

PROOF. Initially we give the construction of a function ζ such as the one described in the lemma. Considers $\Phi \in \mathcal{S}(\mathbb{R})$ such that $\operatorname{supp} \Phi = \{x : -2 \leqq |x| \leqq 2\}$ and $\Phi(x) > 0$ for $-2 < |x| < 2$. Define

$$F(x) = \sum_{j \in \mathbb{Z}} \Phi(x - j).$$

Then for every x, the sum defining $F(x)$ has no more than nine nonzero terms. So, $F(x) > 0$ for every $x \in \mathbb{R}$. Define, for $\xi \in \mathbb{R}$,

$$\zeta(\xi) = \frac{\Phi(\xi)}{F(\xi)}.$$

Then, $\zeta \in C^\infty(\mathbb{R})$ and $\operatorname{supp}\zeta \subseteq [-2,2]$. From the relation

$$F(\xi-k)=\sum_{j\in\mathbb{Z}}\Phi(\xi-j-k)=\sum_{j\in\mathbb{Z}}\Phi(\xi-j)=F(\xi),$$

it follows that

$$\sum_{j\in\mathbb{Z}}\zeta(\xi-j)=\sum_{j\in\mathbb{Z}}\frac{\Phi(\xi-j)}{F(\xi-j)}=\frac{1}{F(\xi)}\sum_{j\in\mathbb{Z}}\Phi(\xi-j)=1.$$

Finally, since $\sum_{j\in\mathbb{Z}}\Phi(\xi-j)\geqq\Phi(\xi)$, it follows that $\zeta(\xi)\leqq 1$ for every $\xi\in\mathbb{R}$.

Now we prove inequality (8.11). Define, for $j\in\mathbb{Z}$, $\zeta_j(x)=\zeta(x-j)$. Also, for any $s\in\mathbb{R}$, let $l_2(H^s(\mathbb{R}))$ denote the Hilbert space of all sequences $\{f_j\}_{j\in\mathbb{Z}}$ such that $f_j\in H^s(\mathbb{R})$ for each j and $\sum_{j\in\mathbb{Z}}\|f_j\|_s^2<\infty$. Consider the linear operator $T: H^s(\mathbb{R})\to l_2(H^s(\mathbb{R}))$ defined by $Tf=\{\zeta_j(x)f(x)\}_{j\in\mathbb{Z}}$. For $f\in H^1(\mathbb{R})$, we have that

$$\|Tf\|^2_{l_2(H^s(\mathbb{R}))}=\sum_{j\in\mathbb{Z}}\|\zeta_j f\|^2\leqq\sum_{j\in\mathbb{Z}}\int_{-\infty}^{\infty}\zeta(x-j)|f(x)|^2dx=\|f\|^2$$

and

$$\sum_{j\in\mathbb{Z}}\int_{-\infty}^{\infty}[\zeta_j'(x)]^2|f(x)|^2dx\leqq\sum_{j\in\mathbb{Z}}\int_{-2}^{2}(\zeta'(x))^2|f(x+j)|^2dx\leqq 9C(\zeta)\|f\|^2$$

where $C(\zeta)>0$. Then T is a bounded linear operator for $s=0$ and $s=1$. By interpolation (see [**56**]) we have that $T: H^{\frac12}(\mathbb{R})\to l_2(H^{\frac12}(\mathbb{R}))$ is bounded. From this fact inequality (8.11) follows. □

LEMMA 8.7. *Suppose $B>0$ and $\delta>0$ are given. Then there exists $\eta=\eta(B,\delta)$ such that if $f\in H^{\frac12}(\mathbb{R})$ with $\|f\|_{\frac12}\leqq B$ and $|f|_{p+2}\geqq\delta$, then*

$$\sup_{y\in\mathbb{R}}\int_{y-2}^{y+2}|f(x)|^{p+2}dx\geqq\eta.$$

PROOF. Let ζ be as in Lemma 8.6. Since $\sum_{j\in\mathbb{Z}}\zeta(x-j)=1$, implying that no more than nine of the terms in the sum are nonzero at any given value of x, it follows that there exists a constant $A_1>0$ such that $\sum_{j\in\mathbb{Z}}\zeta^{p+2}(x-j)\geqq A_1$ for all $x\in\mathbb{R}$. Now, we claim that there exists $j_0\in\mathbb{Z}$ such that

$$\|\zeta_{j_0}f\|^2_{\frac12}\leqq\left(1+A_2|f|_{p+2}^{-p-2}\right)|\zeta_{j_0}f|_{p+2}^{p+2}\tag{8.12}$$

with $A_2=A_0B^2/A_1$. If we assume the contrary, then there exists f which is not identically zero, $\|f\|_{\frac12}\leqq B$, and

$$\|\zeta_j f\|^2_{\frac12}\geqq\left(1+A_2|f|_{p+2}^{-p-2}\right)|\zeta_j f|_{p+2}^{p+2}$$

holds for every $j\in\mathbb{Z}$. Summing over j and applying Lemma 8.6, we obtain

$$\begin{aligned}A_0B^2&\geqq A_0\|f\|^2_{\frac12}\geqq\left(1+A_2|f|_{p+2}^{-p-2}\right)\sum_{j\in\mathbb{Z}}|\zeta_j f|_{p+2}^{p+2}dx\\&=\left(1+A_2|f|_{p+2}^{-p-2}\right)\sum_{j\in\mathbb{Z}}\int_{-\infty}^{\infty}\zeta_j^{p+2}|f|^{p+2}\geqq\left(1+A_2|f|_{p+2}^{-p-2}\right)A_1|f|_{p+2}^{p+2}\\&=A_1|f|_{p+2}^{p+2}+A_1A_2=A_1|f|_{p+2}^{p+2}+A_0B^2,\end{aligned}$$

which contradicts the fact that f is not identically zero.

Then, from (8.12) and the Sobolev embedding, $|g|_{p+2} \leqq A_3\|g\|_{\frac{1}{2}}$, we have that

$$|\zeta_{j_0} f|_{p+2}^2 \leqq A_3^2\|\zeta_{j_0} f\|_{\frac{1}{2}}^2 \leqq A_3^2\Big(1+\frac{A_2}{\delta^{p+2}}\Big)|\zeta_{j_0} f|_{p+2}^{p+2}.$$

Therefore,

$$\int_{j_0-2}^{j_0+2} |f(x)|^{p+2} dx \geqq \int_{-\infty}^{\infty} |\zeta_{j_0} f|^{p+2} dx \geqq \Big[A_3^2\Big(1+\frac{A_2}{\delta^{p+2}}\Big)\Big]^{-\frac{p+2}{p}} \equiv \eta.$$

This completes the proof of the lemma. □

The following lemma shows that the possibility $\alpha = 0$ does not occur here.

THEOREM 8.8 (Ruling out Vanishing). *For every minimizing sequence* $\{\psi_n\}$ *for problem* (P_2), *we have that* $\alpha > 0$.

PROOF. We know that $|\psi_n|_{p+2}^{p+2} \geqq \lambda$ and $\|\psi_n\|_{\frac{1}{2}} \leqq B$ for every $n \in \mathbb{Z}$. Then, Lemma 8.7 implies that there exists $\eta > 0$ such that $Q_n(2) \geqq \eta$ for all n. Then,

$$\alpha = \lim_{\omega\to\infty} Q(\omega) \geqq Q(2) = \lim_{n\to\infty} Q_n(2) \geqq \eta > 0.$$

This completes the proof of the theorem. □

Next we wish to rule out the possibility of *dichotomy*. To do this, we use a procedure which is an analogue to that used in Albert [**8**] for nonlocal equations. Before doing this, we need the following classical Calderón Commutator Theorem.

LEMMA 8.9. *Let* $g \in C^\infty(\mathbb{R})$ *with* $g' \in L^\infty(\mathbb{R})$. *Then* $[\mathcal{H}, g]\partial_x \in \mathcal{B}(L^2(\mathbb{R}); L^2(\mathbb{R}))$ *with*

$$\|[\mathcal{H}, g]\partial_x f\| \leqq C|g'|_\infty \|f\|.$$

PROOF. See Calderón [**76**] or Coifman and Meyer [**86**]. □

Now, we choose $\varphi \in C^\infty(\mathbb{R})$ such that $\varphi \equiv 1$ on $[-1,1]$ and $\varphi(x) = 0$ for $x \notin [-2,2]$. Let $\zeta \in C^\infty(\mathbb{R})$ be such that $\varphi^2 + \zeta^2 = 1$ on $\mathbb{R}$. For each $\omega \in \mathbb{R}$ define

$$\varphi_\omega(x) = \varphi(x/\omega) \quad \text{and} \quad \zeta_\omega(x) = \zeta(x/\omega).$$

LEMMA 8.10. *Let* $\epsilon > 0$, $B > 0$, *and* $r \in \mathbb{R}$. *Suppose* $\{\psi_n\} \subseteq H^{\frac{1}{2}}(\mathbb{R})$ *is such that* $\|\psi_n\|_{\frac{1}{2}} \leqq B$. *Then there exists* $\omega_0 = \omega_0(B,\epsilon) > 0$ *such that for* $\omega \geqq \omega_0$ *and* $n \in \mathbb{N}$,

$$\Big|J(\psi_n) - J(g_n) - J(h_n)\Big| < \epsilon$$

where $g_n(x) = \varphi_\omega(x-r)\psi_n$ *and* $h_n(x) = \zeta_\omega(x-r)\psi_n$.

PROOF. Since J is invariant under translations, we assume $r = 0$, so that $g_n = \varphi_\omega\psi_n$ and $h_n = \zeta_\omega\psi_n$. For all $\omega > 0$, it follows that

$$\begin{aligned} 2J(g_n) = \int_{-\infty}^{\infty} g_n\mathcal{H}g_n' + cg_n^2\, dx &= \int_{-\infty}^{\infty} \psi_n\varphi_\omega\mathcal{H}(\psi_n\varphi_\omega')dx \\ &+ \int_{-\infty}^{\infty} \varphi_\omega^2(c\psi_n^2 + \psi_n\mathcal{H}\psi_n')dx + \int_{-\infty}^{\infty} \psi_n\varphi_\omega[\mathcal{H},\varphi_\omega]\psi_n'\, dx. \end{aligned}$$

Since $|\varphi_\omega|_\infty = 1$ and $|\varphi_\omega'|_\infty \leqq |\varphi'|_\infty/\omega$, it follows from Lemma 8.9 that

$$\begin{aligned} \int_{-\infty}^{\infty} \psi_n\varphi_\omega[\mathcal{H},\varphi_\omega]\psi_n'\, dx \leqq |\varphi_\omega|_\infty\|\psi_n\|\|[\mathcal{H},\varphi_\omega]\psi_n'\| &\leqq \frac{C|\varphi'|_\infty}{\omega}\|\psi_n\|^2 \\ &\leqq \frac{CB^2|\varphi'|_\infty}{\omega} \end{aligned}$$

and

$$\int_{-\infty}^{\infty} \psi_n \varphi_\omega \mathcal{H}(\psi_n \varphi_\omega')dx \leqq |\varphi_\omega|_\infty \|\psi_n\| \|\varphi_\omega' \psi_n\| \leqq \frac{B^2|\varphi'|_\infty}{\omega}$$

where C is independent of r, n, and ω. So, it follows for large ω that

$$\left| J(g_n) - \frac{1}{2}\int_{-\infty}^{\infty} \varphi_\omega^2 (c\psi_n^2 + \psi_n \mathcal{H}\psi_n')dx \right| \leqq \frac{B^2|\varphi'|_\infty}{\omega}(C+1) < \frac{\epsilon}{2}.$$

Similarly, we obtain for large ω that

$$\left| J(h_n) - \frac{1}{2}\int_{-\infty}^{\infty} \zeta_\omega^2 (c\psi_n^2 + \psi_n \mathcal{H}\psi_n')dx \right| \leqq \frac{B^2|\zeta'|_\infty}{\omega}(C+1) < \frac{\epsilon}{2}.$$

So, since $J(\psi_n) = \int_{-\infty}^{\infty} (\varphi_\omega^2 + \zeta_\omega^2)(c\psi_n^2 + \psi_n \mathcal{H}\psi_n')dx$, the lemma follows. □

LEMMA 8.11. *For every $\epsilon > 0$, there exist $N \in \mathbb{N}$ and sequences $\{g_N, g_{N+1}, ...\}$ and $\{h_N, h_{N+1}, ...\}$ in $H^{\frac{1}{2}}(\mathbb{R})$ such that for $j \geqq N$,*

(1) $|K(g_j) - \alpha| < \epsilon$,
(2) $|K(h_j) - (\lambda - \alpha)| < \epsilon$,
(3) $J(\psi_j) \geqq J(g_j) + J(h_j) - \epsilon$.

PROOF. For a given ϵ, we see from the definition of α that there exists $\omega_1 > 0$ such that for $\omega \geqq \omega_1$,

$$\alpha - \epsilon < Q(\omega) \leqq Q(2\omega) \leqq \alpha.$$

Let ω_0 be as in Lemma 8.10 and fix $\omega = \max(\omega_0, \omega_1)$. Since Q_j tends pointwise to Q on $\mathbb{R}^+$, we can find N such that for all $j \geqq N$, $|Q_j(r) - Q(r)| < \epsilon/2$ and $|Q_j(2r) - Q(2r)| < \epsilon$. So, for $j \geqq N$ it follows that

$$\alpha - \epsilon < Q_j(r) \leqq Q_j(2r) < \alpha + \epsilon,$$

and therefore we can find a y_j such that

$$\int_{y_j-\omega}^{y_j+\omega} |\psi_j|^{p+2}\, dx > \alpha - \epsilon \quad \text{and} \quad \int_{y_j-2\omega}^{y_j+2\omega} |\psi_j|^{p+2}\, dx < \alpha + \epsilon. \tag{8.13}$$

Now for each $j \geqq N$ define $g_j(x) = \varphi_\omega(x - y_j)\psi_j$ and $h_j(x) = \zeta_\omega(x - y_j)\psi_j$ (note that g_j's and h_j's belong to $H^{\frac{1}{2}}(\mathbb{R})$). Then from the inequalities

$$\begin{aligned}\int_{-\infty}^{\infty} |g_j(x)|^{p+2}dx &= \int_{-2\omega}^{2\omega} \varphi^{p+2}(x/\omega)|\psi_j(x+y_j)|^{p+2}\, dx \\ &\leqq \int_{-2\omega}^{2\omega} |\psi_j(x+y_j)|^{p+2}\, dx < \alpha + \epsilon\end{aligned}$$

and

$$\begin{aligned}\int_{-\infty}^{\infty} |g_j(x)|^{p+2}dx &= \int_{-\infty}^{\infty} \varphi_\omega^{p+2}(x)|\psi_j(x+y_j)|^{p+2}\, dx \\ &\geqq \int_{-\omega}^{\omega} \varphi^{p+2}(x/\omega)|\psi_j(x+y_j)|^{p+2}\, dx = \int_{y_j-\omega}^{y_j+\omega} |\psi_j|^{p+2}\, dx > \alpha - \epsilon,\end{aligned}$$

we obtain part (1) of the lemma. Since $K(\psi_j) = \lambda$, we similarly have part (2) from the inequalities in (8.13). Finally, by choosing $r = y_j$ in Lemma 8.10, we obtain part (3). This completes the proof of the lemma. □

Now we prove that the dependence of M_λ on λ is sub-linear; hence we obtain the *sub-additivity property* of M_λ. This property will be key for us in ruling out *dichotomy.*

LEMMA 8.12. *For all* $\theta \in (0,\lambda)$,

$$M_\lambda < M_\theta + M_{\lambda-\theta}. \tag{8.14}$$

PROOF. Initially we prove that for all η, λ positive numbers, we have $M_\lambda = (\lambda/\eta)^{\frac{2}{p+2}} M_\eta$. In fact, let $K(\psi) = \lambda$. Then $\varphi \equiv (\eta/\lambda)^{\frac{1}{p+2}}\psi$ satisfies $K(\varphi) = \eta$. So we see that

$$J(\varphi) = (\eta/\lambda)^{\frac{2}{p+2}} J(\psi) \geqq M_\eta \;\Rightarrow\; M_\lambda \geqq (\lambda/\eta)^{\frac{2}{p+2}} M_\eta.$$

The other inequality is similarly obtained. Now, we consider $\theta \in (0,\lambda)$. There exists $\eta \in (0,1)$ such that $\theta = \eta\lambda$. Therefore, since the function $h(\eta) = \eta^{\frac{2}{p+2}} + (1-\eta)^{\frac{2}{p+2}}$ satisfies $h(\eta) > 1$ for all $\eta \in (0,1)$ and $M_\lambda > 0$, one sees that

$$M_\theta + M_{\lambda-\theta} = \left[\eta^{\frac{2}{p+2}} + (1-\eta)^{\frac{2}{p+2}}\right] M_\lambda > M_\lambda.$$

This completes the proof. □

THEOREM 8.13 (Ruling out Dichotomy). *For every minimizing sequence* $\{\psi_n\}$ *for problem* (P_2) *we have* $\alpha = \lambda$.

PROOF. Suppose $0 < \alpha < \lambda$. Then the following must be true:

$$M_\lambda \geqq M_\alpha + M_{\lambda-\alpha}. \tag{8.15}$$

In fact, for a given $\epsilon > 0$, we consider N and sequences $\{g_j\}_{j\geqq N}$ and $\{h_j\}_{j\geqq N}$ in $H^{\frac{1}{2}}(\mathbb{R})$ as in Lemma 8.11. For $j \geqq N$, we set

$$\tilde{g}_j = \frac{\alpha^{1/p+2}}{|g_j|_{p+2}} g_j \quad \text{and} \quad \tilde{h}_j = \frac{(\lambda-\alpha)^{1/p+2}}{|h_j|_{p+2}} h_j.$$

Then $K(\tilde{g}_j) = \alpha$ and $K(\tilde{h}_j) = \lambda - \alpha$. So, it follows that $J(\tilde{g}_j) \geqq M_\alpha$ and $J(\tilde{h}_j) \geqq M_{\lambda-\alpha}$. Therefore,

$$J(g_j) \geqq \frac{|g_j|^2_{p+2}}{\alpha^{2/p+2}} M_\alpha \quad \text{and} \quad J(h_j) \geqq \frac{|h_j|^2_{p+2}}{(\lambda-\alpha)^{2/p+2}} M_{\lambda-\alpha}.$$

Hence from Lemma 8.11, it follows that

$$\begin{aligned} J(\psi_j) &\geqq \frac{|g_j|^2_{p+2}}{\alpha^{2/p+2}} M_\alpha + \frac{|h_j|^2_{p+2}}{(\lambda-\alpha)^{2/p+2}} M_{\lambda-\alpha} - \epsilon \\ &\geqq \frac{(\alpha-\epsilon)^{2/p+2}}{\alpha^{2/p+2}} M_\alpha + \frac{(\lambda-\alpha)^{2/p+2}}{(\lambda-\alpha)^{2/p+2}} M_{\lambda-\alpha} - \epsilon. \end{aligned}$$

Now taking the limit first as $j \to \infty$ (for a fixed ϵ) and then as $\epsilon \to 0$, we obtain (8.15). But if $\alpha \in (0,\lambda)$, then this result would contradict Lemma 8.12. Thus the theorem is proved. □

From Theorem 8.8 and Theorem 8.13 we have ruled out the two possibilities: $\alpha = 0$ and $0 < \alpha < \lambda$. Hence $\alpha = \lambda$ and so the *compactness* alternative of Lions ([**193**], [**194**]) gives the following.

THEOREM 8.14 (Compactness). *Suppose $\alpha = \lambda$. Then there exists a sequence $\{y_k\} \subseteq \mathbb{R}$ with the property that for every $z \in (\frac{\lambda}{2}, \lambda)$, there exists a real number $\omega = \omega(z)$ such that for all n sufficiently large ($n \geqq N(z)$),*

$$\int_{y_n-\omega}^{y_n+\omega} |\psi_n(x)|^{p+2}\,dx > z.$$

PROOF. Let $z \in (\frac{\lambda}{2}, \lambda)$. Then there exists $\epsilon > 0$ such that $z = \lambda - \epsilon$ and so we can find $\omega_1 = \omega_1(\epsilon)$ such that

$$-\frac{\epsilon}{2} < Q(\omega_1) - \lambda < \frac{\epsilon}{2}.$$

Moreover, there exists $N_1(\omega_1, \epsilon) \equiv N_1(z)$ such that for all $n \geqq N_1(z)$, $-\frac{\epsilon}{2} < Q_n(\omega_1) - Q(\omega_1) < \frac{\epsilon}{2}$. Hence, for all $n \geqq N_1(z)$

$$Q_n(\omega_1(z)) = \sup_{y\in\mathbb{R}} \int_{y-\omega_1(z)}^{y+\omega_1(z)} |\psi_n(x)|^{p+2}\,dx > z.$$

So, for each z there exists a sequence $\{y_n(z)\}$ such that

$$\int_{y_n(z)-\omega_1(z)}^{y_n(z)+\omega_1(z)} |\psi_n(x)|^{p+2}\,dx > z.$$

Similarly, for $z = \frac{\lambda}{2}$ we can find $\omega_1 = \omega_1(\frac{\lambda}{2})$ and $N_1(\frac{\lambda}{2})$ and we can define $y_n \equiv y_n(\frac{\lambda}{2})$ such that for all $n \geqq N_1(\frac{\lambda}{2})$,

$$\int_{y_n-\omega_1(\frac{\lambda}{2})}^{y_n+\omega_1(\frac{\lambda}{2})} |\psi_n(x)|^{p+2}\,dx > \frac{\lambda}{2}.$$

Next, we note that the intervals $[y_n - \omega_1(\frac{\lambda}{2}), y_n + \omega_1(\frac{\lambda}{2})]$ and $[y_n(z) - \omega_1(z), y_n(z) + \omega_1(z)]$ must overlap. In fact, otherwise we have

$$\lambda = \int_{-\infty}^{\infty} |\psi_n|^{p+2}\,dx \geqq \int_{y_n(z)-\omega_1(z)}^{y_n(z)+\omega_1(z)} |\psi_n|^{p+2}\,dx + \int_{y_n-\omega_1}^{y_n+\omega_1} |\psi_n|^{p+2}\,dx > z + \frac{\lambda}{2} > \lambda.$$

Therefore, we must have $|y_n - y_n(z)| \leqq \omega_1(z) + \omega_1(\frac{\lambda}{2})$ for every $z \in (\frac{\lambda}{2}, \lambda)$. So, defining $\omega \equiv 2\omega_1(z) + \omega_1(\frac{\lambda}{2})$ and $N(z) \equiv \max(N_1(z), N_1(\frac{\lambda}{2}))$, we obtain that

$$[y_n(z) - \omega_1(z), y_n(z) + \omega_1(z)] \subseteq [y_n - \omega, y_n + \omega].$$

Then, we have that for every $z \in (\frac{\lambda}{2}, \lambda)$, there exists a real number $\omega = \omega(z)$ such that for all $n \geqq N(z)$ we have

$$\int_{y_n-\omega}^{y_n+\omega} |\psi_n(x)|^{p+2}\,dx \geqq \int_{y_n(z)-\omega_1(z)}^{y_n(z)+\omega_1(z)} |\psi_n(x)|^{p+2}\,dx > z.$$

This completes the proof of the theorem. □

Finally we have the proof of Theorem 8.4.

PROOF. Let $\{\psi_n\}$ be a minimizing sequence to M_λ. By using the notation of Theorem 8.14, we define the sequence

$$\widetilde{\psi}_n(x) = \psi_n(x + y_n) \quad \text{for } x \in \mathbb{R}.$$

For each $k \in \mathbb{N}$, let $z = \lambda - \frac{1}{k}$. Then in Theorem 8.14 there exists $\omega_k = \omega(k)$ such that for all $n \geqq N(k)$

$$\int_{-\omega_k}^{\omega_k} |\widetilde{\psi}_n(x)|^{p+2}\,dx = \int_{y_n-\omega_k}^{y_n+\omega_k} |\psi_n(x)|^{p+2}\,dx > \lambda - \frac{1}{k}. \tag{8.16}$$

Now, since $\|\widetilde{\psi}_n\|_{\frac{1}{2}} = \|\psi_n\|_{\frac{1}{2}} \leqq B$, there exits a subsequence of $\{\widetilde{\psi}_n\}$, denoted again by $\{\widetilde{\psi}_n\}$, which converges weakly in $H^{\frac{1}{2}}(\mathbb{R})$ to a limit function ψ. In other words, $\widetilde{\psi}_n \rightharpoonup \psi$ as $n \to \infty$. Therefore, the embedding $H^{\frac{1}{2}}(\mathbb{R}) \hookrightarrow L^{p+2}(\mathbb{R})$ and Fatou's Lemma imply that $\widetilde{\psi}_n \rightharpoonup \psi$ in $L^{p+2}(\mathbb{R})$ and

$$\int_{-\infty}^{\infty} |\psi(x)|^{p+2}dx \leqq \liminf_{n\to\infty} \int_{-\infty}^{\infty} |\widetilde{\psi}_n(x)|^{p+2}dx = \lambda. \tag{8.17}$$

Moreover, since $J(f)$ is a norm equivalent to the standard norm in $H^{\frac{1}{2}}(\mathbb{R})$, we obtain the basic inequality

$$J(\psi) \leqq \liminf_{n\to\infty} J(\widetilde{\psi}_n) = M_\lambda. \tag{8.18}$$

So, from (8.17) and (8.18) we only need to show that $K(\psi) = \lambda$. In fact, let us denote $\varphi_n = \widetilde{\psi}_n|_{(-\omega_k,\omega_k)}$. Using the relation (see Appendix A)

$$\|\varphi_n\|_{H^{\frac{1}{2}}(-\omega_k,\omega_k)} \leqq \|\widetilde{\psi}_n\|_{\frac{1}{2}} \leqq B$$

and the fact that $H^{\frac{1}{2}}(-r,r)$ is reflexive for all $r \in \mathbb{R}$, it follows that there exists a subsequence of $\{\varphi_n\}$ denoted by $\{\varphi_{n_j,k}\}_{j\geqq 1}$ such that $\varphi_{n_j,k} \rightharpoonup f_k$ in $H^{\frac{1}{2}}(-\omega_k,\omega_k)$ as $j \to \infty$. Since the embedding $H^{\frac{1}{2}}(-\omega_k,\omega_k) \hookrightarrow L^{p+2}(-\omega_k,\omega_k)$ is compact, we have that $\varphi_{n_j,k} \to f_k$ in $L^{p+2}(-\omega_k,\omega_k)$ as $j \to \infty$. Now, since $\widetilde{\psi}_{n_j,k} \rightharpoonup \psi$ in $L^{p+2}(-r,r)$ as $j \to \infty$ for all $r \in \mathbb{R}$ and $k \in \mathbb{N}$, it follows that $f_k = \psi|_{(-\omega_k,\omega_k)}$ for all $k \in \mathbb{N}$. So, by using a Cantor diagonalization argument, we can find a subsequence $\{\widetilde{\psi}_{n_m}\}_{m\geqq 1}$ such that

$$\widetilde{\psi}_{n_m} \to \psi \ \text{ in } \ L^{p+2}(-\omega_k,\omega_k), \ \text{ as } m \to \infty, \quad \text{for all } k \in \mathbb{N}.$$

Therefore from (8.16) and (8.17) it follows that

$$\lambda \geqq \int_{-\infty}^{\infty} |\psi(x)|^{p+2}dx \geqq \int_{-\omega_k}^{\omega_k} |\psi(x)|^{p+2}\,dx \geqq \lambda - \frac{1}{k} \quad \text{for all } \ k \in \mathbb{N}.$$

Hence, $K(\psi) = \lambda$. Finally, since $\widetilde{\psi}_{n_k} \rightharpoonup \psi$ in $H^{\frac{1}{2}}(\mathbb{R})$ as $k \to \infty$ and $J(\psi) = \lim_{k\to\infty} J(\widetilde{\psi}_{n_k})$, we see that $\widetilde{\psi}_{n_k} \to \psi$ in $H^{\frac{1}{2}}(\mathbb{R})$ as $k \to \infty$. This proves Theorem 8.4. □

The following two lemmas will imply Theorem 8.3. Here, for $f : \mathbb{R} \to \mathbb{C}$, f^* will represent the *symmetric decreasing rearrangement* of f, which is a decreasing function of $|x|$ (see Appendix A).

LEMMA 8.15. *If $f \in H^{\frac{1}{2}}(\mathbb{R})$, then $|f|$ and $f^* \in H^{\frac{1}{2}}(\mathbb{R})$. We also have*

$$\|D^{\frac{1}{2}} f\|^2 \geqq \|D^{\frac{1}{2}}|f|\|^2 \quad \textit{and} \quad \|D^{\frac{1}{2}} f\|^2 \geqq \|D^{\frac{1}{2}} f^*\|^2.$$

PROOF. The proof is based on Lemmas 3.4 and 3.5 in [**14**]. Initially we show that $|f| \in H^{\frac{1}{2}}(\mathbb{R})$. Let $\nu > 0$, and define the function $N_\nu(x)$ by

$$\widehat{N_\nu}(\xi) = \frac{1}{\nu + |\xi|}\,.$$

Then N_ν is a positive, even, and decreasing function. Moreover, $N_\nu \in L^p(\mathbb{R})$ for every $p \in [1, \infty]$.

Now, if $g = |f|$, then $\langle g, N_\nu * g\rangle \geqq \langle f, N_\nu * f\rangle$ for every $\nu > 0$. Therefore,

$$\int_{-\infty}^{\infty} \frac{1}{\nu + |\xi|} |\widehat{g}(\xi)|^2 \, d\xi = \int_{-\infty}^{\infty} \overline{\widehat{g}(\xi)}\ \widehat{N_\nu * g}(\xi)\, d\xi = \langle g, N_\nu * g\rangle$$
$$\geqq \langle f, N_\nu * f\rangle = \int_{-\infty}^{\infty} \frac{1}{\nu + |\xi|} |\widehat{f}(\xi)|^2 \, d\xi.$$

Since $\int_{\mathbb{R}} |\widehat{g}(\xi)|^2 d\xi = \int_{\mathbb{R}} |\widehat{f}(\xi)|^2 d\xi$, it follows that

$$\int_{-\infty}^{\infty} \nu\Big[1 - \frac{\nu}{\nu + |\xi|}\Big] |\widehat{f}(\xi)|^2 \, d\xi \geqq \int_{-\infty}^{\infty} \nu\Big[1 - \frac{\nu}{\nu + |\xi|}\Big] |\widehat{g}(\xi)|^2 \, d\xi. \tag{8.19}$$

Since $\lim_{\nu\to\infty} \nu\Big[1 - \frac{\nu}{\nu+|\xi|}\Big] = |\xi|$, taking the limit as $\nu \to \infty$ on both sides of (8.19) and using the Monotone Convergence Theorem gives

$$\int_{-\infty}^{\infty} |\xi||\widehat{f}(\xi)|^2 \, d\xi \geqq \int_{-\infty}^{\infty} |\xi||\widehat{g}(\xi)|^2 \, d\xi,$$

which shows that $|f| \in H^{\frac{1}{2}}(\mathbb{R})$ and $\|D^{\frac{1}{2}} f\|^2 \geqq \|D^{\frac{1}{2}}|f|\|^2$.

Let us prove that $f^* \in H^{\frac{1}{2}}(\mathbb{R})$. Since $N_\nu^* = N_\nu$, then for $h = |f|^* \equiv f^*$ the Riesz inequality (see Appendix A) gives

$$\int_{-\infty}^{\infty} h(x)(N_\nu * h)(x)\, dx \geqq \int_{-\infty}^{\infty} f(x)(N_\nu * f)(x)\, dx.$$

Then it follows that

$$\int_{-\infty}^{\infty} \frac{1}{\nu + |\xi|} |\widehat{h}(\xi)|^2 \, d\xi = \int_{-\infty}^{\infty} h(x)(N_\nu * h)(x)\, dx \geqq \int_{-\infty}^{\infty} f(x)(N_\nu * f)(x)\, dx.$$

Also, by Parseval's identity we have $\|\widehat{f^*}\|^2 = \|f^*\|^2 = \|f\|^2 = \|\widehat{f}\|^2$. Hence, by using an analysis similar to that in the preceding proof, we have that $f^* \in H^{\frac{1}{2}}(\mathbb{R})$. □

LEMMA 8.16. *If $\psi \in N_\lambda$, then $|\psi|^* \in G_\lambda$. Moreover, $M_\lambda \geqq I_\lambda$.*

PROOF. Initially we show that $\psi^* \equiv |\psi|^* \in N_\lambda$. In fact, since rearrangement preserves the L^{p+2}-norm, it follows that $K(|\psi|^*) = K(\psi) = \lambda$. So, from Theorem 8.14

$$M_\lambda = J(\psi) \geqq J(|\psi|^*) \geqq M_\lambda.$$

Therefore we have that $|\psi|^* \in N_\lambda$. Since $H(|\psi|^*) = K(\psi) = \lambda$, it follows easily that

$$M_\lambda \geqq I_\lambda. \tag{8.20}$$

Suppose now that $|\psi|^* \notin G_\lambda$. Then there exists $\varphi \in H^{\frac{1}{2}}(\mathbb{R})$ such that $H(\varphi) = \lambda$ and $J(\varphi) < J(|\psi|^*)$. By defining

$$f = \frac{\lambda^{\frac{1}{p+2}}}{|\varphi|_{p+2}} \varphi,$$

we have $K(f) = \lambda$ and

$$J(f) = \frac{\lambda^{\frac{2}{p+2}}}{|\varphi|_{p+2}^2} J(\varphi) \leqq J(\varphi),$$

since $\lambda = H(\varphi) \leqq K(\varphi)$. Therefore,

$$J(|\psi|^*) = M_\lambda \leqq J(f) \leqq J(\varphi) < J(|\psi|^*),$$

which is a contradiction. So, $|\psi|^* \in G_\lambda$. □

Now we have the proof of Theorem 8.3 regarding the existence of solitary wave solutions to the GBO equations.

PROOF. By Lemma 8.16 we see that $G_\lambda \neq \emptyset$. If $\psi \in G_\lambda$, then by the Lagrange multiplier principle [**203**] there exists $\theta \in \mathbb{R}$ such that ψ is a solution of the Euler-Lagrange equation $\delta J(\psi) = \theta \delta H(\psi)$, where $\delta J(\psi)$ and $\delta H(\psi)$ are the Fréchet derivatives of J and H at ψ. Now, since δJ and δH are given as distributions in $H^{-\frac{1}{2}}(\mathbb{R})$, we see that the equality

$$\mathcal{H}\psi' + c\psi = \theta(p+2)\psi^{p+1} \tag{8.21}$$

is true in the sense of distributions. Now, a bootstrapping argument will show that $\psi \in H^\infty(\mathbb{R})$ and so (8.21) holds in the pointwise sense. In fact, since $\psi^{p+1} \in L^2(\mathbb{R})$ ($H^{\frac{1}{2}}(\mathbb{R}) \hookrightarrow L^{2p+2}(\mathbb{R})$), we get from (8.21) that $\mathcal{H}\psi' \in L^2(\mathbb{R})$ and so $\psi' \in L^2(\mathbb{R})$. Then, $\psi \in H^1(\mathbb{R})$. Similarly, since $\psi^{p+1} \in H^1(\mathbb{R})$, then $\psi \in H^2(\mathbb{R})$. Therefore, proceeding in this form, we obtain that $\psi \in H^n(\mathbb{R})$ for all $n \in \mathbb{N}$. Now we see that $\theta > 0$. In fact, since $I_\lambda > 0$, it follows from (8.21) that

$$2I_\lambda = 2J(\psi) = \theta(p+2)H(\psi) = \theta(p+2)\lambda$$

and so θ is positive. Then, by defining η such that $\eta^p = \theta(p+2)(p+1)$, it follows that $\phi \equiv \eta\psi$ satisfies the solitary wave equation

$$\mathcal{H}\phi' + c\phi = \frac{1}{p+1}\phi^{p+1}. \tag{8.22}$$

Therefore, by choosing $\psi = |\varphi|^*$ with $\varphi \in N_\lambda$, we have that $\phi(x)$, defined above, is a nonnegative, smooth, even, decreasing function of $|x|$, solitary wave solution of the GBO equations. Finally, we see that ϕ is strictly positive. In fact, by taking the Fourier transform on both sides of (8.22) we see that

$$\widehat{\phi}(\xi) = \frac{1}{1+p}\,\frac{1}{|\xi|+c}\,\widehat{\phi^{p+1}}(\xi) = \frac{1}{1+p}\,\widehat{N_c}(\xi)\widehat{\phi^{p+1}}(\xi) = \frac{1}{1+p}\,(N_c * \phi^{p+1})^\wedge(\xi).$$

So, ϕ satisfies the convolution equation

$$\phi(x) = \frac{1}{1+p}\,N_c * \phi^{p+1}(x) = \frac{1}{1+p}\int_{-\infty}^{\infty} N_c(x-y)\phi^{p+1}(y)dy.$$

Therefore, since $N_c > 0$, we see that if $\phi(x_0) = 0$, then $\phi \equiv 0$, which is a contradiction. The proof of Theorem 8.3 is complete. □

Next we will establish some results associated with the structure of the set of minimizers G_λ for I_λ.

PROPOSITION 8.17. Consider G_λ, the set of minimizers for I_λ. Then we have

(1) If $\psi \in G_\lambda$, then $\psi^* \in G_\lambda$.
(2) $G_\lambda \subseteq N_\lambda$.
(3) $I_\lambda = M_\lambda$.

PROOF. (1) Let $\psi \in G_\lambda$. Since $\psi^p, \psi^2 \in L^2(\mathbb{R})$, it follows from the properties of the symmetric decreasing rearrangement that

$$0 < \lambda = \int_{-\infty}^{\infty} \psi^p \psi^2 dx \leqq \int_{-\infty}^{\infty} (\psi^p)^* (\psi^2)^* dx = \int_{-\infty}^{\infty} (\psi^*)^{p+2} dx \equiv \mu.$$

The idea now is to show that $\lambda = \mu$ and $J(\psi) = J(\psi^*)$. Indeed, consider $\varphi \in G_\mu$ and define $\rho = (\lambda/\mu)^{\frac{1}{p+2}}$. Then, $H(\rho^{-1}\psi) = \mu$ and $H(\rho\varphi) = \lambda$. So, we have the inequalities

$$J(\rho^{-1}\psi) = \rho^{-2} J(\psi) \geqq J(\varphi)$$

and

$$J(\rho\varphi) = \rho^2 J(\varphi) \geqq J(\psi).$$

Hence, we obtain that

$$J(\varphi) \geqq \rho^{-2} J(\psi) \geqq J(\varphi) \Rightarrow J(\psi) = \rho^2 J(\varphi).$$

Therefore, Lemma 8.15 and $\lambda \leqq \mu$ imply that

$$J(\psi^*) \leqq J(\psi) = \rho^2 J(\varphi) \leqq J(\varphi) \leqq J(\psi^*).$$

It follows that $J(\psi^*) = J(\psi)$, and the equality $\rho^2 J(\varphi) = J(\varphi)$ implies that $\lambda = \mu$. This completes the proof of item (1).

(3) By Lemma 8.16 we only need to show that $I_\lambda \geqq M_\lambda$. Indeed, if $\psi \in G_\lambda$, then by item (1) we get from $\psi^* \equiv |\psi|^* \geqq 0$ that $K(|\psi|^*) = H(|\psi|^*) = H(\psi) = \lambda$, and so

$$I_\lambda = J(\psi) = J(|\psi|^*) \geqq M_\lambda.$$

(2) Let $\psi \in G_\lambda$. Then from item (1) and properties of the symmetric decreasing rearrangement it follows that $\lambda = H(\psi) = H(\psi^*) = K(\psi^*) = K(\psi)$. Moreover, from item (3) we get

$$J(\psi) = I_\lambda = M_\lambda.$$

So, $\psi \in N_\lambda$. This proves the proposition. □

The following theorem will be important in our stability theory of solitary wave solutions for the GBO equations.

THEOREM 8.18. *Let $\lambda > 0$. Then*

(1) *Every minimizing sequence to I_λ converges, up to a translation, in $H^{\frac{1}{2}}(\mathbb{R})$ to an element of G_λ.*
(2) *Let $\{\varphi_n\}$ be a minimizing sequence to I_λ. Then we have*
    (a) $\lim_{n\to\infty} \inf_{\psi\in G_\lambda, y\in\mathbb{R}} \|\varphi_n(\cdot + y) - \psi\|_{\frac{1}{2}} = 0.$
    (b) $\lim_{n\to\infty} \inf_{\psi\in G_\lambda} \|\varphi_n - \psi\|_{\frac{1}{2}} = 0.$

PROOF. (1) Let $\{\varphi_n\}$ be a minimizing sequence to I_λ. Then, $K(\varphi_n) \geqq H(\varphi_n) = \lambda > 0$. Define

$$\psi_n = \frac{\lambda^{1/(p+2)}}{|\varphi_n|_{p+2}} \varphi_n.$$

So, we have that $K(\psi_n) = \lambda$ and

$$M_\lambda \leqq J(\psi_n) = \frac{\lambda^{2/(p+2)}}{|\varphi_n|_{p+2}^2} J(\varphi_n) \leqq J(\varphi_n). \tag{8.23}$$

Hence, taking the limits on both sides of the inequality in (8.3) and using Proposition 8.17, we obtain that $\{\psi_n\}$ is a minimizing sequence to M_λ. Then there is a subsequence $\{\psi_{n_k}\}$ and a sequence of real numbers $\{y_{n_k}\}$ such that

$$\psi_{n_k}(\cdot + y_{n_k}) \to \psi \qquad \text{as } k \to \infty$$

in the $H^{\frac{1}{2}}(\mathbb{R})$-norm with $\psi \in N_\lambda$. Now, by passing to a subsequence if it is necessary, we have that $\{\lambda^{2/(p+2)}/|\varphi_n|_{p+2}^2\}$ converges to a nonnegative real number $L \leqq 1$. But, from the equality

$$M_\lambda = \lim_{n\to\infty} J(\psi_n) = LI_\lambda$$

it follows that $L = 1$.

We will show now that the subsequence $\{\varphi_{n_k}(\cdot + y_{n_k})\}$ converges strongly in $H^{\frac{1}{2}}(\mathbb{R})$ to an element of G_λ. In fact,

$$\lim_{k\to\infty} \varphi_{n_k}(\cdot + y_{n_k}) = \lim_{k\to\infty} \frac{|\varphi_{n_k}|_{p+2}}{\lambda^{1/(p+2)}} \psi_{n_k}(\cdot + y_{n_k}) = \lim_{k\to\infty} \psi_{n_k}(\cdot + y_{n_k}) = \psi,$$

in $H^{\frac{1}{2}}(\mathbb{R})$. So, from the Sobolev Embedding Theorem it follows that $\varphi_{n_k}(\cdot + y_{n_k}) \to \psi$ in $L^{p+2}(\mathbb{R})$ as $k \to \infty$. Therefore,

$$H(\psi) = \lim_{k\to\infty} H(\varphi_{n_k}(\cdot + y_{n_k})) = \lim_{k\to\infty} H(\varphi_{n_k}) = \lambda.$$

Finally, since $J(\psi) = M_\lambda = I_\lambda$, we have that $\psi \in G_\lambda$. This shows item (1).

(2) Suppose that (a) is not true. Then there exists $\{\varphi_{n_k}\}$ such that

$$\eta \equiv \inf_{\psi\in G_\lambda, y\in\mathbb{R}} \|\varphi_{n_k}(\cdot + y) - \psi\|_{\frac{1}{2}} \geqq \epsilon \qquad \text{for all } k.$$

Since $\{\varphi_{n_k}\}$ is a minimizing sequence to I_λ, we have by item (1) that there exists $\{y_{n_k}\} \subseteq \mathbb{R}$ such that (up to a subsequence) $\varphi_{n_k}(\cdot + y_{n_k}) \to \varphi$ in $H^{\frac{1}{2}}(\mathbb{R})$ as $k \to \infty$. So, for k sufficiently large we have

$$\frac{\epsilon}{2} \geqq \|\varphi_{n_k}(\cdot + y_{n_k}) - \varphi\|_{\frac{1}{2}} \geqq \eta \geqq \epsilon,$$

which is a contradiction. This shows (2)(a).

Finally, part (b) follows from the property that if $\psi \in G_\lambda$, then $\psi(\cdot + y) \in G_\lambda$ for all $y \in \mathbb{R}$, and from the equality

$$\inf_{\psi\in G_\lambda} \|\varphi_n - \psi\|_{\frac{1}{2}} = \inf_{\psi\in G_\lambda, y\in\mathbb{R}} \|\varphi_n - \psi(\cdot - y)\|_{\frac{1}{2}} = \inf_{\psi\in G_\lambda, y\in\mathbb{R}} \|\varphi_n(\cdot + y) - \psi\|_{\frac{1}{2}}.$$

This completes the proof of the theorem. □

8.3. Stability of Solitary Waves for the GBO Equations

In this section we will see that the variational characterization of solitary wave solutions for the GBO equations (8.2), established in Section 8.2 and defined in (8.6), can give us information about the problem of stability. In particular, we obtain a new proof about the orbital stability of the solitary wave solutions in (3.6) associated with the BO equation (see [**49**] or Chapter 7, Theorem 7.11). This proof does not need any spectral information.

We start by seeing that there exists a $\lambda > 0$ such that every element in the set of minimizers, G_λ, satisfies equation (8.3).

LEMMA 8.19. *For* $\lambda = [2(p+1)I_1]^{(p+2)/p}$ *in the variational problem* (8.6), *we have that if* $\phi \in G_\lambda$, *then* ϕ *is a solitary wave solution for the GBO equations.*

PROOF. If $\phi \in G_\lambda$, then by the Lagrange multiplier principle, there exists $\theta \in \mathbb{R}$ such that

$$\mathcal{H}\phi' + c\phi = \theta(p+2)\phi^{p+1}. \tag{8.24}$$

By multiplying both sides of (8.24) by ϕ and integrating, we obtain from $J(\phi) = I_\lambda$ and $H(\phi) = \lambda$ that

$$2I_\lambda = \theta(p+2)\lambda.$$

So, using the fact that $I_\lambda = \lambda^{2/(p+2)} I_1$ and the specific value for λ, we obtain

$$\theta(p+2) = 2\lambda^{-\frac{p}{p+2}} I_1 = \frac{1}{p+1}.$$

This completes the proof. □

Next we obtain a characterization of the set G_λ with $\lambda = [2(p+1)I_1]^{(p+2)/p}$. Define the set Υ_c for $c > 0$ as

$$\Upsilon_c = \{\phi \in H^{\frac{1}{2}}(\mathbb{R}) \ : \ H(\phi) = 2(p+1)J(\phi) = [2(p+1)I_1]^{\frac{p+2}{p}}\}. \tag{8.25}$$

Then, $G_\lambda = \Upsilon_c$. Indeed, by Lemma 8.19 we immediately obtain that if $\phi \in G_\lambda$, then

$$2(p+1)J(\phi) = H(\phi) = \lambda = [2(p+1)I_1]^{\frac{p+2}{p}} \ \Rightarrow \phi \in \Upsilon_c.$$

Now suppose that $\phi \in \Upsilon_c$. Then immediately $H(\phi) = \lambda$. Moreover, since

$$I_\lambda = \lambda^{2/(p+2)} I_1 = [2(p+1)]^{2/p} I_1^{(2+p)/p},$$

it follows that

$$J(\phi) = [2(p+1)]^{\frac{2}{p}} I_1^{\frac{2+p}{p}} = I_\lambda \ \Rightarrow \phi \in G_\lambda.$$

Next we establish the main theorem of this section (see Definition 5.3).

THEOREM 8.20. *Let $c > 0$ and $\lambda = [2(p+1)I_1]^{\frac{p+2}{p}}$. Then the set $\Upsilon_c = G_\lambda$ is $H^{\frac{1}{2}}(\mathbb{R})$-stable with regard to the flow of the GBO equations if $p < 2$.*

In order to prove Theorem 8.20, we need some of our previous lemmas. Initially, for any $c > 0$ and **any** ϕ in Υ_c we define the classical function

$$d(c) = E_{\mathrm{GBO}}(\phi) + cF(\phi), \tag{8.26}$$

where

$$E_{\mathrm{GBO}}(u) = \frac{1}{2}\int_{-\infty}^{\infty} \left[u\mathcal{H}u' - \frac{2}{(p+1)(p+2)}u^{p+2}\right] dx$$

and

$$F(u) = \int_{-\infty}^{\infty} |u|^2 \, dx.$$

LEMMA 8.21. *Let $d(c)$ be defined as in (8.26). Then d has the following properties:*

(1) *d is constant on Υ_c.*
(2) *d is differentiable and strictly increasing for $c > 0$ and $p < 2$, satisfying*

$$d(c) = \frac{p}{2(p+1)(p+2)} \, c^{\frac{2}{p}} \int_{-\infty}^{\infty} \psi^{p+2}(x) \, dx \tag{8.27}$$

where $\psi \in H^\infty(\mathbb{R})$ satisfies the equation

$$\mathcal{H}\psi' + \psi = \frac{1}{p+1}\psi^{p+1}. \tag{8.28}$$

(3) *For every* $\phi \in \Upsilon_c$, $d'(c) = F(\phi)$.
(4) $d''(c) > 0 \Leftrightarrow p < 2$.

PROOF. Let $\phi \in \Upsilon_c$. Then

$$\begin{aligned} d(c) &= E_{\mathrm{GBO}}(\phi) + cF(\phi) = J(\phi) - \frac{1}{(p+1)(p+2)} H(\phi) \\ &= \Big[\frac{1}{2(p+1)} - \frac{1}{(p+1)(p+2)}\Big] H(\phi) = \frac{p}{2(p+1)(p+2)} H(\phi) \\ &= \frac{p}{p+2}\, [2(p+1)]^{\frac{2}{p}} I_1^{\frac{p+2}{p}}. \end{aligned} \tag{8.29}$$

Now we prove (8.27). Define ψ as

$$\psi(x) = c^{-\frac{1}{p}}\, \phi(x/c).$$

Then, since ϕ satisfies (8.3), it follows easily that ψ satisfies (8.28). Next, from the relation $H(\phi) = c^{\frac{2}{p}} H(\psi)$ and (8.29) it follows that

$$d(c) = \frac{p}{2(p+1)(p+2)} H(\phi) = \frac{p}{2(p+1)(p+2)}\, c^{\frac{2}{p}} H(\psi).$$

With regard to item (3), we show initially for ψ in (8.28) that

$$F(\psi) = \frac{1}{(p+1)(p+2)} H(\psi).$$

In fact, by multiplying (8.28) by $x\psi'$ and integrating by parts, we obtain

$$\int_{-\infty}^{\infty} x\psi' \mathcal{H}\psi' dx = F(\psi) - \frac{1}{(p+1)(p+2)} H(\psi).$$

But, since $\mathcal{H}(x\psi')(y) = y\mathcal{H}(\psi')(y)$ and $\mathcal{H}^2 = -I$, we then have that

$$\int_{-\infty}^{\infty} x\psi' \mathcal{H}\psi' dx = 0,$$

and so we have our equality. Now, since $F(\psi) = c^{1-\frac{2}{p}} F(\phi)$, we obtain from (8.27) that

$$d'(c) = \frac{1}{(p+1)(p+2)} c^{\frac{2}{p}-1} H(\psi) = c^{\frac{2}{p}-1} c^{1-\frac{2}{p}} F(\phi) = F(\phi).$$

From relation (8.27) we obtain item (4) immediately. This proves the lemma. □

Next, for $\epsilon > 0$ define the following ϵ-neighborhood, $U_\epsilon(\Upsilon_c)$, of set Υ_c:

$$U_\epsilon(\Upsilon_c) = \{\varphi \in H^{\frac{1}{2}}(\mathbb{R}) \;:\; \inf_{\phi \in \Upsilon_c} \|\varphi - \phi\|_{\frac{1}{2}} < \epsilon\}.$$

Then we have the following lemma.

LEMMA 8.22. *Let* $c > 0$. *For* $\phi \in \Upsilon_c$ *consider the relation*

$$d(c) = \frac{p}{2(p+1)(p+2)} H(\phi). \tag{8.30}$$

Then, there exists a small positive number ϵ *and a* C^1 *map* $\rho : U_\epsilon(\Upsilon_c) \to (0, \infty)$ *defined by*

$$\rho(u) = d^{-1}\Big(\frac{p}{2(p+1)(p+2)} H(u)\Big), \tag{8.31}$$

such that $\rho(\phi) = c$ *for every* $\phi \in \Upsilon_c$.

PROOF. Without loss of generality we can set $c = 1$. It follows immediately from (8.25) that Υ_c is a bounded set in $H^{\frac{1}{2}}(\mathbb{R})$. Moreover Υ_c is contained in the sphere of $H^{\frac{1}{2}}(\mathbb{R})$ with center at zero and radius $[2(p+1)]^{\frac{2}{p}} I_1^{\frac{p+2}{p}}$. Let $\eta > 0$ be sufficiently large such that $B(0;\eta) \supseteq \Upsilon_c$, where $B(0;\eta)$ is the open ball in $H^{\frac{1}{2}}(\mathbb{R})$ with center 0 and radius η. Now, since the function $u \to H(u)$ is uniformly continuous on bounded sets, it follows that there exists $\epsilon > 0$ such that if $u, v \in B(0;\eta)$ and $\|u - v\|_{\frac{1}{2}} < 2\epsilon$, then $|H(u) - H(v)| < \eta$. Consider the following neighborhoods of $d(c)$ and Υ_c: $\mathcal{I} = (d(c) - \eta, d(c) + \eta)$ and $U_\epsilon(\Upsilon_c)$, respectively. So, we have that if $u \in U_\epsilon(\Upsilon_c)$, then $H(u) \in \mathcal{I}$. Indeed, from the definition of $U_\epsilon(\Upsilon_c)$ there exists $\phi \in \Upsilon_c$ such that $\|u - \phi\|_{\frac{1}{2}} \leqq \epsilon$ and so $H(u) \in \mathcal{I}$. Therefore, the function ρ in (8.31) is well-defined on $U_\epsilon(\Upsilon_c)$ and satisfies $\rho(\phi) = c$ for all $\phi \in \Upsilon_c$. This proves the lemma. □

Next we establish the key inequality in our study of stability (see (6.3) and (7.22)).

LEMMA 8.23. *Let $c > 0$ and suppose that $d''(c) > 0$, in other words, $p < 2$. Then there exists $\epsilon > 0$ such that for all $u \in U_\epsilon(\Upsilon_c)$ and any $\phi \in \Upsilon_c$,*

$$E_{\text{GBO}}(u) - E_{\text{GBO}}(\phi) + \rho(u)[F(u) - F(\phi)] \geqq \frac{1}{4}\, d''(c)\, |\rho(u) - c|^2, \tag{8.32}$$

where $\rho(u)$ is defined in (8.31).

PROOF. Let ϵ be defined by Lemma 8.22 and let it be sufficiently small. Then it follows that $\rho(U_\epsilon(\Upsilon_c)) \subseteq (c - \delta, c + \delta)$ with $c - \delta > 0$. We denote by J_ϱ the functional

$$J_\varrho(\psi) = \frac{1}{2}\int_{-\infty}^{\infty} [D^{\frac{1}{2}}\psi(x)]^2 + \varrho\psi^2(x)\, dx,$$

and we denote by ϕ_ϱ any element of Υ_ϱ. Then we see that

$$E_{\text{GBO}}(u) + \rho(u)F(u) = J_{\rho(u)}(u) - \frac{1}{(p+1)(p+2)}H(u). \tag{8.33}$$

Since

$$d(\rho(u)) = \frac{p}{2(p+1)(p+2)}H(u)$$

for $u \in U_\epsilon(\Upsilon_c)$ and

$$d(\rho(u)) = \frac{p}{2(p+1)(p+2)}H(\phi_{\rho(u)}),$$

we then have that $H(u) = H(\phi_{\rho(u)})$. Therefore

$$J_{\rho(u)}(u) \geqq J_{\rho(u)}(\phi_{\rho(u)}). \tag{8.34}$$

Then it follows from (8.33), (8.34), (8.29), Lemma 8.21(3), and the Taylor expansion of d at c that

$$\begin{aligned} E_{\text{GBO}}(u) + \rho(u)F(u) &\geqq J_{\rho(u)}(\phi_{\rho(u)}) - \frac{1}{(p+1)(p+2)}H(\phi_{\rho(u)}) \\ &= d(\rho(u)) \geqq d(c) + F(\phi)(\rho(u) - c) + \frac{1}{4}\, d''(c)\, |\rho(u) - c|^2 \\ &= E_{\text{GBO}}(\phi) + \rho(u)F(\phi) + \frac{1}{4}\, d''(c)|\rho(u) - c|^2. \end{aligned}$$

This proves the lemma. □

Now, we will prove our Theorem 8.20 of nonlinear stability of the set Υ_c in $H^{\frac{1}{2}}(\mathbb{R})$.

PROOF THEOREM 8.20. Assume that Υ_c is $H^{\frac{1}{2}}(\mathbb{R})$-unstable with regard to the flow of the GBO equations (see Definition 5.3). Then there is a sequence of initial data $u_k(0) \in U_{1/k}(\Upsilon_C) \cap H^s(\mathbb{R})$, with $s > 3/2$, such that

$$\sup_{0 \leqq t < \infty} \inf_{\phi \in \Upsilon_c} \|u_k(t) - \phi\|_{\frac{1}{2}} \geqq \epsilon,$$

where $u_k(t)$ is the solution of (8.2) with initial data $u_k(0)$ (Theorem 4.2). Then, by continuity in t, we can find for large enough k a time t_k such that

$$\inf_{\phi \in \Upsilon_c} \|u_k(t_k) - \phi\|_{\frac{1}{2}} = \frac{\epsilon}{2}. \tag{8.35}$$

Now, by definition of $U_{1/k}(\Upsilon_c)$ and since E_{GBO} and F are conserved quantities for the GBO equations, we can find $\phi_k \in \Upsilon_c$ such that

$$\begin{aligned}|E_{\mathrm{GBO}}(u_k(t_k)) - E_{\mathrm{GBO}}(\phi_k)| = |E_{\mathrm{GBO}}(u_k(0)) - E_{\mathrm{GBO}}(\phi_k)| \to 0,\\ |F(u_k(t_k)) - F(\phi_k)| = |F(u_k(0)) - F(\phi_k)| \to 0\end{aligned}$$

as $k \to \infty$. Moreover, by choosing k large enough, we have from Lemma 8.23 that

$$E_{\mathrm{GBO}}(u_k(t_k)) - E_{\mathrm{GBO}}(\phi_k) + \rho(u_k(t_k))[F(u_k(t_k)) - F(\phi_k)] \geqq \frac{1}{4}\, d''(c)|\rho(u_k(t_k)) - c|^2.$$

So, since $u_k(t_k)$ is uniformly bounded for k, it follows from the last inequality that $\rho(u_k(t_k)) \to c$ as $k \to \infty$. Hence, by (8.31) and the continuity of d we have

$$\lim_{k \to \infty} H(u_k(t_k)) = \frac{2(p+1)(p+2)}{p}\, d(c). \tag{8.36}$$

On the other hand, by Lemma 8.21(1) it follows that

$$\begin{aligned}J(u_k(t_k)) &= E_{\mathrm{GBO}}(u_k(t_k)) + cF(u_k(t_k)) + \frac{1}{(p+1)(p+2)} H(u_k(t_k))\\ &= d(c) + [E_{\mathrm{GBO}}(u_k(t_k)) - E_{\mathrm{GBO}}(\phi_k)] + c[F(u_k(t_k)) - F(\phi_k)]\\ &\quad + \frac{1}{(p+1)(p+2)} H(u_k(t_k)).\end{aligned}$$

Therefore by (8.36) and (8.29), we obtain

$$\lim_{k \to \infty} J(u_k(t_k)) = \frac{p+2}{p}\, d(c) = [2(p+1)]^{\frac{2}{p}} I_1^{\frac{p+2}{p}}. \tag{8.37}$$

We note from (8.36) that $H(u_k(t_k))$ is positive for k large, and so we define the following sequence in $H^{\frac{1}{2}}(\mathbb{R})$:

$$\psi_k(t_k) = [H(u_k(t_k))]^{-\frac{1}{p+2}} u_k(t_k).$$

Then $\{\psi_k(t_k)\}$ satisfies $H(\psi_k(t_k)) = 1$, and from (8.36), (8.37), and (8.29) it follows that

$$\lim_{k \to \infty} J(\psi_k(t_k)) = \lim_{k \to \infty} [H(u_k(t_k))]^{-\frac{2}{p+2}} J(u_k(t_k)) = I_1.$$

Therefore $\{\psi_k(t_k)\}$ is a minimizing sequence to I_1. Hence from Theorem 8.18 there exists a sequence $\{\varphi_k\} \subseteq G_1$ such that

$$\lim_{k \to \infty} \|\psi_k(t_k) - \varphi_k\|_{\frac{1}{2}} = 0. \tag{8.38}$$

Now, from Lagrange's multiplier principle there are $\theta_k \in \mathbb{R}$ such that

$$\mathcal{H}\varphi_k + c\varphi_k = \theta_k(p+2)\varphi_k^{p+1}$$

and so $2I_1 = \theta_k(p+2)$, which implies $\theta_k = \theta$ for all k. Define $\phi_k = \eta\varphi_k$ with $\eta^p = \theta(p+1)(p+2) = 2(p+1)I_1$. Then every ϕ_k satisfies equation (8.3) and some calculations show that

$$2(p+1)J(\phi_k) = H(\phi_k) = \eta^{p+2},$$

which implies that for every k we have $\phi_k \in \Upsilon_c$. Finally, it follows from (8.29), (8.36), (8.38), and Υ_c being a bounded set in $H^{\frac{1}{2}}(\mathbb{R})$ that the inequality

$$\begin{aligned}\|u_k(t_k) - \phi_k\|_{\frac{1}{2}} &= [H(u_k(t_k))]^{\frac{1}{p+2}} \, \|[H(u_k(t_k))]^{-\frac{1}{p+2}}(u_k(t_k) - \phi_k)\|_{\frac{1}{2}} \\ &\leqq [H(u_k(t_k))]^{\frac{1}{p+2}} \Big[\|\psi_k(t_k) - \eta^{-1}\phi_k\|_{\frac{1}{2}} + \|\phi_k\|_{\frac{1}{2}} |\eta^{-1} \\ &\qquad - [H(u_k(t_k))]^{-\frac{1}{p+2}}|\Big]\end{aligned}$$

implies that $\|u_k(t_k) - \phi_k\|_{\frac{1}{2}} \to 0$ as $k \to \infty$. But according to (8.35) this is a contradiction. So the proof of the theorem is complete. □

By using Theorem 8.20 and the uniqueness of the solitary wave solutions (3.6) for the BO equation (Amick and Toland [**18**], Albert [**7**]), we have the following stability result for the BO equation (2.6).

THEOREM 8.24. *Let ϕ_c be the solitary wave solution* (3.6) *associated to the BO equation. Then the ϕ_c-orbit, Ω_{ϕ_c}, is $H^{\frac{1}{2}}(\mathbb{R})$-stable.*

PROOF. The idea of the proof is to show that Υ_c given by (8.25) satisfies $\Upsilon_c = \Omega_{\phi_c}$. Hence, by using Theorem 8.20 with $p = 1$, we obtain the theorem. So, let $\psi \in \Upsilon_c = G_\lambda$. Then by Lemma 8.19 we have that ψ is a solitary wave solution for the BO equation, namely,

$$\mathcal{H}\psi' + c\psi = \frac{1}{2}\psi^2. \tag{8.39}$$

Due to the uniqueness of the solutions of equation (8.39), modulo translation, we have that there exists $y \in \mathbb{R}$ such that $\psi(x) = \phi_c(x+y)$. Hence, $\psi \in \Omega_{\phi_c}$. Now we show that $\Omega_{\phi_c} \subseteq \Upsilon_c$. In fact, fix $\phi_c(\cdot + y)$ for some $y \in \mathbb{R}$. Then, since $\Upsilon_c \neq \emptyset$, there is a $\varphi \in \Upsilon_c$ that satisfies (8.39). Therefore, we see again that due to uniqueness there exists $\gamma \in \mathbb{R}$ such that $\varphi(x) = \phi_c(x+\gamma)$. Since Υ_c is invariant by translations, we obtain that

$$\varphi(x - \gamma + y) = \phi_c(x+y) \in \Upsilon_c.$$

This completes the proof of the theorem. □

8.4. Comments

We can also show the existence of minimizers to problem (P_1) without using problem (P_2). In fact, a sketch of the proof may be the following: For every minimizing sequence $\{\psi_n\} \subseteq H^{\frac{1}{2}}(\mathbb{R})$ associated with I_λ we apply the Concentration-Compactness Principle with the sequence of nonnegative functions $\{\rho_n\} \subseteq L^1(\mathbb{R})$ defined by

$$\rho_n(x) = |\psi_n(x)|^2 + |J^{\frac{1}{2}}\psi_n(x)|^2,$$

where $J^{\frac{1}{2}}$ is the Bessel potential of order $\frac{1}{2}$ (see Appendix A). Now, we rule out vanishing; this result follows easily from the following lemma (which is a slightly different version of the classical Lions's Lemma I.1 in [**193**]), namely,

- Let $p > 2$ and $s \geqq \frac{1}{2} - \frac{1}{p}$. Suppose $\{f_n\}$ is a bounded sequence in $H^s(\mathbb{R})$. If there is an $R > 0$ for which
$$\lim_{n\to\infty} \sup_{y\in\mathbb{R}} \|f_n\|_{H^s(y-R,y+R)} = 0,$$
then it follows that
$$\lim_{n\to\infty} \int_{-\infty}^{\infty} |f_n(x)|^p \, dx = 0.$$

Now we rule out dichotomy, a result that comes basically from the sub-additivity property of I_λ (see [**34**]), namely,
$$I_\lambda < I_\theta + I_{\lambda-\theta} \qquad \text{for all } \theta \in (0, \lambda)$$
and from the following estimate of the commutator for $f \in L^2(\mathbb{R})$:
$$\|[J^{\frac{1}{2}}, g]f\| \leqq C|g'|_\infty \|f\|$$
where $g \in C^\infty(\mathbb{R})$ with $g' \in L^\infty(\mathbb{R})$ and $C > 0$. So, a *compactness* alternative occurs.

The Concentration-Compactness Principle is a good tool for obtaining the best constant in many Sobolev-type inequalities. Indeed, for determining the best constant $C_{r,\beta}$ in (6.32) for the Gagliardo-Nirenberg-type inequality (6.31), it is sufficient to show that the infimum of the functional
$$J(f) = \frac{\|D^{\beta/2} f\|^{(r-2)/\beta} \|f\|^{(2+r(\beta-1))/\beta}}{|f|_r^r}$$
over $f \in H^{\beta/2}(\mathbb{R})$ is attained. If this is the case, then
$$C_{r,\beta} = \frac{1}{J(\phi)},$$
where ϕ is any element in which J takes its minimum value. It will turn out that any such ϕ must satisfy the equation depicted in (6.33).

To show that the infimum is attained, it considers the following two-parameter family of minimization problems:
$$I(\nu, \tau) = \inf\Big\{ - \int_{-\infty}^{\infty} |f(x)|^r dx : f \in H^{\beta/2}(\mathbb{R}) \text{ with } \|D^{\beta/2} f\|^2 = \nu \text{ and } \|f\|^2 = \tau \Big\},$$
where ν, τ are positive constants. It follows from (6.31) that $I(\nu, \tau) > -\infty$. By homogeneity we have that $I(\nu, \tau) = \nu^a \tau^d I(1, 1)$ where
$$a = \frac{r-2}{2\beta} \quad \text{and} \quad d = \frac{2 + r(\beta - 1)}{2\beta}.$$
Since $I(\nu, \tau) < 0$ and $a + d = \frac{1}{2}r > 1$, it follows from the sub-additive property that
$$I(\nu, \tau) < I(\sigma, \zeta) + I(\nu - \sigma, \tau - \zeta)$$
for $(\sigma, \zeta) \in (0, \nu) \times (0, \tau)$. The sub-additivity of $I(\nu, \tau)$ together with the Concentration-Compactness Principle implies the existence of a $\varphi \in H^{\beta/2}(\mathbb{R}) \cap C^\infty(\mathbb{R})$ such that
$$\|D^{\beta/2}\varphi\|^2 = \nu \;\; \|\varphi\|^2 = \tau \quad \text{and} \quad |\varphi|_r^r = -I(\nu, \tau).$$

Then an appeal to homogeneity assures that

$$J(\varphi) = \inf\{J(f) : f \in H^{\beta/2}(\mathbb{R}) - \{0\}\},$$

where $J'(\varphi) = 0$ and so φ satisfies the equation

$$2aD^\beta\varphi + 2bc_1\varphi = c_2|\varphi|^{r-2}\varphi \tag{8.40}$$

for suitable positive constants c_1, c_2. To determine the value of $C_{r,\beta}$, multiply (8.40) by φ and integrate the result over $\mathbb{R}$ to obtain

$$2a\|D^{\beta/2}\varphi\|^2 = -2bc_1\|\varphi\|^2 + c_2|\varphi|_r^r.$$

Next, multiply (8.40) by $x\varphi'$ and use the equalities $\mathcal{H}(x\varphi'') = x\mathcal{H}(\varphi'')$ and

$$D^{\beta-1}(x\mathcal{H}(\varphi'')) = (\beta - 1)D^\beta\varphi + xD^\beta\varphi'$$

to obtain

$$a(\beta - 1)\|D^{\beta/2}\varphi\|^2 = bc_1\|\varphi\|^2 - \frac{c_2}{r}|\varphi|_r^r.$$

As a consequence,

$$|\varphi|_r^r = \frac{rc_1}{c_2}\|\varphi\|^2 \quad \text{and} \quad \|D^{\beta/2}\varphi\|^2 = \frac{(b-1)c_1}{a(\beta-1)}\|\varphi\|^2.$$

If

$$\Psi(x) = \Big(\frac{c_2}{2bc_1}\Big)^{1/(r-2)}\varphi\Big(\frac{a^{1/\beta}}{(bc_1)^{1/\beta}}x\Big),$$

then Ψ satisfies (6.33) and $C_{r,\beta} = [J(\Psi)]^{-1}$, as mentioned.

CHAPTER 9

More about the Concentration-Compactness Principle

9.1. Introduction

The Concentration-Compactness Principle has a lot of applications to various differential and integro-differential equations in unbounded domains in one or more dimensions. In particular, it allows us to prove the solvability of some problems. Here we only touch upon this subject, and our illustration in Chapter 8, considered an application in the framework of the nonlinear evolution equations, shows this method is an important tool in the study of solitary wave solutions. We refer the reader to Lopes [**195**], [**198**] where a criterion is given for convergence of minimizing sequences for a general range of minimization problems. In particular, it is applied when the basic sub-additivity property is not easy to establish by ruling out dichotomy.

In this chapter we will see other applications of the method to some evolution equations that are of interest in the next chapter.

9.2. Solitary Wave Solutions of Benjamin-Type Equations

The following example of interest is based on the results of Chen and Bona in [**84**], where a theory has been established of the existence of solitary wave solutions of Benjamin-type equations of the form

$$u_t + u^p u_x - \mathcal{L}_i u_x = 0 \quad \text{for } i = 1, 2,$$

where the linear operators $\mathcal{L}_i$ are given by

$$\widehat{\mathcal{L}_1 u}(\xi) = (-l|\xi|^{2r} + |\xi|^{2m})\widehat{u}(\xi) \tag{9.1}$$

where $r \in \mathbb{R}$, $m \in \mathbb{N}$, and $0 \leqq r < m$, and

$$\widehat{\mathcal{L}_2 u}(\xi) = \sum_{j=1}^{k} l_j |\xi|^{2r_j} \widehat{u}(\xi) \tag{9.2}$$

where $l_j \in \mathbb{R}$ with $l_k > 0$ and $0 < r_1 < r_2 < \cdots < r_k$, but r_k is not an integer

In this case the Concentration-Compactness Principle is applied by obtaining solutions of the minimization problems

$$\theta_i(\lambda) = \inf\Big\{ \int_{-\infty}^{\infty} u(c_i + \mathcal{L}_i)u dx : u \in H^{s_i}(\mathbb{R}), \int_{-\infty}^{\infty} u^{p+2} dx = \lambda > 0 \Big\} \tag{9.3}$$

where $s_1 = m$, $s_2 = r_k$, and c_i are chosen such that the linear operators $c_i + \mathcal{L}_i$ are positive. Therefore we obtain the existence of nontrivial solutions ψ_i for the pseudo-differential equations

$$\mathcal{L}_i\psi_i + c_i\psi_i - \frac{1}{p+1}\psi_i^{p+1} = 0.$$

In particular, we obtain the existence of solitary wave solutions for the model

$$u_t + u^p u_x - D^\beta u_x = 0 \tag{9.4}$$

where the homogeneous linear operator D^β is defined for $\beta \geqq 1$ by $\widehat{D^\beta u}(\xi) = |\xi|^\beta \widehat{u}(\xi)$ (see (6.27) and (7.25)).

9.3. Stability of Solitary Wave Solutions: the GKdV Equations

When we apply the Concentration-Compactness Principle to the variational problem (8.1), or more precisely to the minimization problems

$$\theta(\lambda) = \inf\{E_{\mathrm{GKdV}}(\psi) : \psi \in H^1(\mathbb{R}),\ \ F(\psi) = \lambda > 0\},$$

two basic properties are necessary to rule out the possibilities of vanishing and dichotomy. The first one is that the quadratic form associated with E_{GKdV}, namely $\langle L\psi, \psi\rangle$ with $L = -\partial_x^2$, is nonnegative, and so from Section 8.4 it follows that vanishing does not occur here. The second one is based on *homogeneity properties.* We have that for every $\tau > 0$,

$$I_{\tau\lambda} = \tau^{\frac{p+4}{4-p}} I_\lambda,$$

and so it follows that I_λ has the sub-additivity property

$$I_\lambda < I_\eta + I_{\lambda-\eta}, \quad \text{for } \eta \in (0,\lambda) \text{ and } p < 4,$$

which implies that dichotomy does not occur and so the compactness alternative occurs. Therefore every minimizing sequence to $\theta(\lambda)$ will converge, up to translations, in $H^1(\mathbb{R})$ to an element of the set of minimizers for $\theta(\lambda)$, which we will denote by W_λ.

Next, for the benefit of the reader, I provide a sketch of the proof that the set of minimizers associated with the variational problem (8.1), W_λ, forms a stable set in $H^1(\mathbb{R})$ by the flow of the GKdV equations (2.5) if $p < 4$ (see Definition 5.3). Suppose that W_λ is $H^1(\mathbb{R})$-unstable. Then a number $\epsilon > 0$, a sequence $\{f_n\} \subseteq H^1(\mathbb{R})$, and a sequence of times $\{t_n\}$ exist such that

$$\inf_{\varphi \in W_\lambda} \|f_n - \varphi\|_1 < \frac{1}{n}$$

and

$$\inf_{\varphi \in W_\lambda} \|u_n(\cdot, t_n) - \varphi\|_1 \geqq \epsilon \tag{9.5}$$

for all n, where $u_n(x,t)$ solves (2.5) with $u_n(x,0) = f_n$ (see Theorems 4.3 and 4.4). Then, since $f_n \to W_\lambda$ in $H^1(\mathbb{R})$ and for every $\psi \in W_\lambda$ we have that $E_{\mathrm{GKdV}}(\psi) = \theta(\lambda)$ and $F(\psi) = \lambda$, it follows that

$$E_{\mathrm{GKdV}}(f_n) \to \theta(\lambda) \quad \text{and} \quad F(f_n) \to \lambda.$$

Choose $\{\alpha_n\}$ such that $F(\alpha_n f_n) = \lambda$ for all n; thus $\alpha_n \to 1$. Hence, by using the fact that E_{GKdV} and F are *conserved quantities* for the GKdV equations, the sequence $\psi_n = \alpha_n u_n(\cdot, t_n)$ satisfies $F(\psi_n) = F(\alpha_n f_n) = \lambda$ and

$$\lim_{n\to\infty} E_{\mathrm{GKdV}}(\psi_n) = \lim_{n\to\infty} E_{\mathrm{GKdV}}(u_n(\cdot, t_n)) = \lim_{n\to\infty} E_{\mathrm{GKdV}}(f_n) = \theta(\lambda).$$

Therefore $\{\psi_n\}$ is a minimizing sequence for $\theta(\lambda)$. So, there exists a sequence $\{y_n\} \subseteq \mathbb{R}$ such that $\psi_n(\cdot + y_n) \to \psi$ in $H^1(\mathbb{R})$ and $\psi \in W_\lambda$. Hence, it follows that

$$\lim_{n\to\infty} \inf_{\varphi\in W_\lambda} \|\psi_n - \varphi\|_1 = 0.$$

Therefore, for each large n we have that there exists $\varphi_n \in W_\lambda$ such that

$$\|\psi_n - \varphi_n\|_1 < \epsilon/2.$$

Since there exists a positive constant A_1 such that $\|u_n(\cdot, t_n)\|_1 \leqq A_1$ for large n, we obtain using (9.5) that

$$\begin{aligned}\epsilon &\leqq \|u_n(\cdot, t_n) - \varphi_n\|_1 \leqq \|u_n(\cdot, t_n) - \psi_n\|_1 + \|\psi_n - \varphi_n\|_1 \\ &\leqq |1-\alpha_n|\|u_n(\cdot, t_n)\|_1 + \frac{\epsilon}{2} \leqq |1-\alpha_n|A_1 + \frac{\epsilon}{2},\end{aligned}$$

and taking $n \to \infty$ gives $\epsilon \leqq \frac{\epsilon}{2}$, a contradiction. So, W_λ is a stable set in $H^1(\mathbb{R})$ for the flow of the GKdV equations (2.5) if $p < 4$.

Finally, we can see that if $\phi_{c,p}$ denotes the explicit solitary wave solution for the GKdV equations defined in (3.4) and we consider λ in (8.1) as $\lambda_p = \frac{1}{2}\|\phi_{c,p}\|^2$ for $p < 4$, then

$$W_{\lambda_p} = \{\pm\phi_{c,p}(\cdot + y) \;:\; y \in \mathbb{R}\}.$$

This is an example in which the set of minimizers of a variational problem is reduced to either one orbit or two orbits. We refer the reader to Proposition 2.11 in [**8**], where $p = 1$ is done.

9.4. Stability of Solitary Wave Solutions: the Benjamin Equation

We note that if the quadratic form that appears in the functional to be minimized is not positive anywhere, it is not obvious that we can exclude the case of *vanishing* (see [**21**], [**164**], [**198**]). The following variational problem is an example of this situation:

$$\text{(9.6)}\qquad \begin{cases} \text{minimize} \quad V(\psi) = \dfrac{1}{2}\langle L\psi, \psi\rangle - \dfrac{1}{3}\displaystyle\int_{-\infty}^{\infty} \psi^3 dx \\ \text{subject to } F(\psi) = \dfrac{1}{2}\displaystyle\int_{-\infty}^{\infty} \psi^2 dx = \lambda > 0, \end{cases}$$

where the linear operator L is defined by $L = -\partial_x^2 - l\mathcal{H}\partial_x$. Here L does not always have a nonnegative symbol.

For the convenience of the reader, we provide a sketch of the existence of a minimum for the problem (9.6). For $\lambda > 0$ we consider the following family of minimization problems:

$$\text{(9.7)}\qquad I_\lambda = \inf\{V(\psi) : \psi \in H^1(\mathbb{R}) \text{ and } F(\psi) = \lambda\}.$$

LEMMA 9.1. *For all $\lambda > 0$, we have $-\infty < I_\lambda < 0$.*

PROOF. Let $\psi \in H^1(\mathbb{R})$ such that $F(\psi) = \lambda$ and $\int_{-\infty}^{\infty} \psi^3 dx$. Then each $\psi_\theta(x) = \theta^{1/2}\psi(\theta x)$ satisfies $F(\psi_\theta) = \lambda$ and

$$\begin{aligned}V(\psi_\theta) &= \int_{-\infty}^{\infty} \frac{\theta^2}{2}(\psi')^2 - \frac{\theta l}{2}\psi\mathcal{H}\psi' - \frac{\theta^{1/2}}{3}\psi^3 dx \\ &= \frac{\theta^2}{2}\|\psi'\|^2 - \frac{\theta l}{2}\|D^{1/2}\psi\|^2 - \frac{\theta^{1/2}}{3}|\psi|_3^3.\end{aligned}$$

Hence, by taking θ sufficiently small, we get $V(\psi_\theta) < 0$ and therefore I_λ is negative.

To prove that $I_\lambda > -\infty$, it is sufficient to obtain a bound of the form

$$\|\psi'\|^2 \leqq B(\lambda, l) \tag{9.8}$$

with ψ satisfying $F(\psi) = \lambda$ and $V(\psi) \leqq 0$. First note that from the Gagliardo-Nirenberg-type inequality (6.31) and interpolation, we obtain for $A_1 > 0$ that

$$|\psi|_3^3 \leqq A_1\|\psi\|^{5/2}\|\psi'\|^{1/2} \quad \text{and} \quad \|D^{1/2}\psi\| \leqq \|\psi\|\|\psi'\|.$$

Thus,

$$\frac{1}{2}\|\psi'\|^2 \leqq \frac{l}{2}\|D^{1/2}\psi\|^2 + \frac{1}{3}|\psi|_3^3 \leqq \frac{\sqrt{2}}{2}l\lambda^{1/2}\|\psi'\| + A_1\lambda^{5/4}\|\psi'\|^{1/2}.$$

Since the square of $\|\psi'\|$ is bounded by smaller powers, the existence of the bound in (9.8) is obtained. □

From Lemma 9.1 and its proof, it is clear that if $\{\psi_n\}$ is any minimizing sequence for (9.7), then the sequence is bounded in $H^1(\mathbb{R})$.

Now, we give bounds on the quadratic and nonquadratic parts of V. Let

$$Q(\psi) = \frac{1}{2}\int_{-\infty}^{\infty} (\psi')^2 - l\psi\mathcal{H}\psi'\, dx.$$

Then the nonquadratic part of V is $N(\psi) = V(\psi) - Q(\psi)$.

LEMMA 9.2. *If ψ satisfies $F(\psi) = \lambda$, then*

$$Q(\psi) \geqq -\frac{\lambda l^2}{4}.$$

PROOF. From Plancherel's Theorem we obtain

$$Q(\psi) = \frac{1}{2}\int_{-\infty}^{\infty} [\zeta^2 - l|\zeta|]|\widehat{\psi}(\zeta)|^2\, d\zeta \geqq -\frac{l^2}{4}F(\psi),$$

which proves the lemma. □

Now, we find an upper bound for V which is the main ingredient for ruling out the cases of vanishing and dichotomy.

THEOREM 9.3. *There exist constants $A > 0$ and $\sigma > 0$ and an admissible function Φ such that*

$$V(\Phi) \leqq -\frac{\lambda l^2}{4} - A\lambda^\sigma$$

if λ is small enough.

SKETCH OF THE PROOF. For $\beta = \sqrt{l/2}$ we have

$$V(\Phi) + \frac{\lambda l^2}{4} \leqq \int_{-\infty}^{\infty} \frac{1}{2}[(\mathcal{H}\partial_x + \beta^2)\Phi]^2 - \frac{1}{3}\Phi^3 dx \equiv H(\Phi). \tag{9.9}$$

Now, the goal is to show that $H(\Phi)$ can be made negative. The form of (9.9) suggests the following test function $\Phi = c(\Phi_\epsilon + \delta\Psi)$, where for ϵ

$$\begin{cases} \Phi_\epsilon(x) = f(\epsilon x)\cos(\beta^2 x), \quad \Psi(x) = f(\epsilon x), \quad \text{and} \\ f(y) = \dfrac{1}{1+y^2}. \end{cases}$$

The constant c is adjusted to satisfy the constraint $F(\Psi) = \lambda$ and such that the parameters ϵ and δ have the form $\epsilon = \lambda^\beta$ and $\delta = \pm\epsilon^\theta$ (see Angulo [**21**]). So, for $\lambda \to 0$ we can obtain the estimate

$$G(c\Phi_\epsilon + c\delta\Psi) \leqq -A\lambda^\sigma$$

with $A > 0$ and $\sigma = \frac{3}{2} + \beta(\theta + \frac{1}{2})$. This proves the theorem. □

As a consequence of Theorem 9.3, for a very small λ we may restrict our minimization problem (9.7) to the class of admissible functions ψ, such that

$$V(\psi) \leqq -\frac{\lambda l^2}{4} - A\lambda^\sigma.$$

In this class, we have that the nonquadratic part of V satisfies

$$N(\psi) = V(\psi) - Q(\psi) \leqq -A\lambda^\sigma. \tag{9.10}$$

Now we prove that the dependence of I_λ on λ is sub-linear. Hence we obtain the sub-additivity of I_λ with respect to λ, and it will imply that dichotomy cannot occur.

THEOREM 9.4. *For all $\theta > 1$ and λ positive and small, we have*

(1) $I_{\lambda\theta} \leqq \theta I_\lambda - A\lambda^\sigma(\theta^{3/2} - \theta)$,
(2) $I_{\lambda\theta} < \theta I_\lambda$,
(3) $I_\lambda < I_\zeta + I_{\lambda-\zeta}$ *for all* $\zeta \in (0, \lambda)$.

PROOF. We only prove (1) since (2) and (3) follow a standard argument. Choose $\psi \in H^1(\mathbb{R})$ such that $F(\psi) = \lambda$. Then the function $\psi_\theta(x) = \theta^{1/2}\psi(x)$ satisfies $F(\psi_\theta) = \lambda\theta$. Assume that ψ is part of a minimizing sequence for (9.7); in particular, we may assume by Theorem 9.3 and (9.10) that $N(\psi) \leqq -A\lambda^\sigma$ for small λ. Then,

$$V(\psi_\theta) - \theta V(\psi) = (\theta^{3/2} - \theta)N(\psi) \leqq -A(\theta^{3/2} - \theta)\lambda^\sigma.$$

Therefore, $I_{\lambda\theta} \leqq \theta V(\psi) - A(\theta^{3/2} - \theta)\lambda^\sigma$ and so the proof of item (1) is complete. □

Now, we associate with each minimizing sequence $\{\psi_n\}$ for (9.7), the following sequence of nondecreasing functions: $Q_n : [0, \infty) \to [0, \lambda]$ defined by

$$Q_n(\omega) = \frac{1}{2} \sup_{y\in\mathbb{R}} \int_{y-\omega}^{y+\omega} |\psi_n(x)|^2 \, dx.$$

Since $\|\psi_n\|$ remains bounded, $\{Q_n\}$ comprises a uniformly bounded sequence of nondecreasing functions on $[0, \infty)$. A standard argument then implies that $\{Q_n\}$ must have a subsequence, which we denote again by $\{Q_n\}$, that converges pointwise and uniformly on compact sets to a nondecreasing limit function on $[0, \infty)$. Let Q be this limit function, and define

$$\alpha = \lim_{\omega\to\infty} Q(\omega). \tag{9.11}$$

From the assumption that $\|\psi_n\|^2 = \lambda$ it follows that $0 \leqq \alpha \leqq \lambda$. The Concentration-Compactness Principle distinguishes three cases: $\alpha = \lambda$ (*compactness*); $\alpha = 0$ (*vanishing*); and $0 < \alpha < \lambda$ (*dichotomy*).

As we know, the strategy of the principle consists in ruling out vanishing and dichotomy, and so only the case of compactness can occur. Then it will imply that every minimizing sequence is relatively compact, after suitable translations.

In the next theorem, we show that α is positive.

THEOREM 9.5. *If λ is positive and small, then for every minimizing sequence $\{\psi_n\}$, we have $\alpha > 0$.*

PROOF. Suppose $\alpha = 0$. Then

$$Q_n(\omega) = \frac{1}{2} \sup_{y\in\mathbb{R}} \int_{y-\omega}^{y+\omega} |\psi_n(x)|^2\, dx \to 0$$

for every $\omega > 0$. We first claim that ψ_n tends to zero strongly in $L^p(\mathbb{R})$, $p > 2$. In fact, the Sobolev embedding of $W^{1,1}(\omega)$ into $L^{p/2}(\omega)$, the Cauchy-Schwartz inequality, and the estimate $\|\psi_n\|_1 \leqq A_1$ ensure that, for fixed r, there is a constant $A_2 = A(r)$ such that

$$\begin{aligned}\int_{y-\omega}^{y+\omega} |\psi_n(x)|^p\, dx &\leqq A_2(Q_n(\omega) + 2A_1[Q_n(\omega)]^{1/2})^{\frac{p}{2}-1}\\ &\qquad \times \left(\int_{y-\omega}^{y+\omega} |\psi_n|^2 + 2|\psi_n||\psi_n'|dx\right)\\ &= A_2\delta_n^{\frac{p}{2}-1}\int_{y-\omega}^{y+\omega} 2|\psi_n|^2 + |\psi_n'|dx\end{aligned} \tag{9.12}$$

with $\delta_n \to 0$ as $n \to +\infty$. Then, covering $\mathbb{R}$ by intervals of radius r such that any point of $\mathbb{R}$ is contained in at most two intervals, we deduce

$$\int_{-\infty}^{\infty} |\psi_n(x)|^p\, dx \leqq A_3\delta_n^{\frac{p}{2}-1}\|\psi_n\|_1^2 \leqq A_4\delta_n^{\frac{p}{2}-1},$$

and so we conclude the claim.

Therefore,

$$I_\lambda = \lim_{n\to\infty} V(\psi_n) \geqq \liminf_{n\to\infty} Q(\psi_n) \geqq \inf Q = -\frac{\lambda l^2}{4}.$$

On the other hand, by Theorem 9.3 we know that $I_\lambda < -\frac{\lambda l^2}{4}$. This contradiction shows that vanishing cannot occur. Thus the theorem is proved. $\square$

Now from Theorem 9.4 and standard arguments we obtain that $\alpha \in (0, \lambda)$. So, $\alpha = \lambda$ and we get the *compactness* alternative of Lions. Then, the set of minimizers for I_λ,

$$G_\lambda = \{\psi \in H^1(\mathbb{R}) : V(\psi) = I_\lambda \text{ and } F(\psi) = \lambda\},$$

is nonempty for λ small. Moreover, we have the following stability theorem.

THEOREM 9.6. *Let λ be positive and small. Then we have the following.*

(1) *The set G_λ consists of solitary wave solutions ψ for the Benjamin equation satisfying*

$$\psi'' + l\mathcal{H}\psi' + \psi^2 = c\psi, \tag{9.13}$$

where $c > 0$.

(2) *G_λ is stable in $H^1(\mathbb{R})$, in the sense of Definition 5.3, by the flow of the Benjamin equation.*

(3) *The wave speed c in (9.13) satisfies the inequality*

$$\frac{1}{4}l^2 < c \leqq \frac{1}{4}l^2 + Al^{1/2}\lambda^{1/2},$$

where A is a positive constant that does not depend on l and λ.

PROOF. See Theorem 2.12 and Theorem 2.14 in [**21**]. □

REMARK 9.7. Sometimes the property of sub-additivity can be very difficult to show for specific functionals, and so ruling out the case of *dichotomy* can be a big problem. In this case, a new approach has been developed by Lopes in [**195**], [**197**], and [**198**], which gives us sufficient conditions to obtain that every minimizing sequence associated with some variational problems is precompact (modulo translations). The following is an example of this situation:

$$\begin{cases} \text{minimize} \quad Q(\psi) = -\displaystyle\int_{-\infty}^{\infty} \psi(\psi')^2 + F(\psi) + k\psi^2 \, dx \\ \text{subject to } Z(\psi) = \displaystyle\int_{-\infty}^{\infty} \psi^2 + (\psi')^2 \, dx = \lambda > 0, \end{cases}$$

where F is a very general nonlinearity. Here, because of the different behavior of the terms in the constraint Z with respect to dilations, it is very difficult to obtain the sub-additivity of Q. This minimization problem is related to the problem of the stability of *peakons* for the following generalized Camassa-Holm equation (see [**200**]):

$$u_t - u_{txx} = 2u_x u_{xx} + u u_{xxx} - [f(u)/2]_x - k u_x.$$

9.5. Stability of Solitary Wave Solutions: the Fourth-Order Equation

The approach established in Section 8.3 was originally put forward by Levandosky in [**182**]. He studied the stability and instability of fourth-order solitary wave solutions of the following fourth-order wave equation:

$$u_{tt} + \triangle^2 u + u = |u|^{p-1} u$$

where $u = u(x,t)$ for $x \in \mathbb{R}^n$, $t \in \mathbb{R}$, and $p > 1$. As we saw in Section 8.3, this approach depends essentially on the variational characterization of the solitary wave solutions in question and a convexity analysis of the classical function d. Moreover, no *spectral information* was necessary such as is required by the methods established in Chapters 6 and 7.

9.6. Stability of Solitary Wave Solutions: the GKP-I Equations

We consider the Generalized Kadomtsev-Petviashvili Equations (GKP-I)

$$(u_t + (u^{m+1})_x + u_{xxx})_x = \epsilon u_{yy} \tag{9.14}$$

where $u = u(x,y,t)$, $(x,y) \in \mathbb{R}^2$, $t \in \mathbb{R}$, $m = m_1/m_2$, m_1 is an even integer, and m_2 is an odd integer. Many local and global existence theories for GKP-I equations have been established by, for example, Bourgain [**66**], Saut [**248**], Iorio and Leite [**141**], Molinet, Saut, and Tzvekov [**226**] and Ionescu, Kenig, and Tataru [**135**]. Moreover, it was shown by Falkovitch and Turitsyn [**104**], by a virial method, that the GKP-I equations have solutions which blow up for $m \geqq 4$ (see Liu, [**190**], where more recent results about this issue have been obtained).

With regard to the problem of stability and instability of solitary wave solutions of the form $u(x,y,t) = \phi(x - ct, y)$ satisfying (3.15), many conclusive results have been obtained. Indeed, we denote by X the closure of $\partial_x(C_0^\infty(\mathbb{R}^2))$ by the norm

$$\|\partial_x \varphi\|_X = (\|\nabla \varphi\|^2 + \|\partial_x^2 \varphi\|^2)^{1/2},$$

where $\partial_x(C_0^\infty(\mathbb{R}^2))$ denotes the space of functions of the form $\partial_x\varphi$ with $\varphi \in C_0^\infty(\mathbb{R}^2)$, i.e., the space of functions ψ in $C_0^\infty(\mathbb{R}^2)$ such that

$$\int_{-\infty}^{\infty} \psi(x,y)dx = 0$$

for every $y \in \mathbb{R}$. It was shown in de Bouard and Saut ([**95**]) that equation (9.14) does not allow any nontrivial solitary wave solution satisfying $\phi = \partial_x\varphi \in X$ with

$$\phi \in H^1(\mathbb{R}^2) \cap L^\infty_{\text{loc}}(\mathbb{R}^2) \quad \text{and} \quad \partial_x^2\phi, \partial_x\varphi \in L^2_{\text{loc}}(\mathbb{R}^2),$$

if $m \geqq 4$. The existence of solitary wave solutions for GKP-I equations has been shown for $0 < m < 4$ by using variational methods (see Wang, Ablowitz, and Segur [**271**], de Bouard and Saut [**95**], Liu and Wang [**192**]). Moreover, it has been shown that the set of solitary wave solutions, obtained by solving a specific constrained minimization problem, is nonlinearly stable for $0 < m < 4/3$ (see [**94**], [**192**]).

Here we are interested in the approach in [**192**], which uses the same technique for obtaining stability as that established in Section 8.3. We note that with regard to the problem of instability, Wang *et al.* in [**271**] and de Bouard and Saut in [**94**] have shown that if $\frac{4}{3} < m < 4$ and $\phi \in X$ is a cylindrically symmetric solitary wave of the GKP-I equations, then the ϕ-orbit,

$$\Omega_\phi = \{\phi(\cdot + r, \cdot + s) : (r,s) \in \mathbb{R}^2\},$$

is X-unstable by the flow of cylindrically symmetric solutions of the GKP-I equations.

Consider $u(x,y,t) = \phi_c(x - ct, y)$, a solution for equation (9.14) with $\epsilon = 1$. Then $\phi = \phi_c$ satisfies

$$c\,\phi + \partial_x^{-2}\phi_{yy} - \phi_{xx} = \phi^{m+1}, \tag{9.15}$$

where $\partial_x^{-1} = \int_{-\infty}^x$. Define on X the following functionals:

$$E_{\text{GKP-I}}(u) = \frac{1}{2}\int_{\mathbb{R}^2} (u_x)^2 + (\partial_x^{-1}u_y)^2 - \frac{2}{m+2}u^{m+2}\,dxdy,$$

$$F(u) = \frac{1}{2}\int_{\mathbb{R}^2} u^2 dxdy,$$

$$\mathcal{J}_c(u) = \int_{\mathbb{R}^2} (u_x)^2 + (\partial_x^{-1}u_y)^2 + cu^2\,dxdy,$$

$$K(u) = \int_{\mathbb{R}^2} u^{m+2} dxdy.$$

We note that $E_{\text{GKP-I}}$ and F are conservation laws for equations (9.14). Next we have the following theorem of the existence of solitary wave solutions (see [**192**], [**95**]).

THEOREM 9.8. *Let $0 < m < 4$, $c > 0$, and $m = m_1/m_2$ where m_1 is any even integer and m_2 is any odd integer. Then there exists a minimizer φ_c,*

$$\varphi_c \in N = \{u \in X : K(u) = 1\},$$

such that

$$\mathcal{J}_c(\varphi_c) = \inf\{\mathcal{J}_c(u) : u \in N\} \equiv I.$$

Therefore, a solution of (9.15) *exists.*

Now we define the following set of solitary wave solutions for the GKP-I equations with speed $c > 0$, namely,

$$\Theta_c = \{\phi \in X : K(\phi) = \mathcal{I}_c(\phi) = I^{\frac{m+2}{m}}\}.$$

So, if $\phi \in \Theta_c$, then the classical functional $d(c) = E_{\text{GKP-I}}(\phi) + cF(\phi)$ satisfies the relation

$$d(c) = \frac{1}{2}\mathcal{I}_c(\phi) - \frac{1}{m+2}K(\phi) = \frac{m}{2(m+2)}K(\phi) = \frac{m}{2(m+2)}c^{\frac{4-m}{2m}}K(\psi), \tag{9.16}$$

where ψ is a solution of the equation

$$\psi + \partial_x^{-2}\psi_{yy} - \psi_{xx} = \psi^{m+1}.$$

Hence, we have the following stability result (see [**192**]).

THEOREM 9.9 (Nonlinear stability for the GKP-I equations). *Let $0 < m < \frac{4}{3}$ with $m = m_1/m_2$ and $c > 0$. Then Θ_c is X-stable.*

PROOF. The proof is based on the ideas established in Section 8.3. So, we only need to consider when $d''(c) > 0$. From (9.16) it follows immediately that

$$d''(c) = \frac{1}{8m(m+2)}(4-m)(4-3m)c^{\frac{4-5m}{2m}}A$$

where $A = K(\psi) > 0$. Hence, $d''(c) > 0$ if, and only if, $0 < m < \frac{4}{3}$. □

9.7. Comments

We consider the following class of solitary wave solutions:

$$\Lambda_c = \left\{\phi \in H^{\frac{\beta}{2}}(\mathbb{R}) : (p+1)\int_{-\infty}^{\infty} \phi D^\beta \phi + c\phi^2\, dx = K(\phi) = [(p+1)\theta_2(1)]^{p+2/p}\right\},$$

for

$$K(\phi) = \int \phi^{p+2} dx$$

and $c > 0$, associated with equation (9.4). The class Λ_c is determined by the variational problem $\theta_2(\lambda)$ in (9.3) with $\mathcal{L}_2 = D^\beta$ and $\beta \geqq 1$. Then, by considering for every $\phi \in \Lambda_c$ the function $d(c) = E(\phi) + cF(\phi)$, where E is defined in (6.30), we have that

$$d(c) = \frac{p}{2(p+1)(p+2)}c^{\frac{p+2}{p}-\frac{1}{\beta}}K(\psi) \tag{9.17}$$

where ψ satisfies the equation

$$D^\beta\psi + \psi = \frac{1}{p+1}\psi^{p+1}.$$

So, by using the approach established in Section 8.3, we obtain that Λ_c is $H^{\frac{\beta}{2}}(\mathbb{R})$-stable by the flow of the evolution equations (9.4) if $p < 2\beta$ (see Theorem 7.11).

By considering $\beta = 2$ and the uniqueness (up to translations and signal) of solutions of (3.3), it follows from (3.4) that the set Λ_c satisfies

$$\Lambda_c = \{\pm\phi_c(\cdot + y) : y \in \mathbb{R}\},$$

and therefore, ϕ_c is $H^1(\mathbb{R})$-stable by the flow of the GKdV equations (2.5) if $p < 4$ (see Theorem 6.1).

For the set of solitary wave solutions determined by the general minimization problems $\theta_i(\lambda)$ in (9.3), namely,

$$\Lambda_{c,i} = \left\{\phi \in H^{s_i}(\mathbb{R}) : (p+1)\int_{-\infty}^{\infty} [\phi\mathcal{L}_i\phi + c\phi^2]\, dx = K(\phi) = [(p+1)\theta_i(1)]^{p+2/p}\right\}$$

where $s_1 = m$ and $s_2 = r_k$, we do not know what happens with its stability because, in this case, the approach of dilating in the spacial variable does not give us a good expression, like that in (8.27), (9.16), or (9.17) for the associated function, d. But, we conjecture that $\Lambda_{c,1}$ is $H^m(\mathbb{R})$-stable if l is small, $\frac{1}{2} \leqq r < m$, and $p < 4m$. Similarly, we conjecture that $\Lambda_{c,2}$ is $H^{r_k}(\mathbb{R})$-stable if $r_1 \geqq \frac{1}{2}$ and $p < 4r_k$ (see Chapter 10, Section 10.4).

CHAPTER 10

Instability of Solitary Wave Solutions

10.1. Introduction

In Chapters 6, 7, and 8 we saw some basic methods which can give us information about the existence and the nonlinear stability of solitary wave solutions for many basic models of the theory of nonlinear evolution equations. In this chapter we are interested in studying the other side of the coin, namely, the property of instability of travelling wave solutions. As we will see, this issue, as well as the existence of many open problems related to it, is actually of increasing interest to researchers.

The first results on the nonlinear instability of travelling wave solutions to nonlinear evolution equations were concerned with the nonlinear focusing Schrödinger Equations (NLS equations) of the form

$$iu_t + \Delta u + |u|^{2\sigma}u = 0, \tag{10.1}$$

where $u = u(x,t) \in \mathbb{C}$, $x \in \mathbb{R}^n$, and $t \in \mathbb{R}$, which were established initially by Berestycki and Cazenave [**51**], Cazenave [**78**], and Weinstein [**272**] (see [**79**]). The travelling wave solutions studied were the *ground state* solutions:

$$u(x,t) = e^{i\omega t}\phi_\omega(x)$$

where ϕ_ω is a positive, real-valued, radially symmetric solution of the elliptic equation

$$-\Delta\phi_\omega + \omega\phi_\omega - \phi_\omega^{2\sigma+1} = 0. \tag{10.2}$$

It is known for equation (10.2) that when $0 < 2\sigma < \infty$ for $n = 1, 2$ and $0 < 2\sigma < \frac{4}{n-2}$ for $n \geqq 3$, there exists a unique positive radially symmetric solution (see Gidas, Ni, and Nirenberg [**118**] for the symmetric property, Kwong [**174**] for the uniqueness property, and Strauss [**261**], Berestycki and Lions [**53**], [**54**], Berestycki, Gallouet, and Kavian [**52**] for the existence of solution). Also, it is well known that the solutions $u(x,t) = e^{i\omega t}\phi_\omega(x)$ are orbitally stable for every $\omega > 0$ by the flow of equation (10.1) provided $2\sigma < 4/n$ ([**82**], [**274**]). We remember that in this case the orbit generated by the ground state solutions is given by

$$\Omega_{\phi_\omega} = \{e^{i\theta}\phi_\omega(\cdot + y) : \ (\theta, y) \in [0, 2\pi) \times \mathbb{R}^n\}.$$

Moreover, the solutions of the NLS equations can become singular only in the critical case, $2\sigma = 4/n$, and in the supercritical case, $2\sigma > 4/n$ ([**51**], [**274**]). So, we see that the ground state solutions are orbitally stable only in the subcritical case, $2\sigma < 4/n$.

The property of instability of the ground state solutions for the NLS equations is obtained via blow-up mechanics. More precisely, from [**51**] and [**272**] (see [**79**]), we see that for

$$2\sigma = \frac{4}{n}, \qquad \frac{4}{n} < 2\sigma < \frac{4}{n-2}$$

($\frac{4}{n} < 2\sigma < \infty$, if $n = 1, 2$), and $\omega > 0$, the orbit generated by the solution $u(x,t) = e^{i\omega t}\phi_\omega(x)$, Ω_{ϕ_ω}, is $H^1(\mathbb{R}^n)$-unstable, in the sense that there exists a sequence $\{\phi_n\} \subset H^1(\mathbb{R}^n)$ such that

$$\phi_n \to \phi \qquad \text{as } n \to \infty \text{ in } H^1(\mathbb{R}^n)$$

and such that the corresponding maximal solutions u_n of (10.1), with $u_n(0) = \phi_n$, blow up in finite time, which is to say, there exists $T^* = T^*(\phi_n) > 0$ such that

$$\lim_{t \uparrow T^*} \|u_n(t)\|_1 = +\infty.$$

Thus, we see from the above that an understanding of the flow close to an unstable solution is in a certain sense crucial to understanding the behavior and singular formation of solutions associated with nonlinear dispersive evolution equations. An example of this issue has already been established in Chapter 6, Theorem 6.12, where a result of "stability of the blow-up" for the solutions of the class of Korteweg-de Vries equations (6.27) was established for initial data close to an unstable solitary wave solution (see [**29**], [**30**], [**176**], [**215**], [**228**]). We note that in [**29**] a study on the stability of singular solutions of the NLS equations (10.1) in the critical case of $2\sigma = 4/n$ has already been conducted.

Now, in Grillakis, Shatah, and Strauss ([**123**], [**124**]) a set of sufficient conditions (which are similar to those given in Chapter 7) has been established for obtaining the instability of travelling wave solutions for abstract Hamiltonian equations (or systems) of the form

$$u_t = JE'(u(t)), \tag{10.3}$$

where J *should be an onto* skew-symmetric linear operator. So, a direct application of this theory to evolution equations of the form (7.1) (where $J = \partial_x$) is not possible, but it is possible to apply the theory to the NLS equations (10.1). An improve of the instability theory has been made by Bona, Souganidis, and Strauss in [**64**] for the case of $J = \partial_x$ in (10.3), and by Souganidis and Strauss in [**255**] for equations of the form

$$M_0 u_t + \partial_x(f(u)) = 0,$$

where M_0 is a specific pseudo-differential operator of order $\mu \geqq 1$, for instance, $M_0 = 1 - \partial_x^2$, $M_0 = 1 - \mathcal{H}\partial_x$, or more generally, $M_0 = 1 + D^\mu$ ($\mu > 1$).

In particular, if we consider Assumptions 1, 2, and 3 given in Chapter 7, together with some conditions of regularity on the solitary wave solutions ϕ_c of equation (6.27), then it follows from the theory in [**64**] that the ϕ_c-orbit

$$\Omega_{\phi_c} = \{\phi_c(\cdot + y) : \ y \in \mathbb{R}\}$$

will be $H^{\frac{\beta}{2}}(\mathbb{R})$-unstable by the flow of equation (6.27), providing that the basic function d defined by (7.13) satisfies $d''(c) < 0$. Therefore, by using a *scaling argument* we immediately obtain from (7.26) that

$$d''(c) < 0 \Leftrightarrow p > 2\beta. \tag{10.4}$$

Now, we would like to comment on two basic observations that arise from the proof of instability for the ϕ_c-orbit, Ω_{ϕ_c}, associated to equation (6.27) and given in [**64**]. The first one is with regard to the mechanics of obtaining instability: that is, it not clear whether it is produced through some singular behavior of solutions for (7.1), for example, via the blow-up of solutions. The second one is that the analysis given in [**64**] does not give any information about the behavior of the orbit in the *critical case* $p = 2\beta$ (see Section 6.3). We remember that only in the case of the critical Korteweg-de Vries equation (2.4) is it known that the instability of Ω_{ϕ_c} may be produced by the blow-up of solutions (see Section 6.3 or Martel and Merle [**212**] and Merle [**220**]).

Our main purpose in this chapter it is to establish a theory of instability for solitary wave solutions of nonsemilinear evolution equations, whenever these solutions are obtained *via a variational approach.* More precisely, we extend some ideas from Gonçalves [**121**] and Shatah and Strauss [**253**] about the instability of the stationary state for semilinear evolution equations of Schrödinger type. *We note that the method to be established here does not depend on those given by the theories in* [**123**] *and* [**64**], *such as spectral or convexity conditions.*

10.2. Instability of Solitary Wave Solutions: the GB Equations

We will illustrate our method for a generalization of the Benjamin equation (see (2.10)), namely,

$$u_t + (u^p)_x + l\mathcal{H}u_{xx} + u_{xxx} = 0 \tag{10.5}$$

where $p \in \mathbb{N}$, $p \geqq 2$, $l \in \mathbb{R}$ and where $\mathcal{H}$ is the Hilbert transform. The solitary wave solutions for (10.5) will be of the form

$$u(x,t) = \phi(x - ct)$$

where $c > 0$ is a dimensionless wave speed and both $\phi(\xi)$ and its derivatives tend to zero as the variable $\xi = x - ct$ approaches $\pm\infty$. We substitute this form of u into (10.5) and after a few simple manipulations, the equation for ϕ will be

$$\phi'' + l\mathcal{H}\phi' - c\phi = -\phi^p. \tag{10.6}$$

Some known results about equation (10.6) have been established in Chapter 3, Section 3.3, in particular, the existence of solitary wave solutions for $\gamma = \frac{1}{2\sqrt{c}}\, l \in (0,1)$ ([**13**], [**15**], [**21**], [**44**], [**84**]). Stability has been established for $\gamma \sim 0$ by [**44**] and [**13**] in the case when $p = 2$, for $\gamma \sim 1$ by [**21**] with $p = 2$, and for $\gamma \sim 1$ by [**15**] with $p = 3, 4$. Thus, we are interested here in giving some information about the case when $p \geqq 5$.

A very interesting feature of the solutions of equation (10.6) is their algebraic asymptotic decay; namely, by using the theory developed in [**185**], we have that for $l \in (0, 2\sqrt{c})$,

$$\lim_{\xi\to\pm\infty} \xi^2\, \phi(\xi) = C \tag{10.7}$$

where C is a constant (see Theorem 10.3 below).

The solitary wave solutions for the GB equations (10.5) that we are interested in here will be those obtained as minimizers of a variational problem. More precisely, there will exist a $\mu > 0$ such that if ϕ is a minimum point of the minimization problem

$$\Theta(\mu) = \inf\left\{ \int_{-\infty}^{\infty} [(f')^2 - lf\mathcal{H}f' + cf^2]dx \;:\; f \in H^1(\mathbb{R}) \text{ and } \int_{-\infty}^{\infty} f^{p+1}dx = \mu \right\}$$

with $c > \frac{1}{4}l^2$, then ϕ satisfies equation (10.6).

Next, if we consider the orbit generated by this minimum, namely,

$$\Omega_\phi = \{\phi(\cdot + y) \ : \ y \in \mathbb{R}\},$$

then we will prove that *Ω_ϕ is $H^1(\mathbb{R})$-unstable by the flow of the GB equations* (10.5) *when $l > 0$ is small and $p > 5$. Moreover, the trajectories used to exhibit instability have the remarkable property that they are global solutions of the GB equations in time and uniformly bounded in $H^1(\mathbb{R})$* (see Theorem 10.11 and (10.37) below).

With regard to the case $p = 5$ our method does not give information about the problem of instability (except when $l < 0$, we obtain that $p \geqq 5$), but we conjecture that the solitary waves are still unstable. Now, if we consider $l = 0$ in (10.5), our approach immediately recaptures the condition of instability for the solitary wave solutions associated with the GKdV equations (2.5), namely, $p > 5$ in (10.5). On the other hand, our method does not need to give conditions of spectral type or concavity as is needed in [**64**] for obtaining instability.

We start our theory of instability for the GB equations by establishing some information about the Cauchy problem associated with (10.5). This will be sufficient information for our purpose. The proof of local existence of solutions is based on some estimates for oscillatory integrals and on a fixed-point argument, as has been shown recently by Linares and Scialom in [**189**]. The global existence of solutions follows by using the following conservations laws for (10.5):

$$\begin{aligned} E_{\mathrm{GB}}(u) &= \frac{1}{2}\int_{-\infty}^{\infty} \left[(u')^2 - lu\mathcal{H}u' - \frac{2}{p+1}u^{p+1}\right] dx \ \text{ and} \\ F(u) &= \frac{1}{2}\int_{-\infty}^{\infty} u^2 dx. \end{aligned} \tag{10.8}$$

THEOREM 10.1 (Cauchy Problem for the GB Equations). *Let $p \in \mathbb{N}$, $p \geqq 2$. For $u_0 \in H^1(\mathbb{R})$ there exists $T = T(\|u_0\|_{H^1}) > 0$ and a unique solution*

$$u(t) \equiv U(t)u_0 \in C([0,T]; H^1(\mathbb{R}))$$

of (10.5) *with $u(x,0) = u_0(x)$. Moreover, for $T > T_0$ the map $u_0 \to U(t)u_0$ is Lipschitz continuous from $H^1(\mathbb{R})$ to $C([0,T_0]; H^1(\mathbb{R}))$. Finally, $T = +\infty$ if we have the following:*

1. *For $2 \leqq p < 5$, $u_0 \in H^1(\mathbb{R})$.*
2. *For $p = 5$, $\|u_0\|$ has to be not too large.*
3. *For $p > 5$, $\|u_0\|_1$ is small.*

Now, we are interested in the existence of solitary wave solutions for the GB equations as well as in some spatial asymptotic decay of these solutions. We will rely on the results put forward in [**84**] and [**185**]. Initially, we have a result concerning the existence of solitary wave solutions for (10.6). Let $\lambda > 0$ and define the family of minimization problems

$$\Theta(\lambda) = \inf\left\{\int_{-\infty}^{\infty}[(f')^2 - lf\mathcal{H}f' + cf^2]dx \ : \ f \in H^1(\mathbb{R}) \text{ and } \int_{-\infty}^{\infty} f^{p+1}(x)dx = \lambda\right\}. \tag{10.9}$$

Then we have the following theorem.

THEOREM 10.2 (Solitary Wave Solutions for the GB Equations). *Let $p \geqq 2$, $p \in \mathbb{N}$, $c > 0$, and $l \in (0, 2\sqrt{c})$.*

(1) *For every* $\lambda > 0$ *the minimization problem* (10.9) *has a solution and therefore via a rescaling we have a nontrivial solution* ϕ *of equation* (10.6). *Moreover,* $\phi \in H^{\infty}(\mathbb{R})$.

(2) *If the variational problem* (10.9) *has a solution, then there exists* $\mu > 0$ *such that every corresponding minimum point* ϕ *of* $\Theta(\mu)$ *satisfies equation* (10.6). *Moreover,* $\mu = [\Theta(1)]^{\frac{p+1}{p-1}}$.

PROOF. The proof of (1) is based on the Concentration-Compactness Method and we refer the reader to Theorem 4.1 in [**84**]. Part (2) is obvious. □

Attention is turned now to some asymptotic properties and the behavior of solitary wave solutions of the GB equations. This study will be important in our instability theory. Since our results are based essentially on the theory reported in [**185**], we would like to illustrate how this theory is applied in the case of the GB equations. We start by giving an idea of the proof of the limit in (10.7). In fact, let $c > 0$ such that

$$\min_{\xi \in \mathbb{R}} \{\xi^2 - l|\xi| + c\} > 0,$$

so $l \in (0, 2\sqrt{c})$. Then equation (10.6) can be written in the convolutional form

$$\phi(x) = (c + \mathcal{T})^{-1}\phi^p(x) = \int_{-\infty}^{\infty} K(x-y)\phi^p(y)\, dy \tag{10.10}$$

where $T = -\partial_x^2 - l\mathcal{H}\partial_x$ and the kernel K is the inverse Fourier transform of

$$\widehat{K}(\xi) = \frac{1}{\xi^2 - l|\xi| + c}.$$

So, via the Residue Theorem it follows that K is determined explicitly by

$$K(x) = -\frac{2l}{\sqrt{2\pi}} \int_0^{\infty} \frac{y e^{-|x|y}}{(c-y^2)^2 + l^2 y^2} dy + \frac{2\sqrt{2\pi}}{\sqrt{4c - l^2}} e^{-\frac{\sqrt{4c-l^2}}{2}|x|} \cos \frac{l}{2} x. \tag{10.11}$$

Note that $K \in C^{\infty}(\mathbb{R} - \{0\})$. Now, according to the theory in [**185**] the asymptotic properties of ϕ satisfying (10.10) are essentially based on those of the kernel K. In the present context, since $\widehat{K} \in H^s(\mathbb{R})$ for $s > \frac{3}{2}$, and from (10.11),

$$\lim_{x \to \pm\infty} x^2 K(x) = -\frac{2l}{c^2\sqrt{2\pi}}.$$

Then it follows that ϕ satisfies the relation in (10.7). We note that, from (10.10), we immediately get that $\phi \in L^1(\mathbb{R})$.

Now we obtain some basic results of decay associated to ϕ in (10.10).

THEOREM 10.3. *If* ϕ *is a solution of* (10.10), *then*

$$\lim_{|\xi| \to \infty} |\xi|^n \phi^{(n-2)}(\xi) = D_n < \infty \quad \text{and} \quad n \geqq 2.$$

Moreover,

$$\xi\phi', \phi'', \mathcal{H}\phi', \xi\mathcal{H}\phi'', \mathcal{H}\phi''', \xi\phi''', \phi^{(4)} \in L^1(\mathbb{R}).$$

Finally, if ϕ_l *is a solution from* (10.6) *obtained via Theorem 10.2(2), then*

$$\lim_{l \to 0^+} l \int_{-\infty}^{\infty} \phi_l(x)\mathcal{H}\phi_l'(x)\, dx = 0. \tag{10.12}$$

PROOF. Since $\lim_{|x|\to\infty} |x|^n K^{(n-2)}(x) = C_n < \infty$ and $\widehat{K^{(n-2)}} \in H^s(\mathbb{R})$ for $s > \frac{3}{2}$, with $n \geqq 2$, we have from (10.10), (10.11), and [**185**] that

$$\lim_{|x|\to\infty} |x|^n \phi^{(n-2)}(x) = D_n$$

for $n \geqq 2$ and $\phi^{(m)} \in L^1(\mathbb{R})$ for all m. Therefore, it follows that $x\phi', x\phi''' \in L^1(\mathbb{R})$. Now, from (10.6) and the Sobolev Embedding Theorem we have $\mathcal{H}\phi' \in L^1(\mathbb{R})$. Similarly, we show that $\mathcal{H}\phi''' \in L^1(\mathbb{R})$. Finally, from relation

$$xlH\phi'' = cx\phi' - px\phi^{p-1}\phi' - x\phi'''$$

and the previous results we have that $x\mathcal{H}\phi'' \in L^1(\mathbb{R})$.

Now we show (10.12). Since $\xi^2 - l|\xi| + c \geqq c - \frac{l^2}{4}$ and $\xi^2 - l|\xi| + c \geqq (1 - \frac{l^2}{4c})\xi^2$, we have via Fourier transform that

$$V_l(\phi_l) \equiv \int_{-\infty}^{\infty} [(\phi_l')^2 - l\phi_l\mathcal{H}\phi_l' + c\phi_l^2]dx \geqq \frac{1}{2}\min\Big\{c - \frac{l^2}{4}, 1 - \frac{l^2}{4c}\Big\}\|\phi_l\|_1^2 \equiv \beta(l)\|\phi_l\|_1^2.$$

Moreover, if $\Lambda_l = \min\{\Lambda(f) : f \in H^1(\mathbb{R}),\ f \neq 0\}$ where

$$\Lambda(f) = \frac{V_l(f)}{(\int f^{p+1}dx)^{\frac{2}{p+1}}},$$

then a few simple manipulations lead to the relation $\Theta(1) = \Lambda_l$. Thus

$$\Theta(\mu) = \mu^{\frac{2}{p+1}}\Theta(1) = \mu^{\frac{2}{p+1}}\Lambda_l.$$

Therefore, since $\int \phi_l^{p+1}(x)dx = \mu$, we get the estimate

$$\begin{aligned} 0 \leqq l\int_{-\infty}^{\infty} \phi_l(x)\mathcal{H}\phi_l'(x)\,dx &\leqq l\|\phi_l\|_1^2 \leqq \frac{l}{\beta(l)}V_l(\phi_l) = \frac{l}{\beta(l)}\Theta(\mu) \\ &= \frac{l}{\beta(l)}\mu^{\frac{2}{p+1}}\Lambda_l = \frac{l}{\beta(l)}[\Theta(1)]^{\frac{2}{p-1}}\Lambda_l = \frac{l}{\beta(l)}\Lambda_l^{\frac{p+1}{p-1}}. \end{aligned} \tag{10.13}$$

Finally, since the map $l \in [0, 2\sqrt{c}) \to \Lambda_l$ is a decreasing function, we have that $\Lambda_l \leqq \Lambda_0$, and hence from (10.13) we immediately obtain (10.12). □

Now we turn our attention to the time asymptotic behavior of solutions for evolution equation (10.5). In other words, for $u(t)$, a solution of the GB equations (10.5), we estimate how fast its tail grows near infinity with t.

THEOREM 10.4. *Let $p \geqq 2$, $p \in \mathbb{N}$. Assume that $u_0 \in H^1(\mathbb{R}) \cap L^1(\mathbb{R})$ and $\mathcal{H}\partial_x u_0, \partial_x^2 u_0 \in L^1(\mathbb{R})$. If u is a solution of (10.5) which corresponds to u_0, then*

$$\sup_{-\infty<x<\infty} \Big|\int_{-\infty}^{x} u(y,t)\,dy\Big| \leqq C(1 + t^{2/3})$$

for $0 \leqq t < T_0$, where T_0 denotes the existence time for u. The constant C depends only on $\sup_{0\leqq t<T_0}\|u(t)\|_1$, $|u_0|_{L^1}$, $|\mathcal{H}\partial_x u_0|_{L^1}$, and $|\partial_x^2 u_0|_{L^1}$.

SKETCH OF THE PROOF. The proof is based essentially on the following estimate for the free group $U_0(t)$ associated with the linear evolution problem

$$\begin{cases} u_t + l\mathcal{H}u_{xx} + u_{xxx} = 0, \\ u(0) = u_0, \end{cases} \tag{10.14}$$

i.e., for all $t > 0$

$$|U_0(t)u_0|_\infty \leqq C_0(1 + t)^{-1/3}|u_0|_{L^1}.$$

The remainder of the proof follows along the same lines as those of Theorem 4.3 in Souganidis and Strauss [**255**]. □

Next we establish our instability theorem.

THEOREM 10.5 (Instability for the GB Equations). *Let* $0 < l < 2\sqrt{c}$. *Suppose* ϕ_l *solves minimization problem* (10.9) *with* $\lambda \equiv \mu = [\Theta(1)]^{\frac{p+1}{p-1}}$. *If* l *is small and* $p > 5$, *then*

$$\Omega_{\phi_l} = \{\phi_l(\cdot + y) | y \in \mathbb{R}\}$$

is $H^1(\mathbb{R})$*-unstable by the flow of the GB equations.*

The proof of Theorem 10.5 will be based essentially on the criterion of instability given by Theorem 10.11 below. We start by establishing some basic notation (see Chapter 5, Section 5.2). For $\mathcal{O} \subset H^1(\mathbb{R})$ and $\delta > 0$ we define $\mathcal{V}(\mathcal{O}, \delta)$, the δ-neighborhood of $\mathcal{O}$ in $H^1(\mathbb{R})$, by

$$\mathcal{V}(\mathcal{O}, \delta) = \bigcup_{v \in \mathcal{O}} B_\delta(v) = \{g \in H^1(\mathbb{R}) \ : \ \inf_{v \in \mathcal{O}} \|v - g\|_1 < \delta\}$$

where $B_\delta(v) = \{u \in H^1(\mathbb{R}) : \ \|u - v\|_1 < \delta\}$. For each $y \in \mathbb{R}$ we remember that τ_y is the translation operator defined by $\tau_y v = v(\cdot + y)$. If $\Upsilon \subset L^p(\mathbb{R})$, we denote $\Omega_\Upsilon = \{\tau_y v \mid y \in \mathbb{R}, v \in \Upsilon\}$. With this notation we immediately have the following lemma.

LEMMA 10.6. *For each* $\Upsilon \subset H^1(\mathbb{R})$ *and* $\delta > 0$ *we have* $\mathcal{V}(\Omega_\Upsilon, \delta) = \Omega_{\mathcal{V}(\Upsilon,\delta)}$.

Next, we give some basic lemmas related to the ϕ-orbit, Ω_ϕ, when ϕ is a solution of (10.6).

LEMMA 10.7. *There exist* $\epsilon_0 > 0$ *and a unique function*

$$\Lambda : \mathcal{V}(\Omega_\phi, \epsilon_0) \to \mathbb{R},$$

which is a C^2 *functional, such that* $\Lambda(\phi) = 0$ *and such that for all* $v \in \mathcal{V}(\Omega_\phi, \epsilon_0)$ *and all* $w \in \Omega_\phi$, $\|v - \tau_{\Lambda(v)}\phi\| \leqq \|v - w\|$. *Moreover, for* $y \in \mathbb{R}$

$$\begin{cases} \Lambda(\tau_y v) = \Lambda(v) + y, \\ \Lambda'(v) = -\dfrac{1}{\langle v, \phi''(\cdot + \Lambda(v))\rangle}\, \phi'(\cdot + \Lambda(v)). \end{cases} \tag{10.15}$$

In particular, for $v \in \Omega_\phi$ *we have*

$$\langle \Lambda'(v), v\rangle = 0 \quad \text{and} \quad \Lambda'(v) = \frac{1}{\|\phi'\|^2} v'.$$

PROOF. The idea of the proof is to apply the Implicit Function Theorem to the function

$$F(v, y) = \frac{1}{2}\int_{-\infty}^{\infty} |\phi(x + y) - v(x)|^2 dx.$$

We refer the reader to Lemma 3.3 in [**23**]. □

REMARK 10.8. We note that the value $\Lambda(v)$, obtained in Lemma 10.7, can be seen as the "optimal" translation for ϕ to be the best approximation of v to the ϕ-orbit, Ω_ϕ, in the $L^2(\mathbb{R})$-norm .

Now, we define our main vector field in the study of instability. We consider initially a function $\psi \in L^\infty(\mathbb{R})$ such that $\psi' \in H^1(\mathbb{R})$ and ϕ is a nonzero solution of (10.6). Then for $v \in \mathcal{V}(\Omega_\phi, \epsilon_0)$ we define $B_\psi(v)$ by the formula

$$B_\psi(v) \equiv \tau_{\Lambda(v)}\psi' - \frac{\langle v, \tau_{\Lambda(v)}\psi'\rangle}{\langle v, \tau_{\Lambda(v)}\phi''\rangle}\,\tau_{\Lambda(v)}\phi''. \tag{10.16}$$

We note that the vector field B_ψ is an extension of the formula (4.2) in [**64**] as well as of the formula (5.9) in [**121**] (a similar vector field was also used in [**24**]). Now, for $v \in \Omega_\phi$ we can obtain an easy geometric interpretation of $B_\psi(v)$ as being the derivative of the *orthogonal component of* $\tau_{\Lambda(v)}\psi$ *with regard to* $\tau_{\Lambda(v)}\phi'$. In fact, let $y \in \mathbb{R}$ such that $v = \tau_y\phi$. Then from Lemma 10.7 it follows that $\Lambda(v) = y$. Now, even though $\psi \notin L^2(\mathbb{R})$, we have from Theorem 10.3 and integration by parts that

$$\frac{\langle v, \tau_{\Lambda(v)}\psi'\rangle}{\langle v, \tau_{\Lambda(v)}\phi''\rangle}\,\tau_y\phi' = \frac{\langle \tau_y\phi, \tau_y\psi'\rangle}{\langle \tau_y\phi, \tau_y\phi''\rangle}\,\tau_y\phi' = \frac{\langle \tau_y\phi', \tau_y\psi\rangle}{\|\tau_y\phi'\|^2}\,\tau_y\phi' \equiv P_{\shortparallel}.$$

So, defining $Q_\perp \equiv \tau_y\psi - P_{\shortparallel}$, we obtain that $\langle Q_\perp, \tau_y\phi'\rangle = 0$ and $B_\psi(v) = \partial_x Q_\perp$.

The following lemma contains the main properties of B_ψ in (10.16).

LEMMA 10.9. *Let ψ be a given function such that $\psi \in L^\infty(\mathbb{R})$, $\psi', \psi'' \in H^1(\mathbb{R})$, and ϕ is a solution of* (10.6). *Then we have*

$$B_\psi : \mathcal{V}(\Omega_\phi, \epsilon_0) \to H^1(\mathbb{R}) \textit{ is } C^1 \textit{ with a bounded derivative,} \tag{10.17}$$

$$B_\psi \textit{ commutes with translations,} \tag{10.18}$$

$$\langle B_\psi(v), v\rangle = 0 \quad \textit{for all } v \in \mathcal{V}(\Omega_\phi, \epsilon_0), \tag{10.19}$$

$$\textit{if } \langle \phi, \psi'\rangle = 0, \quad \textit{then } B_\psi(\phi) = \psi', \quad \textit{and} \tag{10.20}$$

$$\textit{if } \langle \phi', \psi'\rangle = 0, \quad \textit{then } \langle B_\psi(v), v'\rangle = 0 \quad \textit{for all } v \in \Omega_\phi. \tag{10.21}$$

PROOF. The proofs of statements (10.17), (10.18), and (10.19) follow the same logic as the proof of Lemma 3.5 in [**23**]. Statements (10.20) and (10.21) follow from the equality $\Lambda(\phi) = 0$ and from the basic formula

$$B_\psi(\phi) = \psi' + \frac{\langle \phi, \psi'\rangle}{\|\phi'\|^2}\phi''. \tag{10.22}$$

This completes the proof of the lemma. □

LEMMA 10.10. *Let $p \geqq 2$, $p \in \mathbb{N}$, $0 < l < 2\sqrt{c}$, and let ϕ be a solution of* (10.6). *Assume that $u_0 \in \mathcal{V}(\Omega_\phi, \epsilon_0)$ such that*

$$u_0, \mathcal{H}\partial_x u_0, \partial_x^2 u_0 \in L^1(\mathbb{R}).$$

If $u(t)$ is a solution of (10.5) *which corresponds to the initial data u_0 and $u(t) \in V(\Omega_\phi, \epsilon_0)$ for $t \in [0, T_1]$, then for*

$$\psi(x) = \int_{-\infty}^{x} [\phi(y) + 2y\phi'(y)]dy$$

we have

$$\mathcal{A}_\psi(u(t)) \equiv \int_{-\infty}^{\infty} \psi(x - \Lambda(u(t)))u(x,t)\,dx < \infty, \quad \textit{for all } t \in [0, T_1]. \tag{10.23}$$

PROOF. Let H_e be the Heaviside function and define

$$\gamma = \int_{-\infty}^{\infty} \phi(y) + 2y\phi'(y)dy.$$

Note that from Theorem 10.3 we have $\gamma < \infty$. Then

$$\mathcal{A}_\psi(u(t)) = \int_{-\infty}^{\infty} [\psi(x - \Lambda(u(t))) - \gamma H_e(x - \Lambda(u(t)))]u(x,t)dx + \gamma \int_{\Lambda(u(t))}^{\infty} u(x,t)dx.$$

Then, it follows from Theorem 10.22 that

$$| A_\psi(u(t))| \leqq \|\psi - \gamma H_e\| \|u_0\| + |\gamma| C(1 + t^{2/3}),$$

and so

$$| A_\psi(u(t))| \leqq K_1(1 + t^\theta) \tag{10.24}$$

where $\theta = \frac{2}{3}$ and K_1 is a constant, because $\psi - \gamma H_e \in L^2(\mathbb{R})$. The last affirmation is a consequence of the following argument. For $x < 0$ it follows from Minkowski's inequality that

$$\begin{aligned} \Big(\int_{-\infty}^{0} (\psi(x) - \gamma H_e(x))^2 dx\Big)^{1/2} &= \Big(\int_{-\infty}^{0} \psi^2(x)dx\Big)^{1/2} \\ &\leqq \int_{-\infty}^{0} |\xi|^{1/2}|\phi(\xi)| + 2|\xi|^{3/2}|\phi'(\xi)|d\xi < \infty, \end{aligned} \tag{10.25}$$

where the last integral is finite by Theorem 10.3. A similar inequality holds for $x > 0$. This proves the lemma. □

Now we state our *instability criterion* for solitary wave solutions of the GB equations (10.5). This theorem will give us a sufficient condition to get the instability of the ϕ-orbit, Ω_ϕ.

THEOREM 10.11 (Instability Criterion). *Let $p \geqq 2$, $p \in \mathbb{N}$, and let ϕ be a solution from* (10.6) *obtained via Theorem 10.2(2). If there is a $\psi \in L^\infty(\mathbb{R})$ such that $\psi', \psi'' \in H^1(\mathbb{R}) \cap L^1(\mathbb{R})$, $\mathcal{H}\psi'', \psi''' \in L^1(\mathbb{R})$, and ψ is chosen such that* (10.24) *is true with $0 < \theta < 1$ and*

$$\langle S''(\phi)B_\psi(\phi), B_\psi(\phi)\rangle < 0, \tag{10.26}$$

where $S(u) = E_{\mathrm{GB}}(u) + cF(u)$, then there exist $\epsilon > 0$ and a sequence $\{u_0^j\}$ in $\mathcal{V}(\Omega_\phi, \epsilon)$ satisfying

(1) *$u_0^j \to \phi$ as $j \to \infty$ in $H^1(\mathbb{R})$,*
(2) *for all $j \in \mathbb{N}$, $u(t) = U(t)u_0^j$ is a global solution of* (10.5) *but escapes from $\mathcal{V}(\Omega_\phi, \epsilon)$ in finite time.*

The proof of Theorem 10.11 is based on some ideas of Gonçalves in [**121**] and Shatah and Strauss in [**253**]. We will break the proof into several steps.

(I) *Dynamical system generated by B_ψ.*

For $v_0 \in V(\Omega_\phi, \epsilon_0)$ and B_ψ defined by (10.16), we define the following initial value problem:

$$\begin{cases} \dfrac{d}{ds}v(s) = B_\psi(v(s)), \\ \quad v(0) = v_0. \end{cases} \tag{10.27}$$

Then from (10.17) we have that B_ψ is a uniformly Lipschitz field and therefore the initial value problem (10.27) produces for each $v_0 \in V(\Omega_\phi, \epsilon_0)$ a unique maximal solution, $v \in C^2((-\sigma, \sigma); \mathcal{V}(\Omega_\phi, \epsilon_0))$, where $v(0) = v_0$ and $\sigma = \sigma(v_0) \in (0, \infty]$. Moreover, for each $\epsilon_1 < \epsilon_0$, there exists $\sigma_1 > 0$ such that $\sigma(v_0) \geqq \sigma_1$ for all $v_0 \in \mathcal{V}(\Omega_\phi, \epsilon_1)$: in other words, the solutions may be defined on a uniform interval. Hence, we can define for fixed ϵ_1 and σ_1 the following flow determined for (10.27):

$$\mathcal{U} : (-\sigma_1, \sigma_1) \times \mathcal{V}(\Omega_\phi, \epsilon_1) \to \mathcal{V}(\Omega_\phi, \epsilon_0),$$
$$(s, v_0) \to \mathcal{U}(s)v_0,$$

where $s \to \mathcal{U}(s)v_0$ is the maximal solution of (10.27) with initial data v_0. From Lemma 10.9, $\mathcal{U}$ is a C^1 function, and for $v_0 \in \mathcal{V}(\Omega_\phi, \epsilon_1)$, the function $s \in (-\sigma_1, \sigma_1) \to U(s)v_0$ is C^2. We also have that the flow $s \to \mathcal{U}(s)v_0$ commutes with *translations.* Moreover,

$$\mathcal{U}(s)\phi,\ \mathcal{H}\partial_x \mathcal{U}(s)\phi,\ \partial_x^2 \mathcal{U}(s)\phi \in L^1(\mathbb{R}) \quad \text{for } s \in (-\sigma_1, \sigma_1). \tag{10.28}$$

In fact, from (10.16) and (10.27) we obtain that

$$\mathcal{U}(s)\phi = \phi + \int_0^s \tau_{\Lambda(\mathcal{U}(t)\phi)} \psi' \, dt - \int_0^s F(t) \tau_{\Lambda(\mathcal{U}(t)\phi)} \phi'' \, dt$$

where the map $s \in (-\sigma_1, \sigma_1) \to F(s)$ is a continuous function. Then from Theorem 10.3 and the hypothesis on ψ given in Theorem 10.11 we have (10.28).

(II) *Behavior of the flow $s \to \mathcal{U}(s)\phi$ by the action $S = E_{\mathrm{GB}} + c\,F$.*

For every $v_0 \in \mathcal{V}(\Omega_\phi, \epsilon_1)$ we know that

$$s \in (-\sigma_1, \sigma_1) \to S(\mathcal{U}(s)v_0) \text{ is } C^2.$$

Then applying Taylor's Theorem we have that

$$S(\mathcal{U}(s)v_0) = S(v_0) + P(v_0)s + \frac{1}{2} R(\mathcal{U}(\theta s)v_0)s^2,$$

with $\theta \in (0, 1)$ and P and R functionals defined on $\mathcal{V}(\Omega_\phi, \epsilon)$ by

$$P(v) = (S'(v), B_\psi(v)), \quad R(v) = (S''(v)B_\psi(v), B_\psi(v)) + (S'(v), B'_\psi(v)(B_\psi(v))).$$

Since R and $\mathcal{U}$ are continuous, $S'(\phi) = 0$, and from (10.26), $R(\phi) < 0$, we have the existence of $\epsilon_2 \in (0, \epsilon_1]$ and $\sigma_2 \in (0, \sigma_1]$ such that

$$S(\ U(s)v_0) \leqq S(v_0) + P(v_0)s \quad \text{for } v_0 \in B(\phi; \epsilon_2),\ s \in (-\sigma_2, \sigma_2). \tag{10.29}$$

Since $\mathcal{U}(s)v_0$ commutes with translations, from Lemma 10.6 we obtain the extension to $\mathcal{V}(\Omega_\phi, \epsilon_2)$:

$$\begin{cases} \text{for } v_0 \in \mathcal{V}(\Omega_\phi, \epsilon_2),\ s \in (-\sigma_2, \sigma_2) \\ \qquad S(\mathcal{U}(s)v_0) \leqq S(v_0) + P(v_0)s. \end{cases} \tag{10.30}$$

So, for $v_0 = \mathcal{U}(\tau)\phi$ with $\tau \neq 0$ small enough, we get

$$S(\mathcal{U}(s)\mathcal{U}(\tau)\phi) \leqq S(\mathcal{U}(\tau)\phi) + P(\mathcal{U}(\tau)\phi)s.$$

Hence for $s = -\tau < 0$ we have

$$S(\phi) \leqq S(\mathcal{U}(\tau)\phi) - P(\mathcal{U}(\tau)\phi)\tau. \tag{10.31}$$

Moreover, from (10.26) the map $\tau \to S(\mathcal{U}(\tau)\phi)$ has a strict local maximum at 0 and therefore

$$S(\mathcal{U}(\tau)\phi) < S(\phi) \quad \text{for } \tau \in (-\sigma_2, \sigma_2),\ \tau \neq 0. \tag{10.32}$$

So from (10.31) and (10.32) for some $\sigma_3 \leqq \sigma_2$ we obtain the following basic behavior of the flow $\mathcal{U}(\tau)\phi$:

$$P(\mathcal{U}(\tau)\phi) < 0 \quad \text{for } \tau \in (0, \sigma_3). \tag{10.33}$$

On the other hand, we have for the functional $J_0(v) = \int v^{p+1}dx$ that

$$\langle J_0'(\phi), B_\psi(\phi)\rangle \neq 0. \tag{10.34}$$

Otherwise, $B_\psi(\phi)$ would be tangent to the manifold

$$\mathcal{Q} = \{v \in H^1(\mathbb{R}) :\ J_0(v) = \mu\}$$

and so one would have

$$\langle S''(\phi)B_\psi(\phi), B_\psi(\phi)\rangle \geqq 0$$

since ϕ *minimizes* S *on* $\mathcal{Q}$ by Theorem 10.2(2). But this last inequality is a contradiction to (10.26). So, if we consider the function

$$(v_0, s) \in \mathcal{V}(\Omega_\phi, \epsilon_1) \times (-\sigma_1, \sigma_1) \longrightarrow J_0(\mathcal{U}(s)v_0) = \int_{-\infty}^{\infty} (\mathcal{U}(s)v_0)^{p+1}dx,$$

which is a C^1 functional with $(\phi, 0) \to J_0(\phi) = \mu$, then from (10.34) we get

$$\frac{d}{ds}J_0(\mathcal{U}(s)v_0)\Big|_{(\phi,0)} = \langle J_0'(\mathcal{U}(s)v_0), \frac{d}{ds}\mathcal{U}(s)v_0\rangle\Big|_{(\phi,0)} = \langle J_0'(\phi), B_\psi(\phi)\rangle \neq 0.$$

Therefore from the Implicit Function Theorem, there exist $\epsilon_3 \in (0, \epsilon_2)$ and $\sigma_3 \in (0, \sigma_2]$ such that for all $v_0 \in B_{\epsilon_3}(\phi)$ there exists a unique $s = s(v_0) \in (-\sigma_3, \sigma_3)$ such that $J_0(\mathcal{U}(s)v_0) = \mu$, i.e., $\mathcal{U}(s)v_0 \in G$. So, applying (10.29) to $(v_0, s(v_0)) \in B_{\epsilon_3}(\phi) \times (-\sigma_3, \sigma_3)$ and using the fact that ϕ minimizes S on F, we have

for $v_0 \in B_{\epsilon_3}(\phi)$, there exists $s \in (-\sigma_3, \sigma_3)$ such that $S(\phi) \leqq S(v_0) + P(v_0)s$.

Finally, the following extension follows from the last affirmation, Lemma 10.7 and (10.18) :

$$\begin{cases} \text{for } v_0 \in V(\Omega_\phi, \epsilon_3), \text{ there exists } s \in (-\sigma_3, \sigma_3) \text{ such that} \\ \qquad S(\phi) \leqq S(v_0) + P(v_0)s. \end{cases} \tag{10.35}$$

(III) *Invariant sets and global existence for the flow of the GB equations.*

Initially, since $B_{-\psi}(\phi) = -B_\psi(\phi)$, we can assume from (10.34) that

$$\langle J_0'(\phi), B_\psi(\phi)\rangle < 0.$$

So, if we consider the continuous flow $\tau \to \mathcal{U}(\tau)\phi$, which is solution of (10.27) with initial data ϕ, we will have for small enough $\tau > 0$ that we can find some small δ such that

$$J_0(\,U(\tau)\phi) = J_0(\phi) + \int_0^\tau \langle J_0'(\mathcal{U}(\xi)\phi), B_\psi(\,U(\xi)\phi)\rangle\, d\xi = \mu - \delta < \mu. \tag{10.36}$$

Next, we define the set

$$\mathbb{D} = \{v \in H^1(\mathbb{R}) | S(v) < S(\phi), J_0(v) < \mu\} \cap \{v \in H^1(\mathbb{R}) | P(v) < 0\} \equiv \mathcal{B} \cap P$$

where $P(v) = \langle S'(v), B_\psi(v)\rangle$. Then we have the following:

(1) $\mathcal{U}(\tau)\phi \in \mathbb{D}$ for all $\tau \in (0, \sigma_3)$.
(2) $\mathcal{B}$ is invariant by the flow $u(t)$ of equation (10.5).
(3) The flow $u(t)$ of (10.5) with initial data in $\mathcal{B}$ is global and has a uniformly bounded trajectory in $H^1(\mathbb{R})$.
(4) For $\mathcal{A}_\psi$ defined in (10.23) we have $\partial_t \mathcal{A}_\psi(u(t)) = -P(u(t))$.

In fact, from (10.32), (10.33), and (10.36) we get affirmation (1). Now, we see that $\mathcal{B}$ is invariant by the flow $u(t)$ of the GB equations. Let $u_0 \in \mathcal{B} \subset H^1(\mathbb{R})$. Then from Theorem 10.1 (local existence theorem), there exists $T^* > 0$ such that $u(t) = U(t)u_0 \in H^1(\mathbb{R})$ and this solution satisfies (10.5) for all $t \in [0, T^*)$. So, since S is conserved by the flow $U(t)u_0$, we have that

$$S(U(t)u_0) = S(u_0) < S(\phi).$$

Now, from the continuity of the map $t \in [0, T^*) \to J_0(U(t)u_0)$ and from the minimization of S on $\mathcal{Q}$ by ϕ, we obtain that $J_0(U(t)u_0) < \mu$ for all $t \in [0, T^*)$. This proves (2). Now, we show that $T^* = +\infty$. In fact, without loss of generality we can assume $c = 1$, so combining the conservation of energy and (10.8), we have for $0 < \beta < \frac{1}{2}$ that

$$\begin{aligned}
&\frac{1}{2}\int_{-\infty}^{\infty} |\partial_x U(t)u_0|^2 + |U(t)u_0|^2 dx \\
&= S(U(t)u_0) + \frac{l}{2}\int_{-\infty}^{\infty} U(t)u_0 \mathcal{H}(\partial_x U(t)u_0)dx + \frac{1}{p+1}J_0(U(t)u_0) \\
&< S(\phi) + \frac{l}{2}\|u_0\|\|U(t)u_0\|_1 + \frac{1}{p+1}\mu \\
&< \beta\|U(t)u_0\|_1^2 + \frac{l^2}{4\beta}\|u_0\|^2 + S(\phi) + \frac{1}{p+1}\mu
\end{aligned} \tag{10.37}$$

for all $t \in [0, T^*)$. This shows that the flow $t \to U(t)u_0$ is global for the GB equations and has a bounded trajectory in $H^1(\mathbb{R})$ for every $u_0 \in \mathcal{B}$. Now, we prove (4). Indeed, let $u(t)$ be a solution of (10.5). Then as long as this flow remains in $\mathcal{V}(\Omega_\phi, \epsilon)$ and without loss of generality, we assume that $\mathcal{A}_\psi(u(t)) < \infty$, and we have the equality

$$\begin{aligned}
\partial_t\ A_\psi(u(t)) &= \partial_t \int_{-\infty}^{\infty} \tau_{\Lambda(u(t))}\psi(x)\ u(x,t)\ dx \\
&= \int_{-\infty}^{\infty} \Big(\langle \tau_{\Lambda(u(t))}\psi'(x), u(t)\rangle \Lambda'(u(t)) + \tau_{\Lambda(u(t))}\psi(x)\Big)\frac{\partial u}{\partial t}(x,t)\ dx \\
&= -\Big\langle \langle \tau_{\Lambda(u(t))}\psi'(x), u(t)\rangle \frac{d}{dx}\Lambda'(u(t)) + \tau_{\Lambda(u(t))}\psi'(x), E'(u(t))\Big\rangle \\
&= -\langle B_\psi(u(t)), S'(u(t))\rangle + c\langle B_\psi(u(t)), u(t)\rangle = -P(u(t)),
\end{aligned} \tag{10.38}$$

where in the last equality we have used the relation $\langle B_\psi(v), v\rangle = 0$ for all $v \in \mathcal{V}(\Omega_\phi, \epsilon)$ (see Lemma 10.9).

Now, we are in a position to give the proof of Theorem 10.11.

PROOF OF THEOREM 10.11. Let $\tau_j \in (0, \sigma_3)$ such that $\tau_j \to 0$ as $j \to \infty$, and we consider the sequence of initial data $\{u_0^j\} \subseteq H^1(\mathbb{R})$ defined by

$$u_0^j \equiv \mathcal{U}(\tau_j)\phi.$$

Then it follows that $u_0^j \to \phi$ in $H^1(\mathbb{R})$ as $j \to \infty$. Moreover, since $u_0^j \in \mathbb{D} \subset \mathcal{B}$, we obtain from step (III) that the flow $t \to U(t)u_0^j$ is global for all j.

Finally, to conclude the proof of the theorem, we need only to verify that $t \to U(t)u_0^j$ escapes from $\mathcal{V}(\Omega_\phi, \epsilon_3)$ for some $\epsilon_3 > 0$ and for all $j \in \mathbb{N}$. In fact, let $\epsilon_3 > 0$ be as determined in step (II) and define

$$T_j = \sup\{\tau > 0 \mid U(t)u_0^j \in \mathcal{V}(\Omega_\phi, \epsilon_3) \text{ for all } t \in (0, \tau)\}.$$

Then, it follows from (10.35) that for all $j \in \mathbb{N}$ and $t \in (0, T_j)$ there exists $s = s_j(t) \in (-\sigma_3, \sigma_3)$ satisfying

$$S(\phi) \leqq S(U(t)u_0^j) + P(U(t)u_0^j)s = S(u_0^j) + P(U(t)u_0^j)s. \tag{10.39}$$

Next, since $u_0^j \in \mathbb{D}$, then $t \to U(t)u_0^j \in P$ for $t \in (0, T_j)$. In fact, if we have some time $t_0 \in (0, T_j)$ such that $P(U(t)u_0^j) > 0$, then continuity implies that there is some $t_1 \in (0, T_j)$ satisfying $P(U(t_1)u_0^j) = 0$ and so from (10.39) it follows that $S(\phi) \leqq S(u_0^j)$, but this is a contradiction. Then, we have that P is bounded away from zero and

$$-P(U(t)u_0^j) \geqq \frac{S(\phi) - S(u_0^j)}{\sigma_3} = \eta_j > 0 \qquad \text{for all } t \in (0, T_j). \tag{10.40}$$

Now suppose that for some j, $T_j = +\infty$. Then from the properties obtained by the flow $s \to U(s)\phi$ in step (I) and considering the conditions on ψ such that (10.24) is true, we have from (10.38) and (10.40) that

$$\mathcal{A}_\psi(U(t)u_0^j) \geqq t\eta_j + \mathcal{A}_\psi(u_0^j) \qquad \text{for all } t \in (0, +\infty).$$

So, from (10.24) it follows for $0 < \theta < 1$ that

$$K_1 \geqq \frac{t\eta_j + \mathcal{A}_\psi(u_0^j)}{1 + t^\theta} \qquad \text{for all } t \in (0, +\infty).$$

Therefore $T_j < +\infty$, which means that $u(t) = U(t)u_0^j$ eventually leaves $\mathcal{V}(\Omega_\phi, \epsilon_3)$. Theorem 10.11 is proved. □

We now apply the criterion established in Theorem 10.11 to obtain Theorem 10.5. Therefore, we need only to give conditions to assure inequality (10.26) by choosing the right function ψ. The following lemma contains this basic information.

LEMMA 10.12. *Let $0 < l < 2\sqrt{c}$ and consider ϕ, a solution of* (10.6). *Then for*

$$\psi(x) = \int_{-\infty}^{x} [\phi(y) + 2y\phi'(y)]dy$$

we have the following properties:

(1) $\psi \in L^\infty(\mathbb{R})$, $\psi', \psi'' \in H^1(\mathbb{R}) \cap L^1(\mathbb{R})$, $\mathcal{H}\psi'', \psi''' \in L^1(\mathbb{R})$, *inequality* (10.24) *is true, and* $\langle \psi', \phi \rangle = 0$.

(2) *For* $S = E_{\mathrm{GB}} + cF$,

$$\begin{aligned} &\langle S''(\phi)B_\psi(\phi), B_\psi(\phi)\rangle \\ &\quad = \frac{6p - p^2 - 5}{p+1} \int_{-\infty}^{\infty} \phi^{p+1}(x)\,dx + \frac{l}{2}\int_{-\infty}^{\infty} \phi(x)\mathcal{H}\phi'(x)\,dx. \end{aligned} \tag{10.41}$$

PROOF. From Theorem 10.3 and Lemma 10.10 we immediately obtain item (1). Moreover, since $\langle \psi', \phi \rangle = 0$, it follows from (10.22) that $B_\psi(\phi) = \psi'$. So, we need only to estimate the quantity $\langle S''(\phi)\psi', \psi' \rangle$. In fact, by denoting $\mathcal{M} = S''(\phi)$ and setting $c = 1$, we have that

$$\mathcal{M} = -\frac{d^2}{dx^2} - l\mathcal{H}\frac{d}{dx} - p\phi^{p-1} + 1.$$

Next we show the following identities:

$$\begin{aligned} &(1)\quad \langle \mathcal{M}\phi, \phi \rangle = (1-p)\int_{-\infty}^{\infty} \phi^{p+1}(x)\,dx, \\ &(2)\quad \langle \mathcal{M}\phi, x\phi' \rangle = \frac{p-1}{p+1}\int_{-\infty}^{\infty} \phi^{p+1}(x)\,dx, \quad \text{and} \\ &(3)\quad \langle \mathcal{M}x\phi', x\phi' \rangle = \frac{p-1}{2(p+1)}\int_{-\infty}^{\infty} \phi^{p+1}\,dx + \frac{l}{2}\int_{-\infty}^{\infty} \phi\mathcal{H}\phi'\,dx. \end{aligned} \tag{10.42}$$

The identities (1) and (2) are trivial. In order to prove (3), we have the following estimate from the relations $\mathcal{M}(x\phi') = -2\phi + 2\phi^p + l\mathcal{H}\phi'$ and $\int x\phi'\mathcal{H}\phi'\,dx = 0$ and integration by parts:

$$\langle \mathcal{M}x\phi', x\phi' \rangle = \int_{-\infty}^{\infty} \phi^2 - \frac{2}{p+1}\phi^{p+1}\,dx. \tag{10.43}$$

Now, from (10.6) and integration by parts we get

$$\langle \mathcal{M}x\phi', x\phi' \rangle = -2\int_{-\infty}^{\infty} x\phi'\phi''dx = \int_{-\infty}^{\infty} l\phi\mathcal{H}\phi' + \phi^{p+1} - \phi^2 dx. \tag{10.44}$$

Then from (10.43) and (10.44) we obtain (3). So, the equality in (10.41) follows from (10.42). This proves the lemma. □

With Theorem 10.11 and Lemma 10.12 we immediately have the following proof of Theorem 10.5 about instability to GB equations.

PROOF OF THEOREM 10.5. From Theorem 10.2(2) we have that ϕ_l solves equation (10.6). Since

$$\int_{-\infty}^{\infty} \phi_l^{p+1}(x)dx = \mu > 0 \quad \text{and} \quad \int_{-\infty}^{\infty} \phi_l(x)\mathcal{H}\phi_l'(x)dx > 0,$$

the theorem follows directly from (10.12), (10.41), and Theorem 10.11. □

REMARK 10.13. If we consider $l = 0$ in (10.5), from (10.41) we obtain the instability result for the GKdV equations (2.3) with $p > 4$ (see [**64**]).

REMARK 10.14. If we consider $l < 0$, we obtain the instability of solitary wave solutions for equation (10.5) with $p \geqq 5$.

10.3. Fifth-Order Korteweg-de Vries Equations

The approach established in the last section can be applied with success to a hamiltonian generalized fifth-order Korteweg-de Vries equation of the form

$$u_t + u_{xxxxx} + bu_{xxx} = (G(u, u_x, u_{xx}))_x, \tag{10.45}$$

where we assume that the nonlinear term has the form

$$G(q, r, s) = F_q(q, r) - rF_{qr}(q, r) - sF_{rr}(q, r), \tag{10.46}$$

for some $F(q,r) \in C^3(\mathbb{R}^2)$ which is homogeneous of degree $p+1$ for some $p>1$. That is, we assume

$$F(\lambda q, \lambda r) = \lambda^{p+1} F(q,r) \tag{10.47}$$

for all $\lambda \geqq 0$ and $(q,r) \in \mathbb{R}^2$. Equation (10.45) arises as a model for a variety of physical phenomena (see Kawahara [**155**], Olver [**233**], Craig and Groves [**93**], Benney [**50**], and references therein). Here we are interested in the case when the coefficient b in (10.45) is proportional to $\tau - \frac{1}{3}$, where τ represents the Bond number which is a nondimensionalized surface tension. So, we have that b in (10.45) may be negative, zero, or positive, results which occur in the respective limits $\tau \to \frac{1}{3}^+$, $\tau = \frac{1}{3}$ (no surface tension), and $\tau \to \frac{1}{3}^-$.

By a solitary wave solution for (10.45) we mean a solution of the form $u(x,t) = \varphi(x+ct)$. If $\varphi(\xi)$ and its derivatives tend to zero as the variable $\xi = x + ct$ approaches $\pm\infty$, we see that φ must satisfy the fourth-order ordinary differential equation

$$\varphi_{xxxx} + b\varphi_{xx} + c\varphi = G(\varphi, \varphi_x, \varphi_{xx}). \tag{10.48}$$

For a general function G such as that established in (10.46), Levandosky [**183**] proved the existence of a class of solitary wave solutions of (10.45) satisfying (10.48) provided $c > b_+^2/4$, where $b_+ = \max\{b, 0\}$. He used the Concentration-Compactness Method. More exactly, he considers the functionals $I_{c,b}$ and K in $C^2(H^2(\mathbb{R}); \mathbb{R})$ given by

$$I_{c,b}(f) = \frac{1}{2}\int_{-\infty}^{\infty} [(f_{xx})^2 - b(f_x)^2 + cf^2]\, dx$$

and

$$K(f) = \int_{-\infty}^{\infty} F(f, f_x)\, dx.$$

Define, for $\lambda > 0$ and $c > b_+^2/4$, the family of minimization problems

$$M_c(\lambda) = \inf\{I_{c,b}(f) \ : \ f \in H^2(\mathbb{R}) \text{ and } K(f) = \lambda\}.$$

Then we have the following theorem of existence of solitary wave solutions.

THEOREM 10.15. *Let $p > 1$ and assume that there is some $f \in H^2(\mathbb{R})$ such that $K(f) > 0$ and $c > b_+^2/4$, where $b_+ = \max\{b,0\}$. Then for*

$$\lambda = \left[\frac{M_c(1)}{p+1}\right]^{\frac{p+1}{p-1}},$$

every minimum point φ of $M_c(\lambda)$ satisfies equation (10.48).

So, by making the appropriate changes in the instability theory established above for the GB equations and considering the vector field B_ψ generated by

$$\psi(x) = \int_{-\infty}^{x} [\varphi(y) + 2y\varphi'(y)]dy,$$

we have the following instability result for the φ-orbit $\Omega_\varphi = \{\varphi(\cdot + y) : y \in \mathbb{R}\}$ associated to the fifth-order equation (10.45) (see Angulo [**24**]).

THEOREM 10.16 (Instability for Fifth-Order Equations). *Let $p \geqq 2$ and suppose that $F(q,r)$, which is homogeneous of degree $p+1$, satisfies relation* (10.47).

Consider φ, a solution of (10.48) obtained via Theorem 10.15. Then if $F(\varphi, \varphi')$ is homogeneous in φ' of degree β, with $\beta \in [0, p+1]$, then the conditions

$$\begin{cases} b = 0 \;\; \text{and} \;\; \beta > \dfrac{9-p}{2} \;\; \text{or} \\ b < 0 \;\; \text{and} \;\; \beta \geqq \dfrac{9-p}{2} \;\; \text{or} \\ \beta > \dfrac{9-p}{2}, \;\; b > 0, \;\; \text{and}\, b \;\; \text{small} \end{cases}$$

imply that the orbit Ω_φ is $H^2(\mathbb{R})$-unstable by the flow of (10.45).

REMARK 10.17. Recently Levandosky in [**184**] considered the stability of a class of solitary wave solutions for equations of the form (10.45). He performed a detailed investigation, both analytically and numerically, for a class of homogeneous nonlinearities (for instance, $F(u, u_x) = |u|^\alpha |u_x|^\beta$ where $\alpha + \beta = p + 1 > 2$ and $p + 2\beta < 5$) and determined the regions of wave speeds for which the solitary waves are stable or unstable.

10.4. A Generalized Class of Benjamin Equations

In Section 9.2 we saw that the variational problems $\theta_i(\lambda)$ in (9.3) may be used for obtaining the existence of solitary wave solutions for the following Benjamin-type equations:

$$u_t + u^p u_x - \mathcal{L}_i u_x = 0, \qquad i = 1, 2, \tag{10.49}$$

where $p \geqq 1$ and the linear operators $\mathcal{L}_i$ are given by (9.1) and (9.2), respectively.
More precisely we have the following theorem.

THEOREM 10.18. *Let $p \geqq 1$, $p \in \mathbb{N}$, and let $c > 0$ be a given wave speed. Then we have the following:*

(1) *For $m \in \mathbb{N}$, $0 \leqq r < m$, $l \geqq 0$, and*

$$\frac{r}{m}\Big(\frac{m}{r} - 1\Big)^{\frac{m-r}{m}} l < c^{\frac{m-r}{m}},$$

problem (9.3) with $i = 1$ has a solution. Moreover, there is a $\mu_1 > 0$ such that every minimum point ϕ_1 of $\theta_1(\mu_1)$ satisfies

$$(\mathcal{L}_1 + c)\phi_1 = \frac{1}{p+1}\phi_1^{p+1}.$$

(2) *For l_j, r_j, $j = 1, 2, ..., k$, satisfying $r_k \geqq \frac{1}{2} - \frac{1}{p}$ and*

$$\min_{x \geqq 0} \Big\{ c + \sum_{j=1}^{k} l_j x^{2r_j} \Big\} > 0,$$

problem (9.3) with $i = 2$ has a solution. Moreover, there is a $\mu_2 > 0$ such that every minimum point ϕ_2 of $\theta_2(\mu_2)$ satisfies

$$(\mathcal{L}_2 + c)\phi_2 = \frac{1}{p+1}\phi_2^{p+1}.$$

PROOF. See Chen and Bona [**84**]. □

Therefore, by making the appropriate changes in the instability criterion for the GB equations in Section 10.2 and considering the vector field B_{ψ_i} in (10.16) generated by

$$\psi_i(x) = \int_{-\infty}^{x} [\phi_i(y) + 2y\phi_i'(y)]dy,$$

we have the following instability theorem for equations (10.49) (see Theorem 4.6 in Angulo [**23**]).

THEOREM 10.19. *Let $p \geqq 2$, $p \in \mathbb{N}$, and let ϕ_i be a solitary wave solution of* (10.49) *obtained via Theorem 10.18. Then we have the following:*

(1) *If l is small and $p > 4m$, then Ω_{ϕ_1} is $H^m(\mathbb{R})$-unstable by the flow of equation* (10.49) *with $i = 1$ and $\frac{1}{2} \leqq r < m$. Moreover, the trajectories used to exhibit instability are global and uniformly bounded in $H^m(\mathbb{R})$.*
(2) *If $r_1 \geqq \frac{1}{2}$ and some $r_j \geqq 1$, then Ω_{ϕ_2} is $H^{r_k}(\mathbb{R})$-unstable by the flow of equation* (10.49) *with $i = 2$ provided $p > 4r_k$. Moreover, the trajectories used to exhibit instability are global and uniformly bounded in $H^{r_k}(\mathbb{R})$.*
(3) *If $r_1 \geqq \frac{1}{2}$ and for all j, $r_j < 1$, then Ω_{ϕ_2} is $H^{r_k}(\mathbb{R})$-unstable by the flow of equation* (10.49) *with $i = 2$ provided $p > 4r_k$. Moreover, the trajectories used to exhibit instability are uniformly bounded in $H^{r_k}(\mathbb{R})$.*

10.5. Linear Instability and Nonlinear Instability

Another approach to obtain nonlinear instability for solitary and periodic travelling wave solutions is that of studying initially the "linearized instability" problem (see Section 5.3). More precisely, suppose that our equation may be written in the Hamiltonian form

$$\frac{du}{dt} = JE'(u(t)) \tag{10.50}$$

on a Hilbert space X, where $J : D(J) \to X$ is one-to-one, onto, and skew-symmetric. Moreover, $E : X \to \mathbb{R}$ is a C^2 functional. The Hamiltonian system has two kinds of conserved quantities. One is the energy E. The other is the "charges", Q_ω, which depend on the symmetries or on the group of invariances of the equation. We assume there are travelling waves ϕ_ω which are solutions of (10.50) and are realized as critical points of a constrained energy E. The approach to stability is to linearize around ϕ_ω, and so we obtain the linearized equation

$$\frac{dw}{dt} = J\mathcal{L}w \tag{10.51}$$

where $\mathcal{L}$ is the Hessian of the constrained energy, namely, $\mathcal{L} = E''(\phi_\omega) - Q_\omega(\phi_\omega)$. The spectrum of the generator $J\mathcal{L}$ will give us the information that we need to know stability. The operator $J\mathcal{L}$ is the product of skew-adjoint and self-adjoint operators which do not commute. Moreover its spectrum is symmetric with respect to both axes and most of the spectrum lies on the imaginary axis. The main question is *whether there is any spectrum off the imaginary axis.* If so, one expects instability. If not, one expects stability. For obtaining a capable condition for stability or instability, we look for conditions on the operator $\mathcal{L}$. Because the self-adjoint operator $\mathcal{L}$ usually has some negative spectrum, we need to use the real-valued function

$$d(\omega) = E(\phi_\omega) - Q_\omega(\phi_\omega).$$

By $d''(\omega)$ we mean its Hessian (a symmetric matrix). So, we have from Grillakis *et al.* [**124**] the following.

THEOREM 10.20. (1) ϕ_ω *is stable if the number* $n(\mathcal{L})$ *of negative eigenvalues of* $\mathcal{L}$ *equals the number* $p(d''(\omega))$ *of positive eigenvalues of* $d''(\omega)$. *(It is never less.)*

(2) ϕ_ω *is nonlinearly unstable if* $n(\mathcal{L}) - p(d''(\omega))$ *is an odd positive integer.*

The condition given in item (2) above implies the existence of some nonzero real λ and some nonzero $v \in X$ such that

$$J\mathcal{L}v = \lambda v.$$

Therefore, in this case we get a nonlinear instability result as a consequence of a linearized instability result.

In particular in (10.50), for the case of one-dimensional evolution equations, we consider the group of invariance, T, as being translation $T(g)r = g(\cdot + r)$, $r \in \mathbb{R}$. So, if we take $n(\mathcal{L}) = 1$ and $\mathcal{N}(\mathcal{L}) = [T'(0)\phi_\omega] = [\frac{d}{dx}\phi_\omega]$, then Theorem 10.20 is rephrased in the following way ([**123**]): ϕ_ω is stable if $d''(\omega) > 0$ and it is unstable if $d''(\omega) < 0$.

The instability of travelling wave solutions of nonlinear wave equations has had advances in the past few years. There are advances in the numerical investigation, as well as in the analytical study of spectral stability, linearized instability, and modulation instability.

Next, we focus specifically on the general linearized instability result of Lopes in [**199**] and its application to weakly coupled KdV systems. In fact, we consider the system

$$\begin{cases} u_t = (-u_{xx} + f(u) + \epsilon F_u(u,v))_x, \\ v_t = (-v_{xx} + g(v) + \epsilon F_v(u,v))_x, \end{cases} \tag{10.52}$$

where ϵ is a small real parameter. A travelling wave solution of (10.52) is a solution of the form

$$(u(x,t), v(x,t)) = (\phi_c(x - ct), \psi_c(x - ct));$$

in other words, (ϕ, ψ) satisfies

$$\begin{cases} -\phi_c'' + f(\phi_c) + c\phi_c + \epsilon F_u(\phi_c, \psi_c) = 0, \\ -\psi_c'' + g(\psi_c) + c\psi_c + \epsilon F_v(\phi_c, \psi_c) = 0. \end{cases} \tag{10.53}$$

If we denote by $F_0(u)$ and $G_0(v)$ the primitives of $f(u)$ and $g(v)$, respectively, such that $F_0(0) = G_0(0) = 0$, then (10.52) has the following conserved quantities:

$$V(u,v) = \frac{1}{2}\int_{-\infty}^{\infty} (u_x^2 + v_x^2)\,dx + \int_{-\infty}^{\infty} (F_0(u) + G_0(v) + \epsilon F(u,v))\,dx,$$

$$Q(u,v) = \frac{1}{2}\int_{-\infty}^{\infty} (u^2 + v^2)\,dx.$$

For $\epsilon = 0$ we have the decoupled equations

$$\begin{cases} -\phi_c'' + f(\phi_c) + c\phi_c = 0, \\ -\psi_c'' + g(\psi_c) + c\psi_c = 0. \end{cases} \tag{10.54}$$

We make the following assumptions:

(A_1) $c > 0$, $f, g : \mathbb{R} \to \mathbb{R}$ and $F : \mathbb{R}^2 \to \mathbb{R}$ are C^2 functions with $f'(0) = g'(0) = 0$;

(A_2) the equations in (10.54) have positive homoclinic symmetric solutions for some $c_0 > 0$, ϕ_{c_0} and ψ_{c_0} respectively;

(A_3) for c close to c_0, denoting by $u(c,\cdot)$, and $v(c,\cdot)$ the homoclinic even positive solutions of (10.54), we have

$$\frac{d}{dc}\int_{-\infty}^{\infty} u^2(c,x)dx > 0 \quad \text{and} \quad \frac{d}{dc}\int_{-\infty}^{\infty} v^2(c,x)dx > 0.$$

(A_4) Define $p(b) = \int_{-\infty}^{\infty} F_v(\phi_0(x-b), \psi_0(x))\psi_0'(x)dx$. Suppose that there is a b_0 such that $p(b_0) = 0$ and $p'(b_0) > 0$.

(A_5) F is $C^6(\mathbb{R}^2)$ and F_0, G_0 are $C^6(\mathbb{R})$.

REMARK 10.21. Assumption (A_4) implies that there is an $\epsilon_0 > 0$ and a solution $(\phi(x,\epsilon), \psi(x,\epsilon))$ of (10.53) for $|\epsilon| < \epsilon_0$, such that $(\phi(x,0), \psi(x,0)) = (\phi_0(x), \psi_0(x))$.

From Angulo, Lopes, and Neves [**33**] we have the following nonlinear instability theorem.

THEOREM 10.22. *If* $f(0) = g(0) = 0$ *and* $F_{uu}(0,0) = F_{vv}(0,0) = F_{uv}(0,0) = 0$ *and under assumptions* $(A_2) - (A_5)$, *there is an* $\epsilon_0 > 0$ *such that for* $0 < \epsilon < \epsilon_0$ *the travelling wave* $(\phi(x,\epsilon), \psi(x,\epsilon))$ *for* (10.53) *given by Remark 10.21 is unstable with respect to* (10.52).

REMARK 10.23. If in assumption (A_4) we have that $p'(b_0) < 0$, then the interval of instability is $-\epsilon_0 < \epsilon < 0$.

The proof of Theorem 10.22 is divided into two main parts. The first part is to prove the linearized instability of the persistent wave. The second one is to show that linear instability implies nonlinear instability. For the first part we recall a result about linearized instability.

Let $L : D(L) \subset X \to X$ and $A : D(A) \subset X \to X$ be linear operators satisfying the following conditions:

(1) L is self-adjoint and has two negative eigenvalues and zero as a simple eigenvalue with eigenfunction of the form $A\phi$;
(2) there is an element $\psi \in D(L)$ such that $L\psi = \phi$ and $\langle \psi, \phi \rangle < 0$;
(3) A is skew-adjoint.

We define the subspace X_2 which is the orthogonal to ϕ and to $A\phi$, and we define the operator $\mathcal{F} : D(\mathcal{F}) = D(L) \cap X_2 \subset X_2 \to X_2$ by

$$\mathcal{F}(z) = Lz - \langle Lz, \phi \rangle \frac{\phi}{|\phi|^2} = L(z - \langle Lz, \phi \rangle \frac{\psi}{|\phi|^2}). \tag{10.55}$$

LEMMA 10.24. *The operator* $\mathcal{F}$ *is invertible and it has one negative eigenvalue. Moreover, if* J *is a skew-adjoint injective linear operator, then* JL *has a real positive eigenvalue.*

PROOF. The first part of the lemma is a consequence of a more abstract theorem in [**124**], but instead of verifying its assumptions, we think it is easier to give a direct proof. Suppose first that $\mathcal{F}z = 0$; then from (10.55) we see that for a real number θ we have

$$z - \langle Lz, \phi \rangle \frac{\psi}{|\phi|^2} = \theta A\phi.$$

Taking the scalar product of this last equation with ϕ, we get $\langle Lz, \phi \rangle = 0$ and then $z = \theta A\phi$. This implies $z = 0$, which shows that $\mathcal{F}$ is invertible.

Next, let ϕ_1, ϕ_2 be the eigenfunctions of L corresponding to the negative eigenvalues λ_1, λ_2, respectively. Let α and β be such that $\langle \alpha\phi_1 + \beta\phi_2, \phi\rangle = 0$. Taking into account that $\langle \phi_1, A\phi\rangle = \langle \phi_2, A\phi\rangle = 0$, an elementary calculation shows that

$$\langle \mathcal{F}(\alpha\phi_1 + \beta\phi_2), \alpha\phi_1 + \beta\phi_2\rangle = \lambda_1\alpha^2|\phi_1|^2 + \lambda_2\beta^2|\phi_2|^2 < 0$$

and this implies that $\mathcal{F}$ has at least one negative eigenvalue.

Now suppose there are $z_1, z_2 \in X_2$ and $\mu_1, \mu_2 < 0$ such that $\mathcal{F}z_1 = \mu_1 z_1$ and $\mathcal{F}z_2 = \mu_2 z_2$. Then

$$Lz_1 - \langle Lz_1, \phi\rangle \frac{\phi}{|\phi|^2} = \mu_1 z_1, \quad Lz_2 - \langle Lz_2, \phi\rangle \frac{\phi}{|\phi|^2} = \mu_2 z_2.$$

Therefore

$$\langle Lz_1, z_1\rangle = \mu_1|z_1|^2 < 0, \quad \langle Lz_2, z_2\rangle = \mu_2|z_2|^2 < 0, \quad \langle Lz_1, z_2\rangle = 0,$$

and $\langle Lz_1, \psi\rangle = \langle z_1, L\psi\rangle = \langle z_1, \phi\rangle = 0$, $\langle Lz_2, \psi\rangle = \langle z_2, L\psi\rangle = \langle z_2, \phi\rangle = 0$, $\langle L\psi, \psi\rangle < 0$. Since ψ does not belong to X_2, from this set of inequalities we conclude that the quadratic form $\langle Lu, u\rangle$ is negative definite on a three-dimensional subspace, which implies that L has at least three negative eigenvalues; this contradiction proves the first part of the lemma. The second part of the lemma follows from a linearized instability result given in [**199**], so the lemma is proved. □

For the second part of the proof of Theorem 10.22, we also need the following nonlinear instability result of [**130**].

THEOREM 10.25. *Let $S : \Omega \subset X \to X$ be a $C^{1,\alpha}$ map, $0 < \alpha \leq 1$, defined in an open neighborhood of a fixed point w_0 of S. If there is an element $\lambda \in \sigma(S'(w_0))$ with $|\lambda| > 1$, then w_0 is an unstable fixed point of S.*

PROOF OF THEOREM 10.22. If in the evolution system (10.52) we replace the solution $(u(x,t), v(x,t))$ by $(u(x-ct,t), v(x-ct,t))$, then we obtain

$$\begin{cases} u_t = (-u_{xx} + f(u) + cu + \epsilon F_u(u,v))_x, \\ v_t = (-v_{xx} + g(v) + cv + \epsilon F_v(u,v))_x, \end{cases} \tag{10.56}$$

and the travelling wave is an equilibrium solution of (10.56). Defining $w = (u, v)$, $E(w) = V(w) + cQ(w)$, and $J = \frac{d}{dx}$, we see that (10.56) can be written as $\dot{w} = JE'(w)$. Moreover, the linearized equation at the equilibrium is $\dot{w} = JL(w)$ where $L = T_\epsilon$,

$$T_\epsilon \begin{pmatrix} h \\ k \end{pmatrix} :=$$
$$\begin{pmatrix} -h_{xx} + f'(\phi(x,\epsilon))h + ch + \epsilon F_{uu}(\phi(x,\epsilon), \psi(x,\epsilon))h + \epsilon F_{uv}(\phi(x,\epsilon), \psi(x,\epsilon))k \\ -k_{xx} + g'(\psi(x,\epsilon))k + ck + \epsilon F_{uv}(\phi(x,\epsilon), \psi(x,\epsilon))h + \epsilon F_{vv}(\phi(\epsilon,x), \psi(x,\epsilon))k \end{pmatrix}.$$

First we show that the linearized system is unstable. According to Corollary 2.1 in [**33**], for $\epsilon > 0$ and small the operator T_ϵ has zero as a simple eigenvalue and two negative eigenvalues. Differentiating (10.53) with respect to c and defining $\psi = -(\frac{d}{dc}\phi_c, \frac{d}{dc}\psi_c)$, we get $L\psi = (\phi_c, \psi_c)$; moreover $\langle L\psi, \psi\rangle < 0$ because

$$\frac{d}{dc} Q(\phi(c,\epsilon), \psi(c,\epsilon)) > 0$$

for c close to c_0 and ϵ small. As a consequence of Lemma 10.24 with $A = J = \frac{d}{dx}$ we get that the linearized equation $\dot{w} = JL(w)$ has a positive real eigenvalue and that it is unstable.

In order to show nonlinear instability, for each $w = (u, v)$ we define $S(w)$ as the solution of (10.56) at time $t = 1$ with initial condition w at time $t = 0$. According to Theorem 4.10 in [**33**], the solution of (10.52) is C^2 with respect to the initial condition in the space $H^1(\mathbb{R}) \times H^1(\mathbb{R})$; since translation in x is a linear continuous map in $H^1(\mathbb{R})$, S is a C^2 map defined in a neighborhood of the fixed point $w_0 = (\phi_0, \psi_0)$. Moreover, from the first part we conclude that there is an element $\lambda \in \sigma(S'(w_0))$ with $|\lambda| > 1$. Therefore the instability of the equilibrium solution follows from Theorem 10.25, and Theorem 10.22 is proved. □

REMARK 10.26. By using Theorem 10.22 and dropping some terms, rescaling the variables, and assuming that a certain parameter is small, it is very easy to obtain an instability result for solitary waves associated to the Gear and Grimshaw system (see Section 2.3).

10.6. Comments

For the GKdV equations (2.5), the solitary profile satisfies the scaling relation

$$\phi_c(x) = c^{\frac{1}{p}} \phi_1(c^{\frac{1}{2}} x).$$

The values of the functional $F(u) = \frac{1}{2} \int u^2(x)dx$ in those profiles are given by

$$F(\phi_c) = \frac{1}{2} c^{\frac{4-p}{2p}} \|\phi_1\|^2.$$

So we have that $\frac{d}{dc} F(\phi_c) > 0$ for $p < 4$ and $\frac{d}{dc} F(\phi_c) < 0$ for $p > 4$. Hence, we obtain the stability and instability scenarios associated to the GKdV equations. If $p = 4$, we obtain $\frac{d}{dc} F(\phi_c) = 0$ for all $c > 0$. For this last case, it is stated in [**64**] that the travelling waves ϕ_c are unstable as a consequence of the claim that the set $\{c : \phi_c \text{ is stable}\}$ is open. This claim, however, is left unproved in [**64**]. Moreover, this is not true in general. Indeed, it is demonstrated by the dynamical system in $\mathbb{R}^2$ described in polar coordinates by

$$\dot{\theta} = \sin\, \theta, \qquad \dot{r} = 0.$$

The set of stationary states is the line $y = 0$; the subset of stable stationary points, $x \leqq 0$, is closed. In fact, if $p = 4$ the solitary waves are not only unstable but also exhibit a blow-up behavior. This blow-up is considered in several papers by Martel and Merle [**212**], [**214**], [**215**], [**220**].

Next, we consider the following Generalized Korteweg-de Vries equation in one dimension:

$$u_t = (-u_{xx} + f(u))_x \tag{10.57}$$

where $u = u(x, t)$ and $f \in C^\infty(\mathbb{R})$ is a real-valued function that satisfies $f(0) = f'(0) = 0$. Depending on the nonlinearity f, (10.57) may admit solitary wave solutions of the form $u(x, t) = \phi_c(x - ct)$. Generically, solitary waves exist for speed c from (finite or infinite) intervals of a real line. For an exhaustive study of the existence and uniqueness of solutions for the equation

$$\phi_c'' - c\phi_c = f(\phi_c),$$

we refer the reader to [**54**]. For a particular f, solitary waves with certain speeds are (orbitally) stable, while others are unstable. The study of the stability of the "critical" solitary wave, the one with speeds c on the border of stability and instability regions (c such that $\frac{d}{dc}\|\phi_c\|^2 = 0$), has been recently considered by Comech, Cuccagna, and Pelinovsky in [**90**]. There, the study of the instability (purely nonlinear) of critical travelling waves of the Generalized KdV equation (10.57) under rather generic assumptions is established. Examples of the nonlinearities f that lead to the existence of critical solitary waves are

(1) $f_{p,q}(\xi) = -A\xi^p + B\xi^q$, with $2 < p < q$, $A > 0$, and $B > 0$,
(2) $f_{p,q,r}(\xi) = A\xi^p - B\xi^q + C\xi^r$, with $2 < p < q < r$, $A > 0$, $B > 0$, and $C > 0$.

Part 5

Stability of Periodic Travelling Waves

Poincaré, Jules Henri (1854–1912)

D'ailleurs, ce qui nous rend ces solutions périodiques si précieuses, c'est qu'elles sont, pour ainsi dire, la seule brèche par où nous puissons essayer de pénétrer dans une place jusqu'ici réputée inabordable.

In *Les Méthodes Nouvelles de la Mécanique Céleste*, Gauthiers-Villars, Paris, 1889.

Hilbert, David (1862–1943)

In dealing with mathematical problems, specialization plays, I believe, a still more important part than generalization. Perhaps in most cases where we unsuccessfully seek the answer to a question, the cause of the failure lies in the fact that problems simpler and easier than the one in hand have been either incompletely solved, or not solved at all. Everything depends, then, on finding those easier problems and on solving them by means of devices as perfect as possible and of concepts capable of generalization. This rule is one of the most important levers for overcoming mathematical difficulties; and it seems to me that it is used almost always, though perhaps unconsciously.

In "Mathematical Problems", delivered by Hilbert before the Second International Congress of Mathematicians at Paris in 1900. Bulletin of the American Mathematical Society, vol. 8, 1902, pp. 437–445, 478–479.

CHAPTER 11

Stability of Cnoidal Waves

11.1. Introduction

As we saw in Chapters 3, 6, and 7, the solitary wave solutions are in general single-crested, symmetric, localized travelling waves, whose hyperbolic sech-profile is classic. We also saw that the study of the nonlinear stability properties has experienced a terrific development and refinement in recent years. Moreover, the proofs have been simplified and sufficient conditions have also been obtained to ensure the stability, up to small localized perturbations, in the wave form. Those conditions have been shown to be effective in a variety of circumstances: see for instance [**6**], [**10**], [**42**], [**58**], [**123**], [**124**], and [**276**].

Now, the situation regarding periodic travelling waves is very different. The stability and the existence of explicit formulas for those progressive wave trains have received comparatively little attention. The first study of these wave forms was carried out by Benjamin in [**45**] with regard to the periodic steady solutions called *cnoidal waves*, which were found initially by Korteweg and de Vries in [**168**] for the Korteweg and de Vries equation (KdV)

$$u_t + uu_x + u_{xxx} = 0.$$

Benjamin put forward an approach to the stability of cnoidal waves in the form

$$\varphi(\xi) = \beta_2 + (\beta_3 - \beta_2)\mathrm{cn}^2\left(\sqrt{\frac{\beta_3 - \beta_1}{12}}\xi; k\right),$$

but he did not provide a detailed justification of his assertions. Several aspects of them seem problematic. Recently Angulo, Bona, and Scialom in [**28**] gave a complete theory of the stability of cnoidal waves with mean zero for the KdV equation, which can also be extended to cnoidal wave solutions with any *a priori* mean, via a Galilean transformation associated with the KdV equation (see [**26**]). The approach for obtaining this stability result was that of the modern stability theory of Grillakis, Shatah, and Strauss [**123**] but adapted to the periodic context. In Section 11.2 we will give the basic arguments to obtain these results.

Other new explicit formulas of periodic travelling waves based on the Jacobian elliptic function of the *dnoidal* type, together with their stability, have been obtained recently by Angulo [**25**] in the context of the nonlinear Schrödinger equation (2.11) (see also Galley and Hărăguş [**113**], [**114**]), the modified Korteweg-de Vries equation (3.28) ([**25**]), and the following Hirota-Satsuma systems ([**26**] and [**27**]):

$$\begin{cases} u_t - a(u_{xxx} + 6u_x u) = 2bvv_x, & x, t \in \mathbb{R}, \quad a, b \in \mathbb{R}, \\ v_t + v_{xxx} + 3uv_x = 0. \end{cases}$$

It is remarkable to see that in all these works it was necessary to use the method of quadrature, that is, to put our differential equation in the form

$$[\psi']^2 = F(\psi)$$

to obtain the explicit profile of the solutions. Here F in general represents a polynomial function. Moreover, the necessary spectral information for studying stability or instability was obtained via an elaborate spectral theory for the periodic eigenvalue problem associated to the Jacobi form of *Lamé's equation*, namely,

$$\begin{cases} \frac{d^2}{dx^2}\Psi + [\rho - n(n+1)k^2 sn^2(x;k)]\Psi = 0, \\ \Psi(0) = \Psi(2K(k)), \quad \Psi'(0) = \Psi'(2K(k)), \end{cases}$$

with specific values of $n \in \mathbb{N}$, $sn(\cdot;k)$ representing the Jacobi elliptic function snoidal, $k \in (0,1)$, and K the complete elliptic integral of the first kind:

$$K(k) = \int_0^1 \frac{dt}{\sqrt{(1-t^2)(1-k^2t^2)}}.$$

Recently Angulo and Natali in [**37**] developed a novel method for studying the existence and stability of periodic travelling waves based on the *Poisson Summation Theorem* and the theory of totally positive operators. Their theory was based on the study of the general equation of KdV type:

$$u_t + u^p u_x - Mu_x = 0, \tag{11.1}$$

where $p \geq 1$ is an integer and M is a differential or pseudo-differential operator in the framework of periodic functions. In other words, M is defined as a Fourier multiplier operator by

$$\widehat{Mg}(k) = \alpha(k)\widehat{g}(k), \quad k \in \mathbb{Z},$$

where the symbol α of M is assumed to be a measurable, locally bounded, even function on $\mathbb{R}$, satisfying the conditions

$$A_1|k|^{m_1} \leqq \alpha(k) \leqq A_2(1+|k|)^{m_2}$$

for $m_1 \leqq m_2$, $|k| \geq k_0$, $\alpha(k) > b$ for all $k \in \mathbb{Z}$, and $A_i > 0$. The travelling wave solutions

$$u(x,t) = \varphi_c(x-ct)$$

will have a profile φ_c as a smooth periodic function with an *a priori* fundamental period $2L$, $L > 0$. Hence substituting this form of u above and integrating once (with the constant integration being considered zero), one obtains that $\varphi = \varphi_c$ is a solution of the equation

$$(M+c)\varphi - \frac{1}{p+1}\varphi^{p+1} = 0. \tag{11.2}$$

So we find the associated linear, closed, unbounded, self-adjoint operator $\mathcal{L}$: $D(\mathcal{L}) \to L^2_{\text{per}}([-L,L])$ defined on a dense subspace of $L^2_{\text{per}}([-L,L])$ by

$$\mathcal{L}u = (M+c)u - \varphi^p u.$$

From the theory of compact symmetric operators applied to the periodic eigenvalue problem

$$\mathcal{L}\psi = \lambda\psi,$$

it is possible to see that the spectrum of $\mathcal{L}$ is a countably infinite set of eigenvalues with

$$\lambda_0 \leq \lambda_1 \leq \lambda_2 \leq \lambda_3 \leq \cdots,$$

where $\lambda_n \to \infty$ as $n \to \infty$. In particular we obtain that $\mathcal{L}(d\varphi/dx) = 0$. As is well known, this property of $\mathcal{L}$ is deduced from the invariance by translations of the solutions of the evolution equation.

Now, according to Benjamin, Weinstein, Grillakis, Shatah, and Strauss the conditions that imply stability are

(P_0) there is a nontrivial smooth curve of periodic solutions for (11.2) of the form
$$c \in I \subseteq \mathbb{R} \to \varphi_c \in H^{m_2}_{\text{per}}([-L, L]);$$

(P_1) $\mathcal{L}$ has a unique negative eigenvalue λ, and it is simple;

(P_2) the eigenvalue 0 is simple;

(P_3) $\displaystyle \frac{d}{dc}\int_{-L}^{L} \varphi_c^2(x)dx > 0.$

The problem regarding the existence of a nontrivial smooth curve of periodic solutions in the form required by (P_0) above presents in the periodic setting new and delicate aspects that need to be handled. The possibility of finding explicit solutions will depend naturally on the form of M: if it is a differential operator of the form $M = -\partial_x^2$, the use of the quadrature method (that means writes (11.2) in the form $[\varphi_c']^2 = F(\varphi_c)$) and the theory of elliptic functions have been shown to be main tools. So the solutions will depend of the Jacobian elliptic functions of *snoidal, cnoidal*, and *dnoidal* type. Now, since the period of these functions depends on the complete elliptic integral $K(k)$, we have that the elliptic modulus k will depend on the velocity c and therefore we have that *a priori* the period of φ_c will depend on c. Hence, by using the Implicit Function Theorem, in many cases the desired smooth branch of periodic solutions with a fixed minimal period has been obtained.

We note that the procedure of the quadrature method in general does not work if M is a pseudo-differential operator. But there are cases, such as in the Benjamin-Ono and ILW equations (see (3.7) and (3.9)), where this procedure works. So Angulo and Natali in [**37**], working on this obstacle, have developed a different approach to obtain explicit solutions for specific forms of M and values of p. This approach is based on the classical *Poisson Summation Theorem.* There are at least two important advantages to this approach. The first one is that it can be used for obtaining solutions when M is a pseudo-differential operator, for example in the case of $\mathcal{H}\partial_x$; the other one is that related to computing the integral in (P_3) above. In general, obtaining property (P_3) is very difficult in the periodic case. As we will see, the verification of (P_3) can be very easy to obtain through a combination of the Poisson Summation Theorem and the Parseval Theorem.

With regard to conditions (P_1) and (P_2), the problem is very delicate. The work of Angulo and Natali established a new set of conditions that gives us (P_1) and (P_2). The analysis relies upon the theory of totally positive operators. As we will see below, this new theory leads to a significant simplification of some recent proofs of stability of periodic travelling wave solutions of KdV-type equations.

In this chapter we show many applications of the theory established in Angulo and Natali. We present a new theory for the existence and stability of positive periodic travelling waves for the following nonlinear dispersive equations: the KdV equation, the Benjamin-Ono equation, the critical KdV, and the critical nonlinear Schrödinger equation:

$$iu_t + u_{xx} + |u|^4 u = 0.$$

In Section 11.2 below (Theorem 11.8), the stability results of the orbit generated by the cnoidal waves ϕ_c in (11.14), namely,

$$\Omega_{\phi_c} \equiv \{\phi_c(\cdot + s)\}_{s\in\mathbb{R}},$$

in $H^1_{\rm per}([0, L])$ under the flow of the periodic KdV equation, will be obtained for initial disturbances of ϕ_c having the same period L. It is a conjecture on the part of Benjamin [**45**] that cnoidal waves of minimal period L are unstable under perturbations, for example, of period $2L$ (see Section 11.6).

We start with the theory of existence and stability of cnoidal wave solutions for the KdV equation with mean zero condition. Moreover, we present an extension of Grillakis *et al.*'s theory for the case when the travelling wave solutions *do not have the property of being a critical point.* We note that our stability theory established here for the cnoidal waves ϕ_c is different and shorter than that presented in [**28**].

11.2. Stability of Cnoidal Waves with Mean Zero for KdV Equation

In this section we establish a novel application of the Grillakis *et al.* theory, in [**123**], to the stability of cnoidal wave solutions of mean zero associated with the KdV equation. Initially, we will show that given a fixed period L, a smooth curve of *cnoidal wave* soutions, $c \to \phi_c$, will exist for

$$\phi_c'' + \frac{1}{2}\phi_c - c\phi_c = A_{\phi_c}, \tag{11.3}$$

where c belongs to a specific interval and every solution ϕ_c has the fundamental period L with mean zero on $[0, L]$, namely,

$$\int_0^L \phi_c(x)\, dx = 0.$$

Therefore our integration constant A_{ϕ_c} must be nonzero. Since we need some tools from the theory of elliptic integrals, we refer the reader to Appendix A, where a basic review of this theory can be seen. We also refer the reader to the books of Byrd and Friedman [**75**] and Armitage and Eberlein [**40**]. So, via a quadrature analysis we obtain that ϕ_c has the standard form (see (3.25))

$$\phi_c(\xi) \equiv \phi_c(\xi; \beta_1, \beta_2, \beta_3) = \beta_2 + (\beta_3 - \beta_2)\, cn^2\Big[\sqrt{\frac{\beta_3 - \beta_1}{12}}\,\xi; k\Big] \tag{11.4}$$

where $\beta_1, \beta_2, \beta_3$ are real constants satisfying $\beta_1 < \beta_2 < \beta_3$,

$$k^2 = \frac{\beta_3 - \beta_2}{\beta_3 - \beta_1}, \quad \beta_1 + \beta_2 + \beta_3 = 3c, \quad \text{and} \quad A_{\phi_c} = -\frac{1}{6}\sum_{i<j} \beta_i\beta_j.$$

Moreover, we know that the fundamental period T_{ϕ_c} of ϕ_c is given by

$$T_{\phi_c} = \frac{4\sqrt{3}}{\sqrt{\beta_3 - \beta_1}} K(k).$$

So, it follows that *a priori* the fundamental period depends on the wave speed c. Therefore, this dependence will be a big barrier when we try to find a curve of cnoidal wave solutions with a fixed period L. Our first step in finding a particular such curve will be to show that it is possible to choose constants β_i' such that the cnoidal wave $\phi_1 = \phi_1(\cdot;\beta_i)$ ($c = 1$ in (11.3)) has mean zero on $[0, T_{\phi_1}]$. So we should see that

$$\beta_2 + \frac{\beta_3 - \beta_2}{2K} \int_0^{2K} \mathrm{cn}^2(\xi; k)\, d\xi = 0. \tag{11.5}$$

Now, if we put $E(u) = E(\theta, k)$ with $\sin\theta = \mathrm{sn} u$, then from [**75**] and from the definition of $E(k)$ we get

$$\int_0^K \mathrm{cn}^2(u;k)\,du = \frac{1}{k^2}[E(u) - k'^2 u]\Big|_0^K = \frac{1}{k^2}[E(k) - k'^2 K(k)].$$

Therefore, from the evenness of cn we have that (11.5) gives us the main condition

$$\beta_2 + (\beta_3 - \beta_2)\frac{E(k) - k'^2 K(k)}{k^2 K(k)} = 0. \tag{11.6}$$

Now, by using the fact that

$$(\beta_3 - \beta_2)k'^2 = (\beta_2 - \beta_1)k^2$$

and by using (11.6), we obtain the following equivalent relations:

$$\beta_1 K(k) + (\beta_3 - \beta_1)E(k) = 0 \tag{11.7}$$

and

$$\frac{dK}{dk} = -\frac{\beta_2}{\beta_3 - \beta_2}\frac{k}{K}. \tag{11.8}$$

Next we have our first basic result, which assures the existence of cnoidal wave solutions with mean zero on its fundamental period.

THEOREM 11.1. *For every $\beta_3 > 3$ there are unique β_1, β_2 satisfying $\beta_1 < \beta_2 < 0 < \beta_3$ and $\beta_1 + \beta_2 + \beta_3 = 3$, such that the cnoidal wave $\phi_1(\cdot;\beta_i)$ defined in (11.4) has mean zero over $[0, T_{\phi_1}]$, namely,*

$$\beta_1 K(k) + (\beta_3 - \beta_1)E(k) = 0,$$

where $k^2 = (\beta_3 - \beta_2)/(\beta_3 - \beta_1)$. Moreover, the map $\beta_3 \in (3, \infty) \to \beta_2 \equiv \beta_2(\beta_3)$ is a continuous map.

SKETCH OF THE PROOF. Let $\beta_3 > 3$. Let us denote by $\alpha := \frac{\beta_3 - 3}{2}$, $\beta := \frac{1}{\alpha}$, and I the open interval $(-\alpha, 0)$. Define for $x \in I$ the continuous function

$$J(x) = \frac{K(k(x))}{E(k(x))} + \frac{\beta_3}{3 - \beta_3 - x} - 1,$$

where $k(x) = \sqrt{\frac{\beta_3 - x}{2\beta_3 + x - 3}}$. The idea of the proof is to see that for every $\beta_3 > 3$, there is a unique $\beta_2 := \beta_2(\beta_3) \in I$ such that $J(\beta_2) = 0$. To accomplish this, we define for $x \in I$

$$A(x) \equiv \int_0^1 \frac{1}{\sqrt{1 - t^2}} \frac{-\beta_3 + (\beta_3 - x)t^2}{\sqrt{2\beta_3 - 3 + x - (\beta_3 - x)t^2}}\, dt. \tag{11.9}$$

It is easy to see that $J(x) = 0$ if, and only if, $A(x) = 0$. So, to prove that J has a unique zero in the interval I, it is enough to prove that the function A is strictly decreasing on that interval, because $A(x) \to +\infty$ as $x \to -\alpha$ and $A(0) < 0$. We refer the reader to [**28**] for a complete proof of this fact.

The continuity of the map $\beta_3 \in (3,\infty) \to \beta_2 = \beta_2(\beta_3)$ is a consequence of the Implicit Function Theorem applied to the function $A(x,\beta_3)$ given in the right-hand side of (11.9), and it is a consequence of considering that $\partial_x A(\beta_2(\beta_3),\beta_3) < 0$. $\square$

REMARK 11.2. It follows from Theorem 11.1 that for every $\beta_3 > 3$ there will exist a unique choice $\beta_2 \in (\frac{3-\beta_3}{2}, 0)$ such that $J(\beta_2) = 0$ and the map $\beta_3 \in (3,\infty) \to \beta_2 = \beta_2(\beta_3)$ is continuous. So, if we see the fundamental period T_{ϕ_1} for $\phi_1 = \phi_1(\cdot;\beta_i)$ as a function of β_3, namely,

$$T_{\phi_1}(\beta_3) = \frac{4\sqrt{3}}{\sqrt{2\beta_3 + \beta_2(\beta_3) - 3}}\, K\Big(\sqrt{\frac{\beta_3 - \beta_2(\beta_3)}{2\beta_3 + \beta_2(\beta_3) - 3}}\Big),$$

we infer immediately that $T_{\phi_1}(\beta_3) \to \infty$ as $\beta_3 \to 3$, since $K(k) \to +\infty$ as $k \to 1$. On the other hand, since $J(\beta_2) = 0$ and $\frac{3-\beta_3}{2} < \beta_2 < 0$, we see that

$$\begin{aligned} 0 \leqq T_{\phi_1}(\beta_3) &= 4\sqrt{3}\, \frac{\sqrt{2\beta_3 + \beta_2 - 3}}{\beta_3 + \beta_2 - 3}\, E(k(\beta_2)) \leqq 2\pi\sqrt{3}\, \frac{\sqrt{2\beta_3 + \beta_2 - 3}}{\beta_3 + \beta_2 - 3} \\ &\leqq 4\pi\sqrt{3}\, \frac{\sqrt{2\beta_3 + \beta_2 - 3}}{\beta_3 - 3} < 4\pi\sqrt{3}\, \frac{\sqrt{2\beta_3 - 3}}{\beta_3 - 3}, \end{aligned}$$

and so $T_{\phi_1}(\beta_3) \to 0$ as $\beta_3 \to +\infty$.

Now, we will construct a smooth curve of cnoidal wave solutions with mean zero and a given arbitrary period. We begin by establishing, via the Implicit Function Theorem, the existence of an interval of speed waves for cnoidal waves which will determine the fundamental period of these waves of a continuous form.

LEMMA 11.3. *Let $L > 0$ be a fixed number and consider β_i, $i = 1,2,3$, an arbitrary but fixed trio defined in Theorem 11.1. Define $c_0 = T^2_{\phi_1}/L^2$ where $\phi_1 = \phi_1(\cdot;\beta_i)$. Then we have the following.*

(1) *There exist an interval $I(c_0)$ around c_0, a ball $B(\overrightarrow{\beta})$ around the vector $\overrightarrow{\beta} = (\beta_1,\beta_2,\beta_3)$, and a unique smooth function*

$$\Pi : I(c_0) \mapsto B(\overrightarrow{\beta}) : \ c \mapsto \ (\alpha_1(c), \alpha_2(c), \alpha_3(c))$$

such that $\Pi(c_0) = (\beta_1,\beta_2,\beta_3)$ and $\alpha_i \equiv \alpha_i(c)$ satisfy the relations

$$\frac{4\sqrt{3}}{\sqrt{\alpha_3 - \alpha_1}} K(k) = L\sqrt{c}, \quad \alpha_1 + \alpha_2 + \alpha_3 = 3,$$

$$\alpha_2 + (\alpha_3 - \alpha_2)\frac{E(k) - k'^2 K(k)}{k^2 K(k)} = 0,$$

where

$$k^2 = \frac{\alpha_3 - \alpha_2}{\alpha_3 - \alpha_1}.$$

(2) *The cnoidal wave (11.4), determined by the values α_i, $\phi_1(\cdot;\alpha_i)$, has the fundamental period $L\sqrt{c}$ and mean zero over $[0, L\sqrt{c}]$. Moreover, for all $c \in I(c_0)$*

$$A_{\phi_1(\cdot;\alpha_i)}(c) = \frac{1}{2L\sqrt{c}} \int_0^{L\sqrt{c}} \phi_1^2(x;\alpha_i)dx = -\frac{1}{6}\sum_{i<j} \alpha_i(c)\alpha_j(c).$$

(3) *For all* $c \in I(c_0)$, *the modulus function*

$$k(c) \equiv \frac{\sqrt{\alpha_3(c)-\alpha_2(c)}}{\sqrt{\alpha_3(c)-\alpha_1(c)}}$$

satisfies $\frac{d}{dc}k(c) > 0$.

PROOF. We consider $\Phi : \Omega \to \mathbb{R}^3$ given by

(11.10)

$$\Phi(\alpha_1,\alpha_2,\alpha_3,c) = \Big(\frac{4\sqrt{3}\,K(k)}{\sqrt{\alpha_3-\alpha_1}} - L\sqrt{c}, \sum_{i=1}^{3}\alpha_i - 3, \alpha_2 + (\alpha_3-\alpha_2)\frac{E(k)-k'^2K(k)}{k^2K(k)}\Big)$$
$$= (\Phi_1,\Phi_2,\Phi_3)$$

where $k^2 \equiv (\alpha_3-\alpha_2)/(\alpha_3-\alpha_1)$ and $\Omega = \{(\alpha_1,\alpha_2,\alpha_3,c) \mid \alpha_1 < \alpha_2 < 0 < \alpha_3,\ \alpha_3 > 3 \text{ and } c > 0\ \} \subset \mathbb{R}^4$. So, from Theorem 11.1, $\Phi(\beta_1,\beta_2,\beta_3,c_0) = (0,0,0)$. Next, from the definition of Φ_2 we immediately see that

$$\nabla_{(\alpha_1,\alpha_2,\alpha_3)}\Phi_2(\beta_1,\beta_2,\beta_3,c_0) = (1,1,1).$$

Moreover, it follows that $\frac{\partial}{\partial\alpha_j}\Phi_i(\beta_1,\beta_2,\beta_3,c_0)$ is given by

$$\begin{cases} \dfrac{\partial}{\partial\alpha_1}\Phi_1(\beta_1,\beta_2,\beta_3,c_0) = \dfrac{-\beta_1L\sqrt{c_0}}{2(\beta_3-\beta_1)(\beta_2-\beta_1)}, \\ \dfrac{\partial}{\partial\alpha_2}\Phi_1(\beta_1,\beta_2,\beta_3,c_0) = \dfrac{\beta_2L\sqrt{c_0}}{2(\beta_2-\beta_1)(\beta_3-\beta_2)}, \\ \dfrac{\partial}{\partial\alpha_3}\Phi_1(\beta_1,\beta_2,\beta_3,c_0) = \dfrac{-\beta_3L\sqrt{c_0}}{2(\beta_3-\beta_1)(\beta_3-\beta_2)}. \end{cases}$$

With regard to Φ_3, we see that

$$\Phi_3(\alpha_1,\alpha_2,\alpha_3,c_0) = \alpha_2 + (\alpha_3-\alpha_2)\frac{k'^2}{kK}\frac{dK}{dk}.$$

So, since K satisfies the *hypergeometric differential equation* (A.6) in Appendix A, we obtain from Theorem 11.1 that

$$\nabla_{(\alpha_1,\alpha_3,\alpha_3)}\Phi_3(\beta_1,\beta_2,\beta_3,c_0)$$

$$= \Big(\frac{\beta_2\beta_3-\beta_1\beta_2-\beta_1\beta_3}{2(\beta_2-\beta_1)(\beta_3-\beta_1)}, \frac{\beta_2\beta_3-\beta_1\beta_3+\beta_1\beta_2}{2(\beta_3-\beta_2)(\beta_2-\beta_1)}, \frac{\beta_1\beta_2-\beta_1\beta_3-\beta_2\beta_3}{2(\beta_3-\beta_2)(\beta_3-\beta_1)}\Big).$$

Therefore we obtain that the Jacobian determinant of $\Phi(\cdot,\cdot,\cdot,c)$ satisfies

$$\frac{\partial(\Phi_1,\Phi_2,\Phi_3)}{\partial(\alpha_1,\alpha_3,\alpha_3)}\Big|_{(\beta_1,\beta_2,\beta_3,c_0)} = L\sqrt{c_0}\,\frac{\beta_1\beta_2+\beta_1\beta_3+\beta_2\beta_3}{4(\beta_2-\beta_1)(\beta_3-\beta_2)(\beta_3-\beta_1)}.$$

Now, since the cnoidal wave $\phi_1 \equiv \phi_1(\cdot;\beta_i)$ satisfies (11.3) with $c = 1$ and we have that $-6A_{\phi_1} = \beta_1\beta_2+\beta_2\beta_3+\beta_3\beta_1$, ϕ_1 has mean zero on $[0,T_{\phi_1}]$, and $\phi_1'(0) = \phi_1'(T_{\phi_1})$, we obtain

$$\frac{1}{2}\int_0^{T_{\phi_1}}\phi_1^2(\xi)\,d\xi = A_{\phi_1}T_{\phi_1} = A_{\phi_1}L\sqrt{c_0}.$$

So we get

$$\frac{\partial(\Phi_1,\Phi_2,\Phi_3)}{\partial(\alpha_1,\alpha_3,\alpha_3)}\Big|_{(\beta_1,\beta_2,\beta_3,c_0)} = -\frac{3}{4(\beta_2-\beta_1)(\beta_3-\beta_2)(\beta_3-\beta_1)}\int_0^{T_{\phi_1}}\phi_1^2(\xi)\,d\xi \neq 0.$$

Therefore from the Implicit Function Theorem, there exist $I(c_0)$ and $B(\beta_1, \beta_2, \beta_3)$, neighborhoods of c_0 and $(\beta_1, \beta_2, \beta_3)$, respectively, and a unique smooth function $\Pi : I(c_0) \mapsto B(\beta_1, \beta_2, \beta_3)$ satisfying item (1) in the lemma. Item (2) follows immediately. Item (3) is a consequence of the relations $\Phi(\Pi(c), c) = (0, 0, 0)$,

$$
\begin{aligned}
\frac{d}{dc}\alpha_1(c) &= \frac{(\sum_{i<j}\alpha_i\alpha_j)^{-1}}{c}\Big(\alpha_3^2(\alpha_1-\alpha_2)+\alpha_2^2(\alpha_1-\alpha_3)\Big),\\
\frac{d}{dc}\alpha_2(c) &= \frac{(\sum_{i<j}\alpha_i\alpha_j)^{-1}}{c}\Big(\alpha_1^2(\alpha_2-\alpha_3)+\alpha_3^2(\alpha_2-\alpha_1)\Big),\\
\frac{d}{dc}\alpha_3(c) &= \frac{(\sum_{i<j}\alpha_i\alpha_j)^{-1}}{c}\Big(\alpha_1^2(\alpha_3-\alpha_2)+\alpha_2^2(\alpha_3-\alpha_1)\Big),
\end{aligned}
\tag{11.11}
$$

and the fact that for $c \in I(c)$

$$
\begin{aligned}
\frac{d}{dc}k(c) &= -\frac{2}{\rho(c)\varrho(c)}(\alpha_3-\alpha_2)(\alpha_2-\alpha_1)(\alpha_3-\alpha_1)(\alpha_1+\alpha_2+\alpha_3)\\
&= -\frac{3(\alpha_2-\alpha_1)}{\rho(c)}k(c),
\end{aligned}
$$

where $\rho(c) = c(\alpha_1\alpha_2 + \alpha_1\alpha_3 + \alpha_2\alpha_3)$ and $\varrho(c) = 2(\alpha_3 - \alpha_1)^{3/2}\sqrt{\alpha_3 - \alpha_2}$. This completes the proof of the lemma. □

Now, we establish the existence of a smooth curve of cnoidal wave solutions to the KdV equation with a fixed period L and with mean zero on $[0, L]$.

THEOREM 11.4 (Smooth Curve of Cnoidal Waves of Mean Zero). *Let $L > 0$ be a fixed period and consider $c \in I(c_0)$. Then there exists $\phi_c \in H^n_{\text{per}}([0, L])$ with the fundamental period L, such that $\int_0^L \phi_c(\xi)\, d\xi = 0$ and ϕ_c satisfies the differential equation*

$$
\phi_c''(\xi) + \frac{1}{2}\phi_c^2(\xi) - c\,\phi_c(\xi) = A_{\phi_c} \qquad \textit{for all } \xi \in \mathbb{R}, \tag{11.12}
$$

where

$$
A_{\phi_c} = \frac{1}{2L}\int_0^L \phi_c^2(\xi)\, d\xi = -\frac{c^2}{6}\sum_{i<j}\alpha_i(c)\alpha_j(c). \tag{11.13}
$$

Moreover, ϕ_c has the specific cnoidal profile

$$
\phi_c(\xi) = \beta_2 + (\beta_3 - \beta_2)\,\text{cn}^2\Big[\sqrt{\frac{\beta_3-\beta_1}{12}}\;\xi; k\Big],
$$

and the map $c \in I(c_0) \mapsto \phi_c \in H^n_{\text{per}}([0, L])$ is a C^1 map.

PROOF. Let $\phi_1 \equiv \phi_1(\cdot; \alpha_1, \alpha_2, \alpha_3)$ be the cnoidal wave determined in Lemma 11.3(2) with $\alpha_i \equiv \alpha_i(c)$ for $c \in I(c_0)$, fundamental period $L\sqrt{c}$, and mean zero on $[0, L\sqrt{c}]$, and let it satisfy the differential equation (11.12) when $c = 1$. So, we define

$$
\phi_c(\xi) \equiv c\,\phi_1(\sqrt{c}\,\xi; \alpha_1, \alpha_2, \alpha_3) \qquad \text{for all } \xi \in \mathbb{R}. \tag{11.14}
$$

Then we see that ϕ_c has the fundamental period L and mean zero over $[0, L]$. Moreover, ϕ_c satisfies

$$
\phi_c''(\xi) + \frac{1}{2}\phi_c^2(\xi) - c\phi_c(\xi) = c^2 A_{\phi_1}(c).
$$

We note that ϕ_c has the form (11.4) with $\beta_i = c\alpha_i$ and $\sum_i \beta_i = 3c$. Next, from the properties of ϕ_1 we get the relation

$$A_{\phi_1}(c) = \frac{1}{2L\sqrt{c}} \int_0^{L\sqrt{c}} \phi_1^2(\xi)d\xi = \frac{1}{2Lc^2} \int_0^L \phi_c^2(\xi)d\xi \equiv \frac{1}{c^2} A_{\phi_c}(c),$$

and so ϕ_c satisfies (11.12). Moreover, from Lemma 11.3 we get (11.13). Finally, from the regularity properties of ϕ_1 and α_i' we get the required regularity on the curve $c \mapsto \phi_c$. This completes the proof of the theorem. $\square$

COROLLARY 11.5. Let $L > 0$ be a fixed number. Then there exists a smooth curve $c \in (0, \infty) \to \phi_c \in H^n_{\mathrm{per}}([0, L])$ of solutions for (11.12) of cnoidal wave type, with fundamental period L and satisfying $\int_0^L \phi_c(\xi)\, d\xi = 0$.

PROOF. It follows from Remark 11.2 that $T_{\phi_1}(\beta_3)$ can take any value in the interval $(0, \infty)$ for $\beta_3 > 3$. So, we can choose $c_0 = T^2_{\psi_1}/L^2$ arbitrarily in the interval $(0, \infty)$. So, by Lemma 11.3 and Theorem 11.4, there is an interval $I(c_0)$ around c_0 and a well-defined smooth curve $c \in I(c_0) \mapsto \phi_c \in H^n_{\mathrm{per}}([0, L])$ of cnoidal waves satisfying equation (11.12). By the uniqueness of map Π in Lemma 11.3, we obtain that the map

$$c \in (0, \infty) \mapsto \phi_c \in H^n_{\mathrm{per}}([0, L])$$

is a smooth curve of cnoidal wave solutions to the KdV equation with a fixed period L and mean zero on $[0, L]$. $\square$

Our attention is now turned to establishing the spectral properties required by the Grillakis *et al.* theory (Assumption 3, in Section 7.3) for the linear operator

$$\mathcal{L}_{\mathrm{cn}} = -\frac{d^2}{dx^2} + c - \phi_c,$$

which is associated with the cnoidal wave solution ϕ_c given by Theorem 11.4 and has fundamental period L and wave speed $c > 0$. This study will be based on the periodic eigenvalue problem considered on $[0, L]$, namely,

$$\begin{cases} \mathcal{L}_{\mathrm{cn}}\chi = \lambda\chi, \\ \chi(0) = \chi(L), \quad \chi'(0) = \chi'(L). \end{cases}$$

So, we have the following theorem.

THEOREM 11.6. *Let ϕ_c be the cnoidal wave solution given by Theorem 11.4, and let $c \in (0, \infty)$. Then, the linear operator*

$$\mathcal{L}_{cn} = -\frac{d^2}{dx^2} + c - \phi_c$$

defined on $H^2_{\mathrm{per}}([0, L])$ has exactly its first three eigenvalues simple. The eigenvalue zero is the second one with eigenfunction ϕ_c'. Moreover, the rest of the spectrum is constituted by a discrete set of eigenvalues which are double.

PROOF. See Theorem B.62 in Appendix B. $\square$

We remember that the KdV equation has the following conservation laws in the periodic case:

$$E_{\mathrm{KdV}}(u) = \frac{1}{2} \int_0^L \left[(u')^2 - \frac{1}{3}u^3\right] dx, \qquad F(u) = \frac{1}{2} \int_0^L u^2 dx. \tag{11.15}$$

Our attention is now turned to studying the standard function d defined by

$$d(c) = E_{\mathrm{KdV}}(\phi_c) + cF(\phi_c), \tag{11.16}$$

where the cnoidal wave solution ϕ_c is given by Theorem 11.4 and $c \in (0, \infty)$. So, by differentiating equation (11.3) with regard to c, we have the main relation

$$\mathcal{L}_{\mathrm{cn}}\Big(-\frac{d}{dc}\phi_c\Big) = \phi_c + \frac{d}{dc}A_{\phi_c}.$$

Since ϕ_c satisfies $(E'_{\mathrm{KdV}} + cF')\phi_c = -A_{\phi_c}$, we immediately obtain that

$$d'(c) = (-A_{\phi_c}, \frac{d}{dc}\phi_c) + F(\phi_c) = \frac{1}{2}\int_0^L \phi_c^2(x)\,dx = -\frac{c^2 L}{6}\sum_{i<j}\alpha_i(c)\alpha_j(c),$$

where in the last equality we have used the fact that $\int_0^L \frac{d}{dc}\phi_c\,dx = 0$. So we have the following theorem:

THEOREM 11.7. *For every $c \in (0, \infty)$ we have that $d''(c) > 0$. More precisely, we can obtain the following formulas:*

$$d''(c) = \frac{2304}{L^3}\Big[K(k) - E(k)\Big]E(k)K(k)\frac{dK}{dk}\frac{dk}{dc} > 0$$

(11.17) *or*

$$d''(c) = 3L\frac{B_{\phi_c}}{A_{\phi_c}} > 0,$$

where $6A_{\phi_c} = -\sum_{i<j}\beta_i\beta_j$ and $6B_{\phi_c} = \prod_{i=1}^3 \beta_i$.

PROOF. The first equality in (11.17) is obtained by using nontrivial relations between the complete elliptic integrals K and E. By using Lemma 11.3, we can express the α_i's as a function of the complete elliptic integrals K and E. In fact,

$$\alpha_1 = -\frac{48K}{cL^2}E, \quad \alpha_2 = -\frac{48K}{cL^2}[E - k'^2 K], \quad \text{and} \quad \alpha_3 = \frac{48K}{cL^2}[K - E].$$

Therefore, by using the fact that K and E satisfy specific *hypergeometric differential equations* (see Appendix A), we can see after many calculations (see Theorem 2.6 in [**28**]) that

$$d''(c) = \frac{2304}{L^3}(K - E)EK\frac{dK}{dk}\frac{dk}{dc}.$$

Now, since for $k \in (0, 1)$ we have the properties that $K(k) > E(k)$ and $K(k)$ is strictly increasing and since from Lemma 11.3 we have that $\frac{dk}{dc} > 0$, we immediately obtain that d is a strictly convex function.

Here, for the reader's convenience, we establish a different proof of the convexity property of d (see [**39**]). First we note that for $\frac{d}{dc} =$ " $'$ " we have

$$\begin{aligned}\Big(\sum_{i<j}\alpha_i\alpha_j\Big)' &= \alpha_1'(\alpha_2 + \alpha_3) + \alpha_2'(\alpha_1 + \alpha_3) + \alpha_3'(\alpha_2 + \alpha_1)\\ &= -(\alpha_1\alpha_1' + \alpha_2\alpha_2' + \alpha_3\alpha_3').\end{aligned}$$

Using expressions for α_i' obtained in (11.11), we see that for $i, j, k \in \{1, 2, 3\}$ with $j \neq i$, $j \neq k$, and $k \neq i$,

$$\alpha_i\alpha_i' = \frac{1}{c\sum_{i<j}\alpha_i\alpha_j}\left[\alpha_i^2(\alpha_j^2 + \alpha_k^2) - \alpha_i\alpha_j\alpha_k(\alpha_j + \alpha_k)\right].$$

Then summation over i shows us that

$$\left(\sum_{i<j}\alpha_i\alpha_j\right)' = -\frac{1}{c\sum_{i<j}\alpha_i\alpha_j}\left[-6\alpha_1\alpha_2\alpha_3 + 2\sum_{i<j}\alpha_i^2\alpha_j^2\right].$$

But a direct computation shows that

$$\left(\sum_{i<j}\alpha_i\alpha_j\right)^2 = \left(\sum_{i<j}\alpha_i^2\alpha_j^2\right) + 6\alpha_1\alpha_2\alpha_3.$$

Using this formula in the previous equation, let us conclude that

$$\begin{aligned}\left(\sum_{i<j}\alpha_i\alpha_j\right)' &= -\frac{1}{c\sum_{i<j}\alpha_i\alpha_j}\left[-18\alpha_1\alpha_2\alpha_3 + 2\left(\sum_{i<j}\alpha_i\alpha_j\right)^2\right]\\ &= \frac{18}{c\sum_{i<j}\alpha_i\alpha_j}\prod_{i=1}^{3}\alpha_i - \frac{2}{c}\sum_{i<j}\alpha_i\alpha_j.\end{aligned}$$

Therefore we obtain that

$$d''(c) = -\frac{3cL}{\sum_{i<j}\alpha_i\alpha_j}\prod_{i=1}^{3}\alpha_i,$$

which is strictly positive. □

The theory of nonlinear stability for the cnoidal wave solution obtained in Theorem 11.4 for the KdV equation will be developed along the following lines. The approach will be based on the theory developed in [**123**]. Since some points of this method need to be carefully specified in the periodic case, we will establish the main changes in the theory established initially in Section 7.2

THEOREM 11.8 (Nonlinear Stability of Cnoidal Wave Solutions). *Let $L > 0$ and $c \in (0,\infty)$. Let us consider the cnoidal wave ϕ_c given by Theorem 11.4. Then, the orbit*

$$\Omega_{\phi_c} = \{\phi_c(\cdot + s) \mid s \in \mathbb{R}\}$$

is $H^1_{\mathrm{per}}([0,L])$-stable with regard to the flow of the KdV equation.

For the proof of Theorem 11.8, we first need to make some changes in Lemmas 7.7, 7.8, and 7.9 in Chapter 7. The first one is an extension of Lemma 7.7 to the periodic case and its proof is similar to that given above (see [**123**]). Denoting U_ϵ as

$$U_\epsilon = \{u \in H^1_{\mathrm{per}}([0,L]) \mid \inf_{s\in\mathbb{R}} \|u - \tau_s\phi_c\|_1 < \epsilon\},$$

we have the following lemma.

LEMMA 11.9. *There exist $\epsilon > 0$ and a C^1 map $\alpha : U_\epsilon \to \mathbb{R}/L$ such that*

(1) $\langle u(\cdot + \alpha(u)), \phi_c'\rangle = 0$,
(2) $\alpha(u(\cdot + r)) = \alpha(u) - r$ *modulo the period,*
(3) $\alpha(\phi_c) = 0$,

for all $u \in U_\epsilon$ and $r \in \mathbb{R}$.

Now, using the basic relation

$$d''(c) = \langle \phi_c, \frac{d}{dc}\phi_c\rangle = \langle \phi_c + \frac{d}{dc}A_{\phi_c}, \frac{d}{dc}\phi_c\rangle = -\langle \mathcal{L}_{\rm cn}\frac{d}{dc}\phi_c, \frac{d}{dc}\phi_c\rangle,$$

we obtain the following result, which is similar to Lemma 7.8.

LEMMA 11.10. *Let $c \in (0,\infty)$ and $d''(c) > 0$. Define*

$$\mathcal{A} = \{\psi \in H^1_{\rm per}([0,L])| \int_0^L \psi(x)dx = 0,\ \langle \psi, \phi_c\rangle = \langle \psi, \phi_c'\rangle = 0,\ \textit{and } \|\psi\| = 1\}.$$

Then

$$\zeta = \inf\{\langle \mathcal{L}_{\rm cn}\psi, \psi\rangle \mid \psi \in \mathcal{A}\} > 0.$$

Therefore, there is a positive constant D such that

$$\langle \mathcal{L}_{\rm cn}\psi, \psi\rangle \geqq D\|\psi\|_1^2$$

for all $\psi \in \mathcal{A}$.

PROOF. The proof follows along the same lines as those of the proof of Lemma 7.8 and uses Theorem 11.6. □

Next, a result similar to Lemma 7.9 is established. This result contains the crucial inequality in our stability theory. Here it is necessary to work in the subspace of functions of mean zero, to obtain this basic inequality.

LEMMA 11.11. *Let $c \in (0,\infty)$ and $d''(c) > 0$. There exist constants $D_1 > 0$ and $\epsilon > 0$ such that*

$$E(u) - E(\phi_c) \geqq D_1\|u(\cdot + \alpha(u)) - \phi_c\|_1^2 \tag{11.18}$$

for all $u \in U_\epsilon$ satisfying $F(u) = F(\phi_c)$ and $\int_0^L u\,dx = 0$.

PROOF. We consider $\epsilon > 0$ given by Lemma 11.10. So for $u \in U_\epsilon$ we write

$$u(\cdot + \alpha(u)) - \phi_c = a\phi_c + \psi$$

where $\langle \psi, \phi_c\rangle = 0$ and $a \in \mathbb{R}$. Note that if u has mean zero, then ψ also has mean zero. Now, according to Taylor's Theorem

$$F(\phi_c) = F(u(\cdot+\alpha(u))) = F(\phi_c) + \langle F'(\phi_c), u(\cdot+\alpha(u)) - \phi_c\rangle + O(\|u(\cdot+\alpha(u)) - \phi_c\|_1^2).$$

Then we have $a = O(\|u(\cdot + \alpha(u)) - \phi_c\|_1^2)$. Now, consider $L(u) = E(u) + cF(u)$. Another of Taylor's expansions gives us

$$L(u) = L(u(\cdot + \alpha(u))) = L(\phi_c) + (L'(\phi_c), v) + \frac{1}{2}(L''(\phi_c)v, v) + o(\|v\|_1^2),$$

where $v \equiv u(\cdot + \alpha(u)) - \phi_c$. Note that v has mean zero, so it follows from the relations $F(u) = F(\phi_c)$, $\langle L'(\phi_c), v\rangle = \langle -A_{\phi_c}, v\rangle = 0$, and $L''(\phi_c) = \mathcal{L}_{\rm cn}$ that

$$\begin{aligned} E(u) - E(\phi_c) &= \frac{1}{2}(\mathcal{L}_{\rm cn}\psi, \psi) + O(a^2) + O(a\|v\|_1) + o(\|v\|_1^2) \\ &= \frac{1}{2}(\mathcal{L}_{\rm cn}\psi, \psi) + o(\|v\|_1^2). \end{aligned}$$

From Lemma 11.9 we have $\langle \psi, \phi_c'\rangle = 0$, and so by Lemma 11.10

$$E(u) - E(\phi_c) \geqq \frac{1}{2}D\|\psi\|_1^2 + o(\|v\|_1^2),$$

for a constant $D > 0$. Therefore, if $\|v\|_1$ is small, it follows that

$$E(u) - E(\phi_c) \geqq \frac{1}{4} D\|v\|_1^2,$$

which proves (11.18). This completes the proof of the lemma. □

REMARK 11.12. We note that the restriction on the space of functions of mean zero is due to the facts that ϕ_c has mean zero and $A_{\phi_c} \neq 0$.

Next, using the convexity of $d(c)$ for $c \in (0, \infty)$ and Lemma 11.11, we have the proof of Theorem 11.8.

PROOF OF THEOREM 11.8. Let $\epsilon > 0$ be given by Lemma 11.9. Since

$$\Omega_{\phi_c} \subseteq \mathcal{V} \equiv \Big\{ f \in H^1_{\mathrm{per}}([0, L]) \mid \int_0^L f(x)\, dx = 0 \Big\},$$

we will prove the stability of the orbit initially in $\mathcal{V}$. In fact, assume that Ω_{ϕ_c} is $\mathcal{V}$-unstable and choose initial data $u_k(0) \in U_{1/k} \cap \mathcal{V}$ such that $u_k(t) \notin U_{\epsilon_0}$ for some $\epsilon_0 > 0$ and some $t \in (0, \infty)$. Here $u_k(t)$ is the solution of the KdV equation with initial data $u_k(0)$ (see Theorem 4.7). Without loss of generality we can assume $\epsilon_0 > \epsilon$. Since

$$H(u) = \int_0^L u\, dx$$

is also a conservation law for the KdV equation, we obtain $H(u_k(t)) = H(u_k(0)) = 0$ for all t. Now, by continuity in t, we can choose the first t_k, with k sufficiently large, such that

$$\inf_{y \in \mathbb{R}} \|u_k(t_k) - \phi_c(\cdot + y)\|_1 = \frac{\epsilon}{2}. \tag{11.19}$$

Since E and F in (11.15) are continuous functionals over the space $H^1_{\mathrm{per}}([0, L])$, then we get

$$E(u_k(t_k)) = E(u_k(0)) \to E(\phi_c), \quad F(u_k(t_k)) \to F(\phi_c)$$

as $k \to \infty$. Next choose $w_k \in U_\epsilon \cap \mathcal{V}$ so that $F(w_k) = F(\phi_c)$ and $\|w_k - u_k(t_k)\|_1 \to 0$. Then by Lemma 11.11

$$0 \leftarrow E(w_k) - E(\phi_c) \geqq D_1\|w_k(\cdot + \alpha(w_k)) - \phi_c\|_1^2 = D_1\|w_k - \phi_c(\cdot - \alpha(w_k))\|_1^2,$$

and therefore

$$\|u_k(t_k) - \phi_c(\cdot - \alpha(w_k))\|_1 \to 0.$$

Since this contradicts (11.19), we must have that Ω_{ϕ_c} is $\mathcal{V}$-stable.

Now we consider the general case. Let $u_0 \in U_\delta$ (where δ is given by the $\mathcal{V}$-stability) with $\theta = \int_0^L u_0(x)dx$, and consider $u(t) \in C([0, \infty); H^1_{\mathrm{per}}([0, L]))$, the solution of the KdV equation with initial data u_0 given by Theorem 4.7. Then by defining

$$v(x, t) = u\Big(x + \frac{\theta}{L}t, t\Big) - \frac{\theta}{L},$$

we have that v satisfies the KdV initial value problem

$$\begin{cases} v_t + v v_x + v_{xxx} = 0, \\ v(0) = u_0 - \dfrac{\theta}{L}, \end{cases}$$

where $v_0 \equiv v(0)$ has mean zero over $[0, L]$. Therefore, using Poincaré's inequality for $H^1([0, L])$,

$$\int_0^L |h(x)|^2\,dx - \frac{1}{L}\left|\int_0^L h(x)\,dx\right|^2 \leqq C(L)\int_0^L |h'(x)|^2\,dx \tag{11.20}$$

for $C(L) = 4\pi^2/L^2$, we have for every $y \in \mathbb{R}$ that

$$\int_0^L |v_0 - \tau_y\phi_c|^2\,dx \leqq C(L)\int_0^L |u_0' - \tau_y\phi_c'|^2\,dx,$$

and so

$$\inf_{y\in\mathbb{R}} \|v_0 - \tau_y\phi_c\|_1^2 \leqq [C(L)+1]\inf_{y\in\mathbb{R}} \|u_0' - \tau_y\phi_c'\|^2 < [C(L)+1]\delta^2.$$

Therefore, using the fact that Ω_{ϕ_c} is $\mathcal{V}$-stable, we see that $v(t) \in U_\epsilon \cap \mathcal{V}$ for all $t \geqq 0$. Next, we will see that $u(t) \in U_\epsilon$ for all $t \geqq 0$. In fact, using inequality (11.20), we have for all $y \in \mathbb{R}$ and $t \geqq 0$

$$\int_0^L |u(t) - \tau_y\phi_c|^2\,dx \leqq C(L)\int_0^L |u'(t) - \tau_y\phi_c'|^2\,dx + \frac{1}{L}\theta^2.$$

So, from (11.20) and $u_0 \in U_\delta$, we obtain the following estimate for $|\theta|$. If we let $y \in \mathbb{R}$, then

$$\begin{aligned}\frac{1}{L}\theta^2 &\leqq 2\int_0^L |v_0 - \tau_y\phi_c|^2 + |u_0 - \tau_y\phi_c|^2\,dx\\ &\leqq 2[(1+C(L)]\int_0^L |u_0' - \tau_y\phi_c'|^2 + |u_0 - \tau_y\phi_c|^2\,dx \leqq 2[1+C(L)]\delta^2.\end{aligned}$$

Therefore, using the fact that $v(t) \in U_\epsilon$, which implies, in particular, that

$$\inf_{y\in\mathbb{R}} \|u'(t) - \tau_y\phi_c'\|^2 < \epsilon^2,$$

we obtain that

$$\inf_{y\in\mathbb{R}} \|u(t) - \tau_y\phi_c\|_1^2 \leqq [1+C(L)][\epsilon^2 + 2\delta^2],$$

for all $t \geqq 0$. This completes the proof of Theorem 11.8. □

11.3. Stability of Constant Solutions for the KdV Equation

Unlike the situation that arises for solitary waves where the natural physically relevant assumption is that $\phi(\xi) \to 0$ as $\xi \to \pm\infty$ (or, equivalently, $\int \phi(x)dx = 0$), for which the only trivial solution is $\phi(\xi) \equiv 0$, the cnoidal wave problem gives two trivial solutions. This is easily appreciated from (11.3), for example. If a constant solution taking the value λ is sought, then one sees that λ must satisfy a quadratic equation. If we let $\psi = \phi_c + r$ where $r + c = \sqrt{c^2 + 2A_{\phi_c}}$, then ψ satisfies (11.3) with $A_{\phi_c} = 0$. Note that according to the formulas above (11.4) for c and A_{φ_c} in terms of $\beta_1, \beta_2, \beta_3$,

$$c^2 + 2A_{\phi_c} = \frac{1}{18}\Big[(\beta_2-\beta_1)^2 + (\beta_3-\beta_1)^2 + (\beta_3-\beta_2)^2\Big] > 0,$$

so this is a real-valued transformation. Thus, for ψ, the two constant solutions are $\psi = \phi_0 \equiv 0$ and $\psi = \psi_0 \equiv 2c$.

The state ϕ_0 is clearly stable in H^1_{per}. This follows immediately from the time independence of E_{KdV} and F in (11.15). Indeed, if $\|\varphi_0 - u_0\|_1 = \|u_0\|_1 \le \delta$, then

$$\int_0^L (u-\varphi_0)^2 dx = \int_0^L u^2 dx = \int_0^L u_0^2\, dx \le \delta^2$$

and

$$\begin{aligned}\int_0^L (u_x - \varphi_{0_x})^2\, dx &= \int_0^L u_x^2\, dx = \frac{1}{3}\int_0^L u^3\, dx + \int_0^L \left(u_{0_x}^2 - \frac{1}{3}u_0^3\right) dx \\ &\le \frac{1}{3}\|u\|_\infty \delta^2 + \delta^2 + \delta^3 \\ &\le \frac{1}{3}\delta^2 \left[\frac{1}{L}\|u\|^2 + \|u\|\|u_x\|\right]^{\frac{1}{2}} + \delta^2 + \delta^3 \\ &\le \frac{1}{3}\delta^2 \left[\frac{1}{L}\delta^2 + \delta\|u_x\|\right]^{\frac{1}{2}} + \delta^2 + \delta^3.\end{aligned}$$

These two inequalities imply the previously mentioned stability result. Note that in the second inequality use has been made of the elementary result

$$\sup_{0\le x\le L} |h(x)|^2 \le \frac{1}{L}\int_0^L h^2\, dx + \left(\int_0^L h^2\, dx \int_0^L h_x^2\, dx\right)^{\frac{1}{2}}, \tag{11.21}$$

which will be needed again later.

To study stability of the constant state ψ_0, we rephrase Benjamin's argument as follows. Let $u = \psi_0 + h$ where $h \in H^1_{\text{per}}$ and suppose $F(\psi_0) = F(u_0)$ for the moment. Then, by invariance of F,

$$0 = F(u) - F(\psi_0) = \frac{1}{2}\int_0^L (2\psi_0 h + h^2)\, dx. \tag{11.22}$$

As a consequence, it follows for $E = E_{\text{KdV}}$ that

$$\begin{aligned}\Delta E &= E(u) - E(\psi_0) = E(\psi_0 + h) - E(\psi_0) \\ &= \frac{1}{2}\int_0^L [(h_x)^2 + (c-\psi_0)h^2]\, dx - \frac{1}{6}\int_0^L h^3\, dx \\ &= \frac{1}{2}\int_0^L [(h_x)^2 - ch^2]\, dx - \frac{1}{6}\int_0^L h^3\, dx.\end{aligned} \tag{11.23}$$

A lower bound for the conditional second variation $\delta^2 E$ of E, the quadratic portion of the right-hand side of (11.23), is obtained as follows. Define

$$h^\perp = h - \frac{1}{L}\int_0^L h\, dx \equiv h - \bar{h}. \tag{11.24}$$

Clearly, $\int_0^L h^\perp dx = 0$, so it follows from Poincaré's inequality (11.20) that

$$\int_0^L (h_x^\perp)^2\, dx \ge \left[\left(\frac{2\pi}{L}\right)^2\right] \int_0^L (h^\perp)^2\, dx. \tag{11.25}$$

Formula (11.22) implies that

$$\bar{h} = -\frac{1}{4cL}\int_0^L h^2 dx = -\frac{1}{4cL}\|h\|^2. \tag{11.26}$$

It thus follows that

$$\int_0^L h^2 dx = \int_0^L (h^\perp)^2 dx + L\bar{h}^2 = \int_0^L (h^\perp)^2 dx + \frac{1}{16c^2L}\|h\|^4. \tag{11.27}$$

As a consequence of (11.21) and Young's inequality, for any $\lambda < 1$, we have that

$$\begin{aligned}\left|\int_0^L h^3 dx\right| &\leq \sup_{0\leq x\leq L} |h(x)| \int_0^L h^2 dx \\ &\leq \frac{1}{L^{1/2}}\left(\int_0^L h^2 dx\right)^{3/2} + \left(\int_0^L h^2 dx\right)^{5/4}\left(\int_0^L h_x^2\, dx\right)^{1/4} \\ &\leq \frac{1}{L^{1/2}}\left(\int_0^L h^2 dx\right)^{3/2} + \frac{1-\lambda}{4}\int_0^L h_x^2\, dx + \frac{1}{(1-\lambda)^{1/3}}\left(\int_0^L h^2\, dx\right)^{5/3}.\end{aligned} \tag{11.28}$$

We are now in a position to obtain the aforementioned lower bound. Fix $c < (2\pi/L)^2$ and choose $\lambda \in (0,1)$ such that

$$\lambda\left(\frac{2\pi}{L}\right)^2 - c \geq 1 - \lambda. \tag{11.29}$$

Starting with (11.23), argue as follows using (11.24), (11.25), (11.27), and the choice of λ in (11.29):

$$\begin{aligned}\Delta E &= \frac{1}{2}\int_0^L (h_x^2 - ch^2)\, dx - \frac{1}{6}\int_0^L h^3\, dx \\ &\geq \frac{1}{2}(1-\lambda)\int_0^L h_x^2\, dx + \frac{1}{2}\int_0^L [\lambda(h^\perp)_x^2 - c(h^\perp)^2]\, dx - \frac{1}{32cL}\|h\|^4 \\ &\quad - \frac{1}{24}(1-\lambda)\int_0^L h_x^2\, dx - \frac{1}{6L^{\frac{1}{2}}}\|h\|^3 - \frac{1}{6(1-\lambda)^{\frac{1}{3}}}\|h\|^{\frac{10}{3}} \\ &\geq \frac{1}{4}(1-\lambda)\int_0^L h_x^2\, dx + \frac{1}{2}\Big[\lambda\Big(\frac{2\pi}{L}\Big) - c\Big]\int_0^L (h^\perp)^2\, dx \\ &\quad - \frac{1}{6L^{\frac{1}{2}}}\|h\|^3 - \frac{1}{6(1-\lambda)^{\frac{1}{3}}}\|h\|^{\frac{10}{3}} - \frac{1}{32cL}\|h\|^4 \\ &\geq \frac{1}{4}(1-\lambda)\int_0^L (h_x^2 + h^2)\, dx - \frac{1}{6L^{\frac{1}{2}}}\|h\|^3 \\ &\quad - \frac{1}{6(1-\lambda)^{\frac{1}{3}}}\|h\|^{\frac{10}{3}} - \frac{1}{32cL}\|h\|^4.\end{aligned}$$

Proceeding just as in the argument following (6.23) allows the conclusion of stability in H^1_{per}. The side condition $F(u) = F(\psi_0)$ is then easily removed by a simple use of the triangle inequality. Reverting to the original variables $\phi_c = \psi - r$ with $r = \sqrt{c^2 + 2A_{\phi_c}} - c$ yields the following formal conclusion.

THEOREM 11.13. *Let $L > 0$ and $c > 0$ be given. Let ϕ_0 and ψ_0 be the two constant solutions of the cnoidal wave equation (11.3). Then ϕ_0 is stable in H^1_{per}, and ψ_0 is stable in H^1_{per} provided $c < (2\pi/L)^2$.*

REMARK 11.14. Notice that the translation group acts trivially on φ_0 and ψ_0, so stability up to translation is simply stability in this case.

REMARK 11.15. If $c > (2\pi/L)^2$, the second variation can be negative. Indeed, in this case, if $h_0 = \cos(2\pi x/L)$, then h_0 has mean zero on $[0, L]$ and a straightforward calculation reveals that

$$\delta^2 E(h_0) = \frac{1}{2}(\mathcal{L}_0 h_0, h_0) = \frac{1}{2}\left[\left(\frac{2\pi}{L}\right)^2 - c\right]\int_0^L h_0^2\, dx < 0,$$

where $\mathcal{L}_0 f = -f_{xx} - cf$ is the operator appearing in the general stability theory outlined previously. Thus the approach to stability outlined above fails in this case. Even when $c = (2\pi/L)^2$, the kernel of $\mathcal{L}_0$, i.e.,

$$\mathcal{N}(\mathcal{L}_0) = \text{span}\{\sin(2\pi x/L), \cos(2\pi x/L)\},$$

is two dimensional. Thus, in this case, condition (P_2) is not met. Of course, neither of these implies instability, but the general theory of Grillakis *et al.* [**123**] does not apply because the spectrum of $\mathcal{L}_0$ has at least two negative eigenvalues when $c > (2\pi/L)^2$. We also note that the theory in [**124**] cannot be applied. Indeed, it seems likely that the constant solution $\psi_0 \equiv 2c$ loses stability to a branch of cnoidal waves as c crosses $(2\pi/L)^2$.

11.4. Cnoidal Waves for the 1D Benney-Luke Equation

In this section we study the existence and stability of periodic travelling wave solutions of the nonlinear one-dimensional Boussinesq-type equation

$$\Phi_{tt} - \Phi_{xx} + a\Phi_{xxxx} - b\Phi_{xxtt} + \Phi_t\Phi_{xx} + 2\Phi_x\Phi_{xt} = 0, \tag{11.30}$$

where a and b are positive numbers. One can see that this Boussinesq-type equation is a rescaled version of the one-dimensional Benney-Luke equation

$$\Phi_{tt} - \Phi_{xx} + \mu\,(a\Phi_{xxxx} - b\Phi_{xxtt}) + \epsilon\,(\Phi_t\Phi_{xx} + 2\Phi_x\Phi_{xt}) = 0, \tag{11.31}$$

which is derived from the evolution of two-dimensional long water waves with surface tension. In this model, $\Phi(x,t)$ represents the nondimensional velocity potential at the bottom fluid boundary, μ represents the long-wave parameter (dispersion coefficient), ϵ represents the amplitude parameter (nonlinear parameter), and $a - b = \sigma - 1/3$ with σ the Bond number which is associated with surface tension.

An important feature is that the Benney-Luke equation (11.31) reduces to the Korteweg-de Vries equation, when we look for waves evolving slowly in time, more precisely, when we seek for a solution of the form

$$\Phi(x,t) = f(X, \tau),$$

where $X = x - t$ and $\tau = \epsilon t/2$. In this case, after neglecting $O(\epsilon)$ terms, $\eta = f_X$ satisfies the KdV equation

$$\eta_\tau - (\sigma - 1/3)\eta_{XXX} + 3\eta\eta_X = 0.$$

We will study travelling waves for equation (11.30) of the form

$$\Phi(x,t) = \phi_c(x - ct).$$

The profile ϕ_c has to satisfy the equation

$$(c^2 - 1)\phi_c' + (a - bc^2)\phi_c''' - \frac{3c}{2}(\phi_c')^2 = A_0, \tag{11.32}$$

where A_0 is a constant of integration. Next, $\psi \equiv \phi'$ satisfies equation

$$(c^2 - 1)\psi + (a - bc^2)\psi'' - \frac{3c}{2}\psi^2 = A_0. \tag{11.33}$$

Note that for a periodic travelling wave solution ϕ with a specific period L, we have that ψ has mean zero on $[0, L]$; therefore A_0 needs to be nonzero. Moreover, if ψ is a periodic solution with mean zero on $[0, L]$, then $A_0 \neq 0$ and ϕ is periodic of period L.

Next we scale function ψ. Define

$$\varphi(x) = -\beta\psi(\theta x) \quad \text{with } \beta = 3c \text{ and } \theta^2 = a - bc^2. \tag{11.34}$$

We have that φ satisfies the ordinary differential equation

$$\varphi'' + \frac{1}{2}\varphi^2 - (1 - c^2)\varphi = A_\varphi \tag{11.35}$$

with $A_\varphi = -3cA_0$. For $0 < c^2 < 1$, a class of periodic solutions to (11.35), called *cnoidal waves*, was already found in Section 11.2. It may be written in terms of the Jacobi elliptic function as

$$\varphi_c(x) \equiv \varphi(x) = \beta_2 + (\beta_3 - \beta_2)\mathrm{cn}^2\left(\sqrt{\frac{\beta_3 - \beta_1}{12}}x; k\right) \tag{11.36}$$

where

$$\beta_1 < \beta_2 < \beta_3, \quad \beta_1 + \beta_2 + \beta_3 = 3(1 - c^2), \quad \text{and} \quad k^2 = \frac{\beta_3 - \beta_2}{\beta_3 - \beta_1}. \tag{11.37}$$

Then ψ_c has the form

$$\psi_c(x) = -\frac{1}{3c}\left[\beta_2 + (\beta_3 - \beta_2)\mathrm{cn}^2\left(\frac{1}{\sqrt{a - bc^2}}\sqrt{\frac{\beta_3 - \beta_1}{12}}x; k\right)\right]. \tag{11.38}$$

Next we consider the degenerate cases. First, fix the value of c and consider whether or not periodic solutions can persist if $\beta_1 = \beta_2$ or $\beta_2 = \beta_3$. As φ can only take values in the interval $[\beta_2, \beta_3]$, we conclude that the second case leads only to the constant solution $\varphi(x) \equiv \beta_2 = \beta_3$. Indeed, the limit of (11.36) as $\beta_2 \to \beta_3$ is uniform in x and the limit is exactly this constant solution. If, on the other hand, c and β_1 are fixed, say, $\beta_2 \downarrow \beta_1$ and $\beta_3 = 3(1 - c^2) - \beta_2 - \beta_1$, then $k \to 1$, the elliptic function cn converges, uniformly on compact sets, to the hyperbolic function sech, and (11.36) becomes, in that limit,

$$\varphi(x) = \varphi_\infty + \gamma\,\mathrm{sech}^2\left(\sqrt{\frac{\gamma}{12}}x\right)$$

with $\varphi_\infty = \beta_1$ and $\gamma = \beta_3 - \beta_1$. If $\beta_1 = 0$, we obtain

$$\varphi(x) = 3(1 - c^2)\,\mathrm{sech}^2\left(\frac{\sqrt{1 - c^2}}{2}x\right).$$

So, by returning to the original function ψ, we obtain the standard solitary wave solution

$$\psi(x) = -\frac{1 - c^2}{c}\,\mathrm{sech}^2\left(\frac{1}{2}\sqrt{\frac{1 - c^2}{a - bc^2}}\,x\right)$$

of speed c such that $0 < c^2 < \min\{1, a/b\}$ of the Benney-Luke equation.

Next, by returning to the original variable ϕ_c, we obtain after integration and using the formula (see [**75**])

$$\int \mathrm{cn}^2(u; k)\, du = \frac{1}{k^2}\left[\int_0^u \mathrm{dn}^2(x; k)\, dx - (1 - k^2)u\right]$$

that

$$\phi_c(x) = -\frac{\beta_1}{3c}x - \frac{\beta_3 - \beta_2}{3cLk^2}\int_0^{Lx} \mathrm{dn}^2(u;k)\,du + M \tag{11.39}$$

where M is a constant of integration and

$$L = \frac{1}{\sqrt{a-bc^2}}\sqrt{\frac{\beta_3-\beta_1}{12}},$$

where $0 < c^2 < \min\{1, a/b\}$.

As a consequence of the analysis in Section 11.2 we have the following.

THEOREM 11.16. *For every $L > 0$ there are smooth curves*

$$c \in I = (-\sqrt{\min\{1,a/b\}}, \sqrt{\min\{1,a/b\}}\,) \setminus \{0\} \to \psi_c \in H^1_{\mathrm{per}}([0,L]),$$

of solutions of the equation

$$(a-bc^2)\psi_c'' - (1-c^2)\psi_c - \frac{3c}{2}\psi_c^2 = A_{\psi_c}, \tag{11.40}$$

where each ψ_c has fundamental period L and mean zero on $[0,L]$. Moreover, there are smooth curves $c \in I \to \beta_i(c)$, $i = 1,2,3$, such that

$$A_{\psi_c} = \frac{-3c}{2L}\int_0^L \psi_c^2(\xi)\,d\xi = \frac{1}{3c}\frac{1}{6}\sum_{i<j}\beta_i(c)\beta_j(c) \tag{11.41}$$

and ψ_c has the cnoidal formula given in (11.38) *and* (11.37).

Our attention is turned to set the main tools to be used in order to establish the stability of the cnoidal wave solutions given by Theorem 11.16. As is well known, the study of the following periodic eigenvalue problem considered on $[0,L]$ is required in the stability theory:

$$\begin{cases} \mathcal{L}_{\mathrm{cn}}\chi \equiv [-(a-bc^2)\dfrac{d^2}{dx^2} + (1-c^2) + 3c\psi_c]\chi = \lambda\chi, \\ \chi(0) = \chi(L), \quad \chi'(0) = \chi'(L), \end{cases} \tag{11.42}$$

where c is fixed such that $0 < c^2 < \min\{1, a/b\}$. The following result is obtained for this context.

THEOREM 11.17. *The linear operator*

$$\mathcal{L}_{\mathrm{cn}} = -(a-bc^2)\frac{d^2}{dx^2} + (1-c^2) + 3c\psi_c$$

defined on $H^2_{\mathrm{per}}([0,L])$ has exactly its three first eigenvalues $\lambda_0 < \lambda_1 < \lambda_2$ simple. Moreover, ψ_c' is an eigenfunction with eigenvalue $\lambda_1 = 0$. The rest of the spectrum is a discrete set of eigenvalues which are double. The eigenvalues only accumulate at $+\infty$.

The result of stability of the periodic travelling wave solutions ϕ_c for equation (11.30) follows from studying the same problem for the Boussinesq system associated with (11.30), namely,

$$\begin{cases} q_t = r_x, \\ r_t = B^{-1}(q_x - aq_{xxx}) - B^{-1}(rq_x + 2qr_x), \end{cases} \tag{11.43}$$

where $q = \Phi_x$, $r = \Phi_t$, and $B = 1 - b\partial_x^2$.

The Boussinesq system (11.43) can be written as a Hamiltonian system in the new variables

$$(q,p) \equiv (q, Br + \frac{1}{2}q^2)$$

as

$$\text{(11.44)} \qquad \begin{cases} q_t = \partial_x B^{-1}\Big(p - \dfrac{1}{2}q^2\Big), \\ p_t = \partial_x\Big(Aq - rq\Big), \end{cases}$$

with $A = 1 - a\partial_x^2$ and $B = 1 - b\partial_x^2$. This system arises as the Euler-Lagrange equation for the action functional

$$\mathcal{S} = \int_{t_0}^{t_1} \mathcal{Q}\begin{pmatrix} q \\ p \end{pmatrix} dt,$$

where the Lagrangian $\mathcal{Q}$ and the Hamiltonian are given, respectively, by

$$\text{(11.45)} \quad \mathcal{Q}\begin{pmatrix} q \\ p \end{pmatrix} = \frac{1}{2}\int_0^L \Big\{B^{-1}\Big(p - \frac{1}{2}q^2\Big)\Big(p - \frac{1}{2}q^2\Big) - qAq + B^{-1}\Big(p - \frac{1}{2}q^2\Big)q^2\Big\}\, dx$$

and

$$\text{(11.46)} \qquad \mathcal{H}\begin{pmatrix} q \\ p \end{pmatrix} = \frac{1}{2}\int_0^L \Big\{\Big(p - \frac{1}{2}q^2\Big)\, B^{-1}\Big(p - \frac{1}{2}q^2\Big) + qAq\Big\}\, dx.$$

In this way we obtain the canonical Hamiltonian form

$$\text{(11.47)} \qquad \partial_x \mathcal{H}_p = q_t, \quad \partial_x \mathcal{H}_q = p_t$$

and the Hamiltonian system in the variable $V = \begin{pmatrix} q \\ p \end{pmatrix}$ as

$$\text{(11.48)} \qquad V_t = \begin{pmatrix} 0 & \partial_x \\ \partial_x & 0 \end{pmatrix} \mathcal{H}'(V).$$

We observe that the Hamiltonian in (11.45)–(11.46) is formally conserved in time for solutions of system (11.44) since

$$\begin{aligned} \frac{d}{dt}\mathcal{H}(V) &= \int_0^L \{\mathcal{H}_q q_t + \mathcal{H}_p p_t\}\, dx = \int_0^L \{\mathcal{H}_q \partial_x \mathcal{H}_p + \mathcal{H}_p \partial_x \mathcal{H}_q\}\, dx \\ &= \int_0^L \partial_x\{\mathcal{H}_q \mathcal{H}_p\}\, dx. \end{aligned}$$

So, the Hamiltonian

$$\text{(11.49)} \qquad \mathcal{H}\begin{pmatrix} q \\ r \end{pmatrix} = \frac{1}{2}\int_0^L \{rBr + qAq\}\, dx,$$

associated to (11.43), is formally conserved in time. Moreover, since the Hamiltonian is translation-invariant, then by Noether's theorem there is an associated momentum functional $\mathcal{N}$ which is also conserved in time. This functional has the form

$$\text{(11.50)} \qquad \mathcal{N}\begin{pmatrix} q \\ r \end{pmatrix} = \int_0^L \Big(Br + \frac{1}{2}q^2\Big) q\, dx.$$

Next we are interested in finding periodic travelling wave solutions for system (11.43), in other words, solutions of the form $(q,r) = (\psi(x-ct), g(x-ct))$. By substituting, we have that the couple (ψ, g) satisfies the nonlinear system

$$
\text{(11.51)} \qquad \begin{cases} g = -c\psi + A_0, \\ c^2(1-b\partial_x^2)\psi = (1-a\partial_x^2)\psi + \dfrac{3c}{2}\psi^2 - A_0\psi + A_{\psi_c}, \end{cases}
$$

with A_0 and A_{ψ_c} constants of integration. Now, since our approach of stability is based in the context of the stability theory of Grillakis *et al.*, we need to have that (ψ, g) satisfies the equation

$$
\text{(11.52)} \qquad \delta\mathcal{F}\begin{pmatrix}\psi \\ g\end{pmatrix} = \begin{pmatrix}A_\psi \\ 0\end{pmatrix}
$$

with

$$
\text{(11.53)} \qquad \mathcal{F} = \mathcal{H} + c\mathcal{N}.
$$

Therefore it follows from (11.51) that we must have $A_0 = 0$. In other words, we have to solve the system

$$
\text{(11.54)} \qquad \begin{cases} g = -c\psi, \\ (1-c^2)\psi + (bc^2 - a)\psi'' + \dfrac{3c}{2}\psi^2 = A_{\psi_c}. \end{cases}
$$

We note that

$$
V_c = \begin{pmatrix}\psi_c \\ -c\psi_c\end{pmatrix}
$$

cannot be a critical point of the action functional $\mathcal{F}$. This shows the necessity of adapting the Grillakis *et al.* stability result to the present case (see Lemma 11.11). More precisely, we need in the stability theory that

$$
\mathcal{F}'(V_c)\vec{v} = \langle (A_{\psi_c}, 0), \vec{v}\rangle = 0,
$$

for $\vec{v} = (f, g)$. So, we need to have

$$
\text{(11.55)} \qquad f \in \mathcal{W}^1 = \left\{ q \in H^1_{\text{per}}([0,L]) : \int_0^L q(y)\,dy = 0 \right\}.
$$

Now we consider the study of the convexity of the function d defined by

$$
\text{(11.56)} \qquad d(c) = \mathcal{F}\left(\psi_c, -cB\psi_c + \frac{1}{2}\psi_c^2\right)
$$

where the cnoidal wave solution ψ_c is given by Theorem 11.16. If we differentiate (11.56) with respect to c and use the fact that

$$
\int_0^L \frac{d}{dc}\psi_c(x)\,dx = 0,
$$

we get that

$$
\text{(11.57)} \qquad \begin{aligned} d'(c) &= \delta\mathcal{F}(V_c)\left(\frac{d}{dc}V_c\right) + \mathcal{N}(V_c) = \left\langle \mathcal{A}, \frac{d}{dc}\psi_c \right\rangle + \mathcal{N}(V_c) \\ &= -\int_0^L \left(c\psi_c^2 + bc(\psi_c')^2 - \frac{1}{2}\psi_c^3 \right) dx. \end{aligned}
$$

Then from Theorem 4.3 in [**39**] we have the following convexity property of the function d.

THEOREM 11.18. *The function d is a strictly convex function for*

(1) $0 < |c| < 1 < \sqrt{a/b}$ *and*
(2) $0 < |c| < c_* + \theta < \sqrt{a/b} < 1$, *where θ is small and c_* is the unique positive root of the polynomial*

$$P(c) = 12b^2c^6 + (13b^2 - 19ab)c^4 + (9ab - 9a^2)c^2 - 6a^2.$$

Next, we obtain the following stability result associated to system (11.43).

THEOREM 11.19. *Consider c with $0 < c^2 < \min\{1, a/b\}$ and let $\{\psi_c\}$ be the cnoidal wave branch of period L given in Theorem 11.16. Then, for c satisfying the conditions in Theorem 11.18, the orbit*

$$\{(\psi_c(\cdot + s), -c\psi_c(\cdot + s))\}_{s\in\mathbb{R}}$$

is stable in $\mathcal{X} = \mathcal{W}^1 \times H^1_{\text{per}}([0, L])$ with regard to L-periodic perturbations and the flow generated by system (11.43). More precisely, given any $\epsilon > 0$ there is a $\delta = \delta(\epsilon)$ such that if $(q_0, r_0) \in \mathcal{X}$ and

$$\|(q_0, r_0) - (\psi_c, -c\psi_c)\|_{\mathcal{X}} < \delta,$$

then

$$\inf_{s\in\mathbb{R}} \|(q(t), r(t)) - (\psi_c(\cdot + s), -c\psi_c(\cdot + s))\|_{\mathcal{X}} < \epsilon$$

for all t, where $(q(t), r(t))$ is the solution of system (11.43) with initial value (q_0, r_0).

In order to prove the stability of periodic solutions for the Benney-Luke equation, we need to establish an appropriate space for it. We start by defining the natural space

$$\mathcal{Z} = \{w \in C^\infty_{\text{per}}([0, T_0]) : w_x \in H^1_{\text{per}}([0, T_0])\},$$

and we define the equivalence relation on $\mathcal{Z}$ given by $u \sim v$ if and only if $u(x) - v(x) = \theta$ for $x \in \mathbb{R}$ and θ a real constant. Now define the quotient space $\mathcal{Y} = \mathcal{Z}/\sim$ with the norm,

$$\|[y]\|_{\mathcal{Y}} = \|y_x\|_{H^1_{\text{per}}} = \|u_x\|_{H^1_{\text{per}}} \quad \text{for all } u \in [y].$$

Let $(\mathcal{V}, \|\cdot\|_{\mathcal{V}})$ be the closure of $\mathcal{Y}$ with respect to $\|\cdot\|_{\mathcal{Y}}$. In particular, if $[\Phi] \in \mathcal{Y}$,

$$\|[\Phi]\|_{\mathcal{V}} = \|[\Phi]\|_{\mathcal{Y}} = \|\Phi_x\|_{H^1_{\text{per}}}.$$

Hereafter we will identify any equivalence class with a representative. Roughly speaking, the space $\mathcal{V}$ can be viewed as the closure of $\mathcal{Z}$ with respect to the "norm" $\|\cdot\|_{\mathcal{V}}$. Moreover, as we can see, $\mathcal{V}$ can be identified with the space $\mathcal{W}^1$ through of the linear operator ∂_x. So, it will give us an easy way to recover all the stability theory associated to system (11.43) for $\mathcal{V}$. Finally, we note that as (11.30) is invariant by the translation $\Phi \to \Phi + \text{const.}$, $\mathcal{V}$ is the natural space to be considered in our stability study.

Now we have to note that the Benney-Luke equation can be written as the system in the variables Φ and $\Phi_t = r$

$$\begin{pmatrix}\Phi\\ r\end{pmatrix}_t = M_0\begin{pmatrix}\Phi\\ r\end{pmatrix} + \mathcal{G}_0\begin{pmatrix}\Phi\\ r\end{pmatrix}, \tag{11.58}$$

where operator M_0 and $\mathcal{G}_0$ are given by

$$M_0 = \begin{pmatrix} 0 & I \\ \partial_x^2 B^{-1} A & 0 \end{pmatrix} \quad \text{and} \quad \mathcal{G}_0 \begin{pmatrix} \Phi \\ r \end{pmatrix} = \begin{pmatrix} 0 \\ -B^{-1}(r\Phi_{xx} + 2\Phi_x r_x) \end{pmatrix}.$$

Finally, we easily derive the following stability result from Theorem 11.19.

THEOREM 11.20. *Consider c with $0 < c^2 < \min\{1, a/b\}$ and let $\{\phi_c\}$ be such that $\phi_c' = \psi_c$ where ψ_c is the cnoidal wave of period L given in Theorem 11.16. Then, for c satisfying the conditions in Theorem 11.18, the orbit*

$$\{(\phi_c(\cdot + s), -c\phi_c'(\cdot + s))\}_{s\in\mathbb{R}}$$

is stable in $\mathcal{V} \times H^1_{\text{per}}([0, L])$ with regard to L-periodic perturbations and the flow generated by the Benney-Luke equation (11.30)–(11.58). More precisely, given any $\epsilon > 0$ there is a $\delta = \delta(\epsilon)$ such that if $(u_0, r_0) \in \mathcal{V} \times H^1_{\text{per}}([0, L])$ and

$$\|(u_0, r_0) - (\phi_c, -c\phi_c')\|_{\mathcal{V}\times H^1_{\text{per}}} < \delta,$$

then

$$\inf_{s\in\mathbb{R}} \|(\Phi(t), r(t)) - (\phi_c(\cdot + s), -c\phi_c'(\cdot + s))\|_{\mathcal{V}\times H^1_{per}} < \epsilon$$

for all t, where $(\Phi(t), r(t))$ is the mild solution for (11.58) with initial value (u_0, r_0).

11.5. Angulo and Natali's Stability Approach

Next we establish the basic points of the theory developed by Angulo and Natali in [**37**] to study the stability of positive, even, periodic travelling wave solutions associated to the general dispersive equation (11.1).

We start by studying the linear operator $\mathcal{L} : D(\mathcal{L}) \to L^2_{\text{per}}([-L, L])$ defined on a dense subspace of $L^2_{\text{per}}([-L, L])$ by

$$\mathcal{L}u = (M + c)u - \varphi^p u. \tag{11.59}$$

We will suppose that $c > -b$, where b satisfies $\alpha(k) > b$ for all $k \in \mathbb{Z}$. With this condition $M+c$ represents a positive operator. Then, by using the spectral theorem for compact and self-adjoint operators, we have the following characterization of the spectrum of $\mathcal{L}$.

THEOREM 11.21. *The operator $\mathcal{L}$ in (11.59) is a closed, unbounded, self-adjoint operator on $L^2_{\text{per}}([0, 2L])$ whose spectrum consists of an enumerable (infinite) set of eigenvalues $\{\lambda_k\}_{k=0}^\infty$ satisfying*

$$\lambda_0 \le \lambda_1 \le \lambda_2 \le \dots$$

and $\lambda_k \to \infty$ as $k \to \infty$. In particular, $\mathcal{L}$ has 0 as an eigenvalue with eigenfunction $\frac{d}{dx}\varphi_c$.

PROOF. Clearly $\mathcal{L}$ defined on $H^{m_2}_{2L}$ is a closed, unbounded, and self-adjoint operator in $L^2_{\text{per}}([0, 2L])$. Let us prove that the spectrum of $\mathcal{T} := M + c$ is a countably infinite set of eigenvalues with

$$\gamma_0 \le \gamma_1 \le \gamma_2 \le \gamma_3 \le \dots, \tag{11.60}$$

where $\gamma_n \to \infty$ as $n \to \infty$. In fact, let $R_c = (M + c)^{-1}$, whose symbol via the Fourier transform is given by $(c + \alpha(k))^{-1}$ for $k \in \mathbb{Z}$. Since $(c + \alpha(k))^{-1} \in \ell^2(\mathbb{Z})$, we see that there is a unique $G_c \in L^2_{\text{per}}([0,2L])$ such that $\widehat{G_c}(k) = (c+\alpha(k))^{-1}$, and because of this, we have the action

$$R_c f(x) = \frac{1}{2L}\int_{-L}^{L} G_c(x-y)f(y)dy,$$

defined for $f \in L^2_{\text{per}}([0,2L])$. Since $[0,2L]$ is a bounded set and we can prove that the kernel $\widetilde{G_\mu}(x,y) := G_\mu(x-y) \in L^2([0,2L]\times[0,2L])$, we obtain that R_c is a Hilbert-Schmidt operator on $L^2_{\text{per}}([0,2L])$ (see Kato [**153**]). Therefore R_c is a compact operator on $L^2_{\text{per}}([0,2L])$ for all $c>0$ (here we assumed without loss of generality that $b=0$), and so we obtain (11.60).

Next, we will show that there is a μ_1 (large enough) such that

$$\mathcal{M} = (\mathcal{L}+\mu_1)^{-1}$$

exists and is a bounded, positive, compact, and self-adjoint operator. In fact, first of all, it is easy to see that $\mathcal{L}$ is limited below; that is, if $f \in D(\mathcal{L})$, we have

$$\langle \mathcal{L}f, f\rangle \geq -\beta\langle f, f\rangle,$$

where $\beta = ||\varphi_c||_{L^\infty_{\text{per}}} + c$. Then, we can choose a μ_1 such that $\mathcal{L}+\mu_1>0$; that is, $\mathcal{M}$ is positive. We denote $\mu_1 := \mu$ only for convenience. Let ν be a positive number such that $\nu+\varphi_c - c>0$ and $\nu+\mu>0$. Thus for $\mu>0$ we have $f=(\mathcal{L}+\mu)g$ if and only if

$$(I-M)g = \Upsilon f$$

where

$$Mg = R_{\mu+\nu}[(\nu+\varphi_c - c)g], \quad \Upsilon = R_{\nu+\mu},$$

and we denote $h = \nu+\varphi_c-c$. Next, from the Parseval Theorem, it follows that

$$||Mg||_{L^2_{\text{per}}} \leq \sup_{k\in\mathbb{Z}}\left\{\frac{1}{\alpha(k)+\nu+\mu}\right\}||h||_{L^\infty_{\text{per}}}||g||_{L^2_{\text{per}}}.$$

Thus, we can choose μ such that $||M||_{B(L^2_{\text{per}})}<1$ and $\mathcal{L}+\mu>0$. Then, $I-M$ is invertible. We see that $g=(I-M)^{-1}\Upsilon f$ and write $\mathcal{M} = (\mathcal{L}+\mu)^{-1} = (I-M)^{-1}\Upsilon$. Since Υ is a compact operator and $(I-M)^{-1}\in B(L^2_{\text{per}})$, we have that $\mathcal{M}$ is a compact operator. Then, there is an orthonormal basis $\{\varphi_k\}_{k=0}^\infty$ of L^2_{per} made of eigenfunctions of $\mathcal{M}$ with nonzero eigenvalues $\{\mu_k\}_{k=0}^\infty$ satisfying

$$\mu_1 \geq \mu_2 \geq \mu_3 \geq \cdots > 0$$

and $\mu_k \to 0$ as $k\to\infty$. Since $\mathcal{M}\varphi_k = \mu_k\varphi_k \in D(\mathcal{L}+\mu)$, we see that

$$\mathcal{L}\varphi_k = \left(\frac{1}{\mu_k}-\mu\right)\varphi_k := \lambda_k\varphi_k.$$

Thus, there is a sequence of eigenvalues of $\mathcal{L}$, $\{\lambda_k\}_{k=0}^\infty$, satisfying

$$\lambda_0 \leq \lambda_1 \leq \lambda_2 \leq \cdots$$

and $\lambda_k\to\infty$ as $k\to\infty$. This argument shows the desired results. □

Next, for all $\theta\geq 0$, we consider the operator $S_\theta : \ell^2(\mathbb{Z}) \to \ell^2(\mathbb{Z})$ defined by

$$S_\theta\beta(n) = \frac{1}{\omega_\theta(n)}\sum_{j=-\infty}^{\infty} K(n-j)\beta_j = \frac{1}{\omega_\theta(n)}(K*\beta)_n,$$

where $\omega_\theta(n) = \alpha(n) + \theta + c$, $K(n) = \widehat{\varphi_c^p}(n)$, $n \in \mathbb{Z}$. Since $\omega_\theta(n) > 0$ for all $n \in \mathbb{Z}$, it follows that the space X defined by

$$X = \left\{ \beta \in \ell^2(\mathbb{Z}) : \|\beta\|_{X,\theta} := \left(\sum_{n=-\infty}^{\infty} |\beta_n|^2 \omega_\theta(n) \right)^{\frac{1}{2}} < \infty \right\}$$

is a Hilbert space with norm $||\beta||_{X,\theta}$ and corresponding inner product

$$\langle \gamma, \beta \rangle_{X,\theta} = \sum_{n=-\infty}^{\infty} \gamma_n \overline{\beta_n} \omega_\theta(n).$$

THEOREM 11.22.

(1) *If $\alpha \in \ell^2$ is an eigensequence of S_θ for a nonzero eigenvalue, then $\alpha \in X$.*
(2) *The restriction of S_θ to X is a compact, self-adjoint operator on X with respect to the norm $||\cdot||_{X,\theta}$.*
(3) *Suppose $\theta \geq 0$. Then, 1 is an eigenvalue of S_θ (as an operator of X) if and only if $-\theta$ is an eigenvalue of $\mathfrak{L}$ (as an operator of L^2_{per}). Furthermore, both eigenvalues have the same multiplicity.*
(4) *For every $\theta \geq 0$, S_θ has a family of eigensequences $\{\psi_{i,\theta}\}_{i=0}^\infty$ forming an orthonormal basis of X with respect to the norm $||\cdot||_{X,\theta}$. The eigensequences correspond to real eigenvalues $\{\lambda_i(\theta)\}_{i=0}^\infty$ whose only possible accumulation point is zero. Moreover,*

$$|\lambda_0(\theta)| \geq |\lambda_1(\theta)| \geq |\lambda_2(\theta)| \geq \cdots.$$

PROOF. See [**37**]. □

DEFINITION 11.23. We say that a sequence $\alpha = (\alpha_n)_{n \in \mathbb{Z}} \subseteq \mathbb{R}$ is in the class $PF(2)$ discrete if

(1) $\alpha_n > 0$, for all $n \in \mathbb{Z}$,

(2) $\alpha_{n_1-m_1}\alpha_{n_2-m_2} - \alpha_{n_1-m_2}\alpha_{n_2-m_1} > 0$ for $n_1 < n_2$ and $m_1 < m_2$.

The definition above is a discrete form of the continuous form of the definition which appears in Section 7.5. The main result of this section is

THEOREM 11.24 (Angulo and Natali). *Let φ_c be an even positive solution of (11.2). Suppose that $\widehat{\varphi_c} > 0$ and $K = \widehat{\varphi_c^p} \in PF(2)$ discrete. Then $\mathfrak{L}$ in (11.59) has exactly one unique negative eigenvalue which is simple, and zero is a simple eigenvalue with eigenfunction $\frac{d}{dx}\varphi_c$.*

SKETCH OF THE PROOF. The basic steps of the proof are
Step I. Since S_θ is a compact, self-adjoint operator on X, it follows that

$$\lambda_0(\theta) = \pm \sup_{||\alpha||_X=1} |\langle S_\theta \alpha, \alpha \rangle_X|. \tag{11.61}$$

Let $\psi(\theta) := \psi$ be an eigensequence of S_θ corresponding to $\lambda_0(\theta) := \lambda_0$. We will show that ψ is one-signed, that is, either $\psi(n) \leq 0$ or $\psi(n) \geq 0$. By contradiction, suppose ψ takes both negative and positive values. Since by hypotheses the kernel

$K = (K(n)) = (\widehat{\varphi_c^p}(n))$ *is positive*, we have

$$S_\theta|\psi|(n) = \frac{1}{\omega_\theta(n)} \sum_{j=-\infty}^{\infty} K(n-j)\psi^+(j) + \frac{1}{\omega_\theta(n)} \sum_{j=-\infty}^{\infty} K(n-j)\psi^-(j)$$

$$> \left| \frac{1}{\omega_\theta(n)} \sum_{j=-\infty}^{\infty} K(n-j)\psi^+(j) - \frac{1}{\omega_\theta(n)} \sum_{j=-\infty}^{\infty} K(n-j)\psi^-(j) \right|,$$

where ψ^+ and ψ^- are the positive and negative parts of ψ, respectively. Therefore

$$\begin{aligned}\langle S_\theta(|\psi|), |\psi|\rangle_{X,\theta} &= \sum_{n=-\infty}^{\infty} S_\theta|\psi|(n)|\psi(n)|\omega_\theta(n) > \sum_{n=-\infty}^{\infty} |\lambda_0||\psi(n)|^2\omega_\theta(n)\\ &= |\lambda_0|||\psi||^2_{X,\theta}.\end{aligned}$$

Hence, if we assume that $||\psi||_X = 1$, we obtain

$$\langle S_\theta(|\psi|), |\psi|\rangle_X > |\lambda_0|,$$

which contradicts (11.61). Then, there is an eigensequence ψ_0 which is nonnegative. Since K *is a positive sequence* and $S_\theta(\psi_0) = \lambda_0\psi_0$, we have $\psi_0(n) > 0$ $\forall n \in \mathbb{Z}$. Since ψ_0 cannot be orthogonal to any nontrivial one-signed eigensequence in X, we have that λ_0 is a simple eigenvalue. Notice that the preceding argument also shows that $-\lambda_0$ cannot be an eigenvalue of S_θ; therefore it follows that $|\lambda_1| < \lambda_0$.

Step II. The next step will be to study the behavior of the eigenvalue $\lambda_1(\theta)$. In fact, we consider the set of indices

$$\triangle = \{(n_1, n_2) \in \mathbb{Z} \times \mathbb{Z};\ n_1 < n_2\}.$$

Denoting $\overline{n} = (n_1, n_2)$ and $\overline{m} = (m_1, m_2)$, for $\overline{n}$, $\overline{m} \in \triangle$ we define the following sequence:

$$K_2(\overline{n}, \overline{m}) := K(n_1 - m_1)K(n_2 - m_2) - K(n_1 - m_2)K(n_2 - m_1).$$

By hypothesis $K \in PF(2)$ discrete; hence we have $K_2 > 0$. Let $\ell^2(\triangle)$ be defined as

$$\ell^2(\triangle) = \left\{ \alpha = (\alpha_{\overline{n}})_{\overline{n}\in\triangle} : \sum\sum\nolimits_{\triangle} |\alpha_{\overline{n}}|^2 := \sum_{n_1\in\mathbb{N}} \sum_{\substack{n_1<n_2\\ n_2\in\mathbb{Z}}} |\alpha(n_1, n_2)|^2 < +\infty \right\},$$

and define the operator $S_{2,\theta} : \ell^2(\triangle) \to \ell^2(\triangle)$ by

$$S_{2,\theta}g(\overline{n}) = \sum\sum\nolimits_{\triangle} G_{2,\theta}(\overline{n}, \overline{m})g(\overline{m}),$$

where

$$G_{2,\theta}(\overline{n}, \overline{m}) = \frac{K_2(\overline{n}, \overline{m})}{\omega_\theta(n_1)\omega_\theta(n_2)}.$$

We also consider the space

$$W = \left\{ \alpha \in \ell^2(\triangle) : ||\alpha||_{W,\theta} := \left(\sum\sum\nolimits_{\triangle} |\alpha(\overline{n})|^2\omega_\theta(n_1)\omega_\theta(n_2) \right)^{\frac{1}{2}} < \infty \right\}.$$

Then W is a Hilbert space with norm $||\cdot||_{W,\theta}$ given above and with inner product

$$\langle \alpha^1, \alpha^2 \rangle_{W,\theta} = \sum\sum\nolimits_{\triangle} \alpha^1(\overline{n})\overline{\alpha^2(\overline{n})}\omega_\theta(n_1)\omega_\theta(n_2).$$

Now, we have the following comments:

(1) We can show, analogous to Theorem 11.22, that $S_{2,\theta}\big|_W$ is a self-adjoint, compact operator. Therefore, the eigenvalues associated to it can be enumerated in order of decreasing absolute value; that is,

$$|\mu_0(\theta)| \geq |\mu_1(\theta)| \geq |\mu_2(\theta)| \geq \cdots.$$

(2) We also obtain that $\mu_0(\theta) := \mu_0$ is positive and simple and $|\mu_1| < \mu_0$.

To study operator $S_{2,\theta}$, we need the following.

DEFINITION 11.25. Let α^1, $\alpha^2 \in \ell^2(\mathbb{Z})$. We define the wedge product $\alpha^1 \wedge \alpha^2$ in $\triangle$ by

$$(\alpha^1 \wedge \alpha^2)(n_1, n_2) = \alpha^1(n_1)\alpha^2(n_2) - \alpha^1(n_2)\alpha^2(n_1).$$

We have the following results from Definition 11.25.

LEMMA 11.26. *Let* $A = \{\alpha^1 \wedge \alpha^2 : \text{ for } \alpha^1,\ \alpha^2 \in X,\ \alpha^1 \wedge \alpha^2 \in \ell^2(\triangle)\}$. *Then* A *is dense in* W.

PROOF. See [**6**], [**10**]. □

LEMMA 11.27. *Let* α^1, $\alpha^2 \in \ell^2(\mathbb{Z})$. *Then* $S_{2,\theta}(\alpha^1 \wedge \alpha^2) = S_\theta\alpha^1 \wedge S_\theta\alpha^2$.

PROOF. See [**6**], [**10**]. □

LEMMA 11.28. *For* $\mu_0(\theta)$ *we have the following.*

(1) $\mu_0(\theta) = \lambda_0(\theta)\lambda_1(\theta)$. *Then, we can conclude that* $\lambda_1 > 0$.
(2) λ_1 *is simple.*

PROOF. See [**37**]. □

Step III. This is the final step. For $i = 0, 1$, we have that the curve $\theta \to \lambda_i(\theta)$ satisfies

$$\frac{d}{d\theta}\lambda_i(\theta) < 0.$$

Moreover,

$$\lim_{\theta \to \infty} \lambda_0(\theta) = 0.$$

Now, since $\widehat{\varphi_c} > 0$, we have that $\lambda_1(0) = 1$. Therefore, since

$$\lambda_0(0) > \lambda_1(0) = 1,$$

it follows that there is a unique $\theta_0 \in (0, +\infty)$ such that $\lambda_0(\theta_0) = 1$. Then, from Theorem 11.22 we obtain that $\kappa \equiv -\theta_0$ is a negative eigenvalue of $\mathcal{L}$ which is simple. Now, for $i \geq 2$ and $\theta > 0$ we have that

$$\lambda_i(\theta) \leq \lambda_1(\theta) < \lambda_1(0) = 1,$$

so 1 cannot be an eigenvalue of S_θ for all $\theta \in (0, +\infty)\setminus\{\theta_0\}$, since 1 is an eigenvalue only for $\theta = 0$ and $\theta = \theta_0$. Then we obtain that $\mathcal{L}$ has a unique negative eigenvalue which is simple. Finally, since $\lambda_1(0) = 1$ and λ_1 is a simple eigenvalue, it follows that $\theta = 0$ is a simple eigenvalue of $\mathcal{L}$ by Theorem 11.22. This proves the theorem. □

REMARK 11.29. We have the following main comments:

(1) In Theorem 11.24, the Fourier transform needed to be evaluated in the minimal period for the solution φ_c. In fact, let L be this minimal period, and we evaluate the Fourier transform of φ_c as being of period $2L$. Then
$$\widehat{\varphi_c}(k) = \frac{1}{2L}\int_{-L}^{L} \varphi_c(x) e^{-\frac{ik\pi x}{L}}\,dx = \frac{1}{2L}\int_{-L}^{L} \varphi_c(x+L) e^{-\frac{ik\pi x}{L}}\,dx$$
$$= \frac{1}{2L}\int_{0}^{2L} \varphi_c(y) e^{-\frac{ik\pi y}{L}} e^{i\pi k}\,dy = (-1)^k \widehat{\varphi_c}(k).$$
Then, for k odd we have $\widehat{\varphi_c}(k) = 0$. Thus, we cannot apply our theory if the minimal period is not fixed.

(2) Another fact that necessitates special mention is the condition that the solution φ_c needs to be a positive solution. In fact, if φ_c has at least one zero on the interval $[0, 2L]$, then the theory cannot be applied. Suppose without loss of generality that $\varphi_c(0) = 0$. Then since φ_c is smooth, it follows that
$$\varphi_c(0) = \sum_{n=-\infty}^{+\infty} \widehat{\varphi_c}(n) = 0.$$
In other words a Fourier coefficient of φ_c must be negative and Theorem 11.24 cannot be applied.

11.5.1. Stability of Positive Cnoidal Waves for the KdV Equation. Next we establish a stability theory of positive periodic travelling wave solutions for the KdV equation based on the Poisson Summation Theorem and Theorem 11.24.

(1) *Existence of positive cnoidal waves.* We start by obtaining our cnoidal waves via the Poisson Summation Theorem (see Section 3.4). We consider initially the solitary wave solutions $\phi_\omega(x) = 3\omega \text{sech}^2\left(\frac{\sqrt{\omega}}{2}x\right)$ associated with the equation
$$\phi_\omega'' + \frac{1}{2}\phi_\omega^2 - c\phi_\omega = 0.$$
So, since the Fourier transform of ϕ_ω is given by
$$\widehat{\phi}_\omega(x) = \frac{12\pi x}{\sinh(\frac{\pi x}{\sqrt{\omega}})},$$
then from the Poisson Summation Theorem we obtain the following periodic function of minimal period L:
$$\psi_\omega(\xi) = \frac{12\pi}{L^2}\sum_{n=-\infty}^{\infty} n\text{cosech}\Big(\frac{\pi n}{\sqrt{\omega}L}\Big) e^{\frac{2\pi i n\xi}{L}}$$
$$= \frac{12\sqrt{\omega}}{L} + \frac{12\pi}{L^2}\sum_{n\neq 0} \text{cosech}\Big(\frac{\pi n}{\sqrt{\omega}L}\Big) e^{\frac{2\pi i n\xi}{L}}.$$
Since $\omega > 0$ is arbitrary, consider $\omega := \omega(k)$ such that
$$\sqrt{\omega(k)} = \frac{K(k)}{K(k')L},$$

$k \in (0,1)$ and $k'^2 = 1 - k^2$. Then, we obtain

$$\psi_{\omega(k)}(\xi) = 12\frac{\sqrt{\omega}}{L} + \frac{12\pi}{L^2}\sum_{n\neq 0} n\text{cosech}\Big(\frac{\pi n K'}{K}\Big)e^{\frac{2\pi i n\xi}{L}}$$

$$= \frac{12\sqrt{\omega}}{L} + \frac{24\pi}{L^2}\sum_{n=1}^{+\infty} n\text{cosech}\Big(\frac{\pi n K'}{K}\Big)\cos\Big(\frac{2\pi n\xi}{L}\Big).$$

Now, we consider the Fourier expansion of dn^2 (see [**75**]), that is,

$$K^2\left(\text{dn}^2\left(\frac{2K\xi}{L};k\right) - \frac{E}{K}\right) = 2\pi\sum_{n=1}^{+\infty}\frac{nq^n}{1-q^{2n}}\cos\left(\frac{2\pi n\xi}{L}\right),$$

where $q = e^{-\left(\frac{\pi K'}{K}\right)}$. We can conclude that

$$\frac{q^n}{1-q^{2n}} = \frac{1}{2}\text{cosech}\Big(\frac{n\pi K'}{K}\Big).$$

Then, we get from above that for $k \in (0,1)$

$$\psi_{\omega(k)}(\xi) = \frac{12\sqrt{\omega(k)}}{L} + \frac{48K^2}{L^2}\Big(\text{dn}^2\Big(\frac{2K\xi}{L};k\Big) - \frac{E}{K}\Big). \tag{11.62}$$

Next, because of equality (11.62), we consider

$$\varphi_c(\xi) = a + b\Big(\text{dn}^2(dx;k) - \frac{E}{K}\Big),$$

a periodic travelling wave solution for the KdV equation of period L. Then, the following nonlinear system is obtained:

$$\begin{cases} \dfrac{b^2}{2} - 6d^2 b = 0, \quad 4bd^2(1+k'^2) + ab - b^2\dfrac{E}{K} - cb = 0, \\ \dfrac{a^2}{2} - \dfrac{abE}{K} + \dfrac{b^2}{2}\Big(\dfrac{E}{K}\Big)^2 - ac - cb\dfrac{E}{K} - 2bd^2k'^2 = 0. \end{cases}$$

Since φ_c is periodic of period L, it follows that $d = \dfrac{2K(k)}{L}$. Then, from the first equation of the system above we have that $b = \dfrac{48K^2}{L^2}$. Substituting those values into the second equation, we get

$$c = \frac{16K}{L^2}\Big[(1+k'^2)K - 3E\Big] + a. \tag{11.63}$$

Next we obtain the value of a. From the third equation above and the value of c in (11.63) we have the quadratic equation in terms of a:

$$a^2 + \frac{32K^2}{L^2}\Big[(1+k'^2)K - 3E\Big]a - \frac{(1+k'^2)1536K^3E}{L^4} + \frac{768K^4k'^2}{L^4} + \frac{2304K^2E^2}{L^4} = 0,$$

whose positive solution is

$$a = -\frac{16K}{L^2}\Big[(1+k'^2)K - 3E\Big] + \frac{16K^2}{L^2}\sqrt{1-k^2+k^4}.$$

Thus, the value of c is

$$c = \frac{16K^2}{L^2}\sqrt{1-k^2+k^4}.$$

Hence for $k \in (0,1)$ we have that $c \in \left(\frac{4\pi^2}{L^2}, +\infty\right)$. Therefore, writing φ_c in a convenient form, in terms of cn^2, we obtain

$$\varphi_c(\xi) = \frac{16K^2}{L^2}\left[\sqrt{1-k^2+k^4}+1-2k^2\right] + \frac{48K^2k^2}{L^2}\mathrm{cn}^2\left(\frac{2K}{L}\xi; k\right).$$

We can see that this formula is the same as the one that was obtained by Angulo in [**27**] and it can be rewritten as

$$\varphi_c(\xi) = \beta_2 + (\beta_3 - \beta_2)\mathrm{cn}^2\left(\sqrt{\frac{\beta_3-\beta_1}{12}}\xi; k\right), \tag{11.64}$$

where

$$\beta_2 = \frac{16K^2}{L^2}\left[\sqrt{1-k^2+k^4}+1-2k^2\right],$$
$$\beta_3 = \frac{16K^2}{L^2}\left[\sqrt{1-k^2+k^4}+1+k^2\right],$$

and β_1 is such that $\beta_3 - \beta_1 = 48K^2/L^2$.

(2) *Existence of a smooth curve of positive cnoidal waves.* Using the Implicit Function Theorem, we can obtain a smooth curve of positive cnoidal waves of the form given in (11.64), i.e.,

$$c \in \left(\frac{4\pi^2}{L^2}, +\infty\right) \to \varphi_c \in H^n_{\mathrm{per}}([0,L])$$

such that $k := k(c)$ is a strictly increasing smooth function (see Angulo [**27**]). Moreover, we can determine that for $k \in (0,1)$ there is a unique $c \in (4\pi^2/L^2, +\infty)$ such that $k(c) = k$. Therefore the function $\omega(k)$ defined above can be expressed as a function of c, $\omega = \omega(k(c))$, and is a strictly increasing function. Then, since

$$K(k)/K(k') \in (0, +\infty),$$

it follows that for $c \in (4\pi^2/L^2, +\infty)$ we obtain $\omega(k(c)) \in (0, +\infty)$. Therefore the mapping

$$c \in \left(\frac{4\pi^2}{L^2}, +\infty\right) \to \psi_{\omega(k(c))} \in H^n_{\mathrm{per}}([0,L])$$

is a smooth curve.

(3) *Fourier coefficient of φ_c.* Now we obtain the Fourier coefficient of φ_c in (11.64) which we will use below. We note that

$$\psi_{\omega(k(c))} - \varphi_c = \frac{12\sqrt{\omega(k(c))}}{L} - a(k(c)),$$

where for $k = k(c)$

$$a(k) = \frac{16K^2}{L^2}\left[\sqrt{1-k^2+k^4}+2-k^2+3\frac{E}{K}\right].$$

Thus, for

$$s(k(c)) \equiv a(k(c)) - \frac{12\sqrt{\omega(k(c))}}{L},$$

we can write $\varphi_c(\xi) \equiv s(k(c)) + \psi_{\omega(k(c))}(\xi)$. Therefore, we immediately obtain that the Fourier coefficients of φ_c are

$$\widehat{\varphi_c}(n) = \begin{cases} a(k), & n = 0, \\ \dfrac{12\pi}{L^2} n \operatorname{cosech}\Big(\dfrac{\pi n}{\sqrt{\omega(k)}L}\Big), & n \neq 0. \end{cases} \tag{11.65}$$

REMARK 11.30. After some calculations, it is possible to see that $s(k)$ is a positive function defined on $(0,1)$, and using Maple's software, we see that $s(k)$ does not have any roots on the extremes of the interval $(0,1)$. Also, it is possible to determine that the function $a(k)$ is a positive strictly increasing function.

(4) *Spectral properties.* Next we determine the properties (P_1) and (P_2) for the linear operator

$$\mathcal{L}_1 = -\frac{d^2}{dx^2} + c - \varphi_c,$$

where φ_c is the positive cnoidal wave obtained above. To obtain this information, we need to verify the conditions in Theorem 8.14 with $p = 1$. Initially it follows from (11.65) that the Fourier transform of φ_c is strictly positive. Now that $\widehat{\varphi_c} \in PF(2)$, we proceed in the following way. Since $s(k) > 0$ and the function $f : \mathbb{R} \to \mathbb{R}$ defined by

$$f(x) = \frac{12\pi}{L^2} x \operatorname{cosech}\Big(\frac{\pi x}{\sqrt{\omega}L}\Big)$$

belongs to $PF(2)$ in the continuous case (see Theorem 7.16 in Chapter 7), we will obtain that $\widehat{\varphi_c}$ belongs to $PF(2)$ in the discrete case. Indeed, since

$$a(k) > \frac{12\sqrt{\omega(k)}}{L} = f(0) > f(x), \qquad x \neq 0,$$

we can redefine f as a smooth function $h : \mathbb{R} \to \mathbb{R}$ such that $h(0) = a(k)$ and $h(x) \equiv f(x)$ on $(-\infty, -1] \cup [1, +\infty)$. On the interval $(-1, 1)$ we "complete" f in a differentiable way, such that h belongs to $PF(2)$ continuous. Therefore, the sequence to be obtained (if we look only at the set of integers) will be $h(n) = \widehat{\varphi_c}(n)$. Therefore we have that $\mathcal{L}_1$ has exactly one negative eigenvalue and it is simple, zero is a simple eigenvalue with eigenfunction φ_c', and the rest of the spectrum is an enumerable (infinite) set of eigenvalues converging to $+\infty$.

(5) *Convexity property.* Next we see that

$$I = \frac{d}{dc} \int_0^L \varphi_c^2(\xi)\, d\xi > 0.$$

In looking at this point, we will see the power of using the Poisson Summation Theorem in our approach. In fact, according to the Parseval Theorem, it follows that $I = L\frac{d}{dc}\|\widehat{\varphi_c}\|^2_{\ell^2}$. Hence,

$$\begin{aligned} \frac{d}{dc}\|\widehat{\varphi_c}\|^2_{\ell^2} &= C_1 a(k) \frac{da}{dk}\frac{dk}{dc} \\ &\quad + \frac{C_2}{\sqrt{\omega(k)^3}} \frac{d\omega}{dk}\frac{dk}{dc} \sum_{\substack{n=-\infty \\ n\neq 0}}^{+\infty} \underbrace{n^3 \operatorname{cosech}^2\Big(\frac{\pi n}{L\sqrt{\omega(k)}}\Big) \coth\Big(\frac{\pi n}{L\sqrt{\omega(k)}}\Big)}_{b_n}, \end{aligned}$$

where $C_1 := C_1(L)$ and $C_2 := C_2(L)$ are positive constants. Next, we need only to show that the quantities $\dfrac{da}{dk}$ and $\dfrac{d\omega}{dk}$ are positive because $k := k(c)$ is a strictly increasing function and $(b_n)_{n\in\mathbb{Z}}$ is obviously a positive sequence. Hence, we have

$$\frac{d\omega}{dk} = 2\frac{K(\frac{dK}{dk}K' - K\frac{dK'}{dk})}{K'^3}.$$

Since $\frac{dK}{dk} > 0$ and $\frac{dK'}{dk} < 0$, we get that $\frac{d\omega}{dk} > 0$. By making use of a similar argument, we can also show that $\frac{da}{dk} > 0$ because for

$$a(k) = \frac{16K^2(k)}{L^2}\left[\sqrt{1-k^2+k^4} + 2 - k^2\right] + 48\frac{E(k)K(k)}{L^2},$$

we have that

$$p(k) = \frac{16K^2(k)}{L^2}\left[\sqrt{1-k^2+k^4} + 2 - k^2\right]$$

and $q(k) = 48\frac{E(k)K(k)}{L^2}$ are positive and strictly increasing functions. Therefore $I > 0$ and the positive cnoidal waves φ_c, for every $c \in (4\pi^2/L^2, +\infty)$, are stable in $H^1_{\text{per}}([0,L])$ by the periodic flow of the KdV equation.

11.5.2. Stability of Periodic Waves for the BO Equation. The Poisson Summation Theorem and Theorem 11.24 can be applied with success to obtain the nonlinear stability of the periodic travelling waves in (3.53), with regard to the periodic flow of the Benjamin-Ono equation (2.6) ([**37**]). We note that in this case Definition 11.23 needs to be changed to the following one:

DEFINITION 11.31. We say that a sequence $\alpha = (\alpha_n)_{n\in\mathbb{Z}} \subseteq \mathbb{R}$ is in the class $PF(2)$ discrete if

(1) $\alpha_n > 0$, for all $n \in \mathbb{Z}$,
(2) $\alpha_{n_1-m_1}\alpha_{n_2-m_2} - \alpha_{n_1-m_2}\alpha_{n_2-m_1} \geqq 0$ for $n_1 < n_2$ and $m_1 < m_2$,
(3) strict inequality holds in (2) whenever the intervals (n_1, n_2) and (m_1, m_2) intersect.

From Section 3.4, we have that the Fourier coefficients of ϕ_c in (3.53) are given by

$$\widehat{\phi_c}(n) = \frac{2\pi}{L}e^{-\gamma|n|}, \quad n \in \mathbb{Z}.$$

Now, from Section 7.5 we obtain that the function $f(x) = e^{-\gamma|x|}$ belongs to $PF(2)$ class in the continuous case, and so $\{\widehat{\phi_c}(n)\}$ is in $PF(2)$ discrete in the sense of Definition 11.31.

For the calculus of $\frac{d}{dc}||\phi_c||$, it follows from the Parseval Theorem that

$$\begin{aligned} I = -\frac{1}{2}\frac{d}{dc}||\phi_c||^2_{L^2_{2L}} &= -\frac{1}{2}\frac{d}{dc}||\widehat{\phi_c}||^2_{\ell^2}\cdot 2L = -\frac{1}{2}\frac{d}{dc}\left(\frac{4\pi^2}{L^2}\sum_{n=-\infty}^{\infty} e^{-2\gamma|n|}\right)\cdot 2L \\ &= -\frac{4\pi^3}{c^2L^3}\left(\frac{1}{1-\left(\frac{\pi}{cL}\right)^2}\right)\left(\sum_{n=-\infty}^{\infty}|n|e^{-2\gamma|n|}\right)\cdot 2L. \end{aligned}$$

Therefore, the orbit generated by the periodic traveling wave ϕ_c is stable by the periodic flow of the BO equation provided $c > \frac{\pi}{L}$.

11.5.3. Stability of Constant Solutions for the BO Equation. To complete the investigation about periodic travelling wave solutions for the BO equation, we will study the stability of the constant solutions. Hence, if $\varphi_c(x) \equiv \tau$ is a constant solution to the BO equation, we have $\varphi_c \equiv 2c$ and $\varphi_c \equiv 0$. Here we consider only the first case.

THEOREM 11.32. *Let $L > 0$ and $c > 0$ be given. Let $\psi_0 \equiv 2c$ be a nontrivial constant solution of equation (2.6). Then ψ_0 is stable in $H^{\frac{1}{2}}_{\text{per}}([-L, L])$ provided $c < \frac{\pi}{L}$.*

The proof of this theorem follows the same ideas established in Section 11.3 using this time the following nonlocal Poincaré-Wirtinger-type inequality.

THEOREM 11.33. *Suppose $f \in H^{\frac{1}{2}}_{\text{per}}([-L, L])$ such that $\int_{-L}^{L} f(x)dx = 0$; then for $D = \mathcal{H}\partial_x$*

$$\int_{-L}^{L} |D^{\frac{1}{2}} f(x)|^2 dx \geq \frac{\pi}{L} \int_{-L}^{L} |f(x)|^2 dx.$$

PROOF. Let f be in $\mathcal{P}$ and consider the Fourier expansion of f given by

$$f(x) = \sum_{n=-\infty}^{+\infty} a_n e^{\frac{in\pi x}{L}} = \sum_{n=-\infty}^{+\infty} \widehat{f}(n) e^{\frac{in\pi x}{L}},$$

where $\widehat{f}(n) = \dfrac{1}{2L} \displaystyle\int_{-L}^{L} f(x) e^{-\frac{in\pi x}{L}} dx$, with $\widehat{f}(0) = 0$. Then,

$$\mathcal{H}\partial_x f(x) = \frac{\pi}{L} \sum_{n=-\infty}^{+\infty} a_n |n| e^{\frac{in\pi x}{L}}.$$

From the Parseval Theorem we obtain

$$\int_{-L}^{L} f(x)\mathcal{H}\partial_x f(x)dx = 2L \sum_{n=-\infty}^{+\infty} \frac{\pi}{L} |a_n|^2 |n| = 2L \sum_{n\neq 0} \frac{\pi}{L} |a_n|^2 |n|$$

and

$$\int_{-L}^{L} f^2(x)dx = 2L \sum_{n\neq 0} |a_n|^2.$$

Therefore,

$$\int_{-L}^{L} f(x)\mathcal{H}\partial_x f(x)dx = 2L \sum_{n\neq 0} \frac{\pi}{L} |a_n|^2 |n| \geq \frac{\pi}{L} \int_{-L}^{L} f^2(x)dx.$$

So, using density arguments, we can show that the inequality above occurs for $f \in H^{\frac{1}{2}}_{\text{per}}([-L, L])$. This completes the proof of the inequality. □

11.5.4. Stability and Instability of Periodic Travelling Waves for the Critical KdV Equation. Using the quadrature method, in Example 3.15 we showed the existence of a family of periodic travelling wave solutions associated to the critical KdV equation

$$u_t + u_{xxx} + (u^5)_x = 0. \tag{11.66}$$

More precisely, there is a smooth curve $c \in \left(\frac{\pi^2}{L^2}, +\infty\right) \to \varphi_c$, where

$$\varphi_c(z) = \frac{\sqrt{\eta_3}\,\mathrm{dn}(\frac{4}{3g}z; k)}{\sqrt{1 + \beta^2 \mathrm{sn}^2(\frac{4}{3g}z; k)}}$$

with η_3, g, and k smooth functions depending on c.

It was shown in [**38**] that there is a unique (threshold) value of the velocity c such that this value separates two different global scenarios of the evolution of a localized initial perturbation of φ_c, with respect to the periodic flow generated by (11.66). More precisely, we have that

$$\text{(11.67)} \qquad \begin{aligned} &(1) \quad \varphi_c \text{ is orbitally stable for } c \in \left(\frac{\pi^2}{L^2}, \frac{r(k_0)}{L^2}\right), \\ &(2) \quad \varphi_c \text{ is orbitally unstable for } c \in \left(\frac{r(k_0)}{L^2}, +\infty\right) \end{aligned}$$

where $r(k) = 4K^2(k)\sqrt{k^4 - k^2 + 1}$ and $k_0 \approx 0.3823174965$.

The stability result in (11.67) is deduced from the approach described in Section 11.3 applied to the operator

$$\text{(11.68)} \qquad \mathcal{L}_{\mathrm{CKdV}} = -\frac{d^2}{dx^2} + c - 5\varphi_c^4.$$

The instability result is based on an extension to the periodic setting of arguments found in Bona, Souganidis, and Strauss [**64**]. We recall that this last approach gives the instability of the solitary wave solutions (3.4) by the flow of the GKdV equation when $p > 4$, because

$$\frac{d}{dc}\int_{-\infty}^{\infty} \phi_c^2(x)dx < 0$$

for those powers. Since for $p = 4$ in (3.4) the associated solitary wave ϕ_c satisfies

$$\text{(11.69)} \qquad \int_{-\infty}^{\infty} |\phi_c(\xi)|^2\, d\xi = \int_{-\infty}^{\infty} |\phi_1(\xi)|^2\, d\xi \qquad \text{for every } c > 0,$$

we have that $\frac{d}{dc}\int_{-\infty}^{\infty} \phi_c^2(x)dx = 0$ and so the approach in [**64**] cannot be used in this case (see [**214**] and [**220**] for instability results in this power). In [**38**] it is shown that the condition

$$\text{(11.70)} \qquad \frac{d}{dc}\int_0^L \varphi_c^2(x)dx < 0$$

still implies instability consequences in the case of periodic traveling wave solutions.

11.5.5. Stability and Instability of Periodic Travelling Waves for the Critical NLS Equation. Next we consider the periodic critical nonlinear Schrödinger equation (CNLS),

$$\text{(11.71)} \qquad iv_t + v_{xx} + |v|^4 v = 0$$

with $v = v(x,t) \in \mathbb{C}$, $x \in [0, L]$, and $t \in \mathbb{R}$. We start by considering solutions of the form

$$v_\omega(x,t) = e^{i\omega t}\varphi_\omega(x),$$

with φ_ω a real L-periodic function. If we substitute this kind of solution in (11.71), we obtain the ordinary differential equation

$$\text{(11.72)} \qquad \varphi_\omega'' + \varphi_\omega^5 - \omega\varphi_\omega = 0,$$

which is the same as the differential equation that determines the periodic travelling wave solution for the critical KdV equation.

An immediate consequence of the CKdV stability approach developed in the last subsection is the stability/instability result associated to the branch $\omega \to \varphi_\omega$. Indeed, we can apply the theory in [**124**] directly since (11.71) has the Hamiltonian structure

$$\frac{d\mathbf{v}(t)}{dt} = JG'(\mathbf{v}(t)), \tag{11.73}$$

where $\mathbf{v} = (\operatorname{Re} v, \operatorname{Im} v)^t$,

$$J = \begin{pmatrix} 0 & 1 \\ -1 & 0 \end{pmatrix}$$

is skew-symmetric, one-to-one, and onto, and

$$G(\mathbf{v}) = \int_0^L \left[\frac{1}{2}|\mathbf{v}'|^2 - \frac{1}{6}|\mathbf{v}|^6\right] dx \tag{11.74}$$

is a conservation law for (11.71). Moreover, the spectral information required in [**124**] for the linear matrix operator

$$\mathcal{L}_{\text{CNLS}} = \begin{pmatrix} -\frac{d^2}{dx^2} + \omega - 5\varphi_\omega^4 & 0 \\ 0 & -\frac{d^2}{dx^2} + \omega - \varphi_\omega^4 \end{pmatrix} \tag{11.75}$$

is deduced from that for (11.68) and from the property that the linear operator

$$\mathcal{L}_I := -\frac{d^2}{dx^2} + \omega - \varphi_\omega^4$$

is a nonnegative operator with zero a simple eigenvalue.

The notion of stability in the case of nonlinear Schrödinger equations is different from that for KdV-type equations because we are working with two symmetries: translations and rotations. So, by defining the orbit generated by φ_ω as

$$\mathcal{O}_{\varphi_\omega} = \{e^{i\theta}\varphi_\omega(\cdot + y) : \quad (y, \theta) \in \mathbb{R} \times [0, 2\pi)\}, \tag{11.76}$$

we need to use the stability Definition 5.4.

We recall that in the case of the ground state solutions associated to (11.71) (φ_ω is positive with $\lim_{|x|\to\infty} \varphi_\omega(x) = 0$), the well-known instability result of these solutions was obtained via blow-up mechanics (see Berestycki and Cazenave [**51**] and Weinstein [**272**]).

Next we have the following result whose proof comes immediately from the theory for the CKdV equation presented above and from the general theory of Grillakis *et al.* in [**123**] and [**124**]. We start with the local well-posedness theory associated with (11.71) (see [**65**]).

THEOREM 11.34. *Let $s \geq 1$. For each $u_0 \in H^s_{\text{per}}$ there is a $T = T(\|u_0\|_{H^1_{\text{per}}})$ and a unique solution $u \in C([0,T]; H^s_{\text{per}})$ of the CNLS equation* (11.71). *Moreover, the correspondence $u_0 \mapsto u$ is a continuous function between the approriate spaces.*

If $s = 1$ and $\|u_0\|_{L^2_{\text{per}}} << 1$, then T can be chosen as $T = +\infty$.

THEOREM 11.35. *The orbit $\mathcal{O}_{\varphi_\omega}$ is stable in $H^1_{\text{per}}([0, L])$ by the flow of the critical nonlinear Schrödinger equation for*

$$\omega \in \left(\frac{\pi^2}{L^2}, \frac{r(k_0)}{L^2}\right)$$

and it is unstable for

$$\omega \in \left(\frac{r(k_0)}{L^2}, +\infty \right),$$

where $r(k) \equiv 4K^2(k)\sqrt{k^4 - k^2 + 1}$ *and* $k_0 \approx 0.3823174965$.

COROLLARY 11.36. For $v_0 \in H^1_{\rm per}([0, L])$ sufficiently close to the orbit $\mathcal{O}_{\varphi_\omega}$ with

$$\omega \in \left(\frac{\pi^2}{L^2}, \frac{r(k_0)}{L^2} \right),$$

we see that the solution $v = v(x,t)$ of (11.71) with $v(x,0) = v_0$ belongs to the class $C_b(\mathbb{R}; H^1_{\rm per})$.

11.6. Comments

The stability result in $H^1_{\rm per}([0, L])$ of the orbit $\Omega_{\phi_c} \equiv \{\phi_c(\cdot + s)\}_{s\in\mathbb{R}}$, generated by the cnoidal wave ϕ_c in (11.4) and studied in [**28**], under the flow of the periodic KdV equation, was obtained for initial disturbances of ϕ_c having the same period L. It is a conjecture going back to Benjamin [**45**] that cnoidal waves of minimal period L are unstable for perturbations, for example, of period $2L$. He provided some ideas that point to the instability of the cnoidal waves with respect to this class of perturbations. Currently, nonrigorous results have been found, and the problem remains open. Similar problems for the periodic travelling wave solutions (3.31) and (3.53), associated to the mKdV and BO equations, respectively, are also open. Some evidence in favor of this scenario was recently shown by Angulo in [**25**] in the case of the NLS equation

$$iu_t + u_{xx} + |u|^2 u = 0.$$

More precisely, it was found that the orbit $\{e^{i\theta}\phi_c : \theta \in \mathbb{R}\}$ generated by the periodic travelling wave solutions of the form $u(x,t) = e^{ict}\phi_c(x)$, with ϕ_c having the dnoidal profile (3.31) and with a prescribed minimal period $L > 0$, is $H^1_{\rm per}([0, 2L])$-unstable with regard to the periodic flow of the NLS equation.

The next natural question arising in our exposition is to decide if the travelling waves $u_i(x,t) = e^{ict}\phi_{c,i}(x)$, $i = 1, 2$, determined by the cnoidal waves $\phi_{c,i}$ in Example 3.14, are stable in $H^1_{\rm per}([0, L])$ for the mKdV or NLS equations. From the following two lemmas (see [**25**]), it follows immediately that the abstract theories established by Grillakis in [**122**] and Grillakis, Shatah, and Strauss in [**124**] do not give any information about the stability or instability problems. For example in the case of the NLS we have the following.

LEMMA 11.37. *Let* $L > 0$ *and let* $\phi_{c,i}$, $i = 1, 2$, *be the cnoidal wave solutions given by Example 3.14. Then the linear operators*

$$\mathcal{L}_i = -\frac{d^2}{dx^2} + c - 3\phi_{c,i}^2, \quad i = 1, 2,$$

defined on $H^2_{\rm per}([0, L])$, *have exactly two negative eigenvalues which are simple. The eigenvalue zero is the third one, which is simple with eigenfunction* $\phi'_{c,i}$. *Moreover, the remainder of the spectrum is constituted by a discrete set of eigenvalues.*

LEMMA 11.38. *Let $\phi_{c,i}$, $i = 1, 2$, be the cnoidal wave solutions given by Example 3.14. Then the linear operators*

$$\mathcal{T}_i = -\frac{d^2}{dx^2} + c - \phi_{c,i}^2, \quad i = 1, 2,$$

defined on $H^2_{\text{per}}([0, L])$, have exactly one negative eigenvalue which is simple, the eigenvalue zero is also simple, and the remainder of the spectrum is constituted by a discrete set of eigenvalues converging to infinity.

Then we have the following arguments:

(1) It follows from [**124**] that one of the principal pieces of information for obtaining an instability result for the solutions $u_i(x, t) = e^{ict}\phi_{c,i}(x)$ is based on accounting for the number of the negative eigenvalues associated to the linear operator (the linearized Hamiltonian)

$$H_{c,i} = \begin{pmatrix} \mathcal{L}_i & 0 \\ 0 & \mathcal{T}_i \end{pmatrix}$$

and on the sign of the quantity

$$d(c) = \frac{d}{dc}\|\phi_{c,i}\|^2.$$

In fact, if $\text{n}(H_{c,i})$ denotes the number of negative eigenvalues of $H_{c,i}$ and $\text{sgn}(d)$ denotes the sign of d, it follows from the Instability Theorem in [**124**] that the orbit generated by u_i will be unstable if

$$\text{n}(H_{c,i}) - \text{sgn}(d) \text{ is odd.}$$

In our case this condition is not satisfied. Indeed, from the two lemmas above we have that for every $i = 1, 2$, $\text{n}(H_{c,i}) = 3$. Now, from the explicit form of $\phi_{c,i}$ in the Example 3.14 we obtain that

$$\begin{aligned}\int_0^L \phi_{c,i}^2(\xi)\, d\xi &= \frac{32K(k_i)}{L}\Big[E(k_i) - (1 - k_i^2)K(k_i)\Big] \\ &= \frac{32}{L}k_i(1 - k_i^2)K(k_i)\frac{dK(k_i)}{dk_i}.\end{aligned}$$

Hence, since $k_i'(c) > 0$ for each $i = 1, 2$, it follows after some calculations that

$$\frac{d}{dc}\int_0^L \phi_{c,i}^2(\xi)\, d\xi = \frac{32}{L}k_i'(c)\Big[k_i(1 - k_i^2)\Big(\frac{dK}{dk_i}\Big)^2 + k_iK^2\Big] > 0.$$

So, $\text{n}(H_{c,i}) - \text{sgn}(d) = 2$.

(2) It follows from [**122**] that to obtain a nonlinear instability result for the orbit

$$\Omega_{\phi_{c,i}} = \{e^{iy}\phi_{c,i} : y \in \mathbb{R}\},$$

it is sufficient to show that the linearization of the NLS equation around the orbit $\Omega_{\phi_{c,i}}$ has the zero solution unstable. So, it is necessary to study the number of real eigenvalues of the linear operators

$$A_i = \begin{pmatrix} 0 & \mathcal{T}_i \\ -\mathcal{L}_i & 0 \end{pmatrix}.$$

The main theorems in [**122**] (Theorems 1.2 and 1.3) are based on the linear operators $R_i \equiv P_i\mathcal{L}_iP_i$, where P_i is the orthogonal projection on $[\ker \mathcal{T}_i]^{\perp}$.

More precisely, if $I(A_i)$ = *the number of pairs of nonzero real eigenvalues of* A_i, then we have

(a) if $|n(R_i) - n(\mathcal{T}_i)| = m > 0$, then $I(A_i) \geqq m$,

(b) if $n(R_i) = n(\mathcal{T}_i)$ and

$$\{f \in L^2_{\text{per}} : (R_i f, f) < 0\} \cap \{f \in L^2_{\text{per}} : (\mathcal{T}_i^{-1} f, f) < 0\} = \emptyset,$$

then $I(A_i) \geqq 1$.

We will see that the conditions in (a) and (b) are not satisfied in the case of $\phi_{c,i}$. In fact, since $d''(c) > 0$, we obtain that

$$\mathrm{n}(R_i) = \mathrm{n}(\mathcal{L}_i) - 1$$

and so by the lemmas above, we have that $\mathrm{n}(R_i) = 1 = \mathrm{n}(\mathcal{T}_i)$. Next, since $\chi_{0,1}(x) = \mathrm{dn}(\gamma x) \in [\ker \mathcal{T}_1]^\perp$ ($\gamma = \sqrt{b^2 - c}$) and $b^2 > 2c$, it follows that

$$\begin{aligned}(R_1\chi_{0,1}, \chi_{0,1}) &= (\mathcal{L}_1\chi_{0,1}, \chi_{0,1}) \\ &= -2b^2 \int_0^L \mathrm{cn}^2(\gamma x)\mathrm{dn}^2(\gamma x)dx + \left(c - \frac{b^2}{2}\right) \int_0^L \mathrm{dn}^2(\gamma x)dx < 0.\end{aligned}$$

Moreover, since $\mathcal{T}_1\chi_{0,1} = \lambda_{0,1}\chi_{0,1}$ with $\lambda_{0,1} < 0$, it follows that

$$(\mathcal{T}_1^{-1}\chi_{0,1}, \chi_{0,1}) < 0.$$

So the criterion in (b) cannot be applied in the case of the solution u_1. Similarly, we can obtain the same conclusion in the case of u_2.

The stability theories established in this chapter have been developed in the space H^1_{per} or $H^{\frac{1}{2}}_{\text{per}}$, which arise naturally in the context of two invariants associated to the equation flow. Better control of high frequency components than would be afforded by being small in these Sobolev norms can be obtained if one works in the higher-order classes H^k_{per}, $k = 2, 3, \ldots$. An analysis presented in favor of obtaining stability in these smaller spaces has been obtained in [**12**], considering the solitary wave solutions for the KdV equation. In the case of the cnoidal wave solutions found in Section 11.2, we have the following result (see [**28**]).

THEOREM 11.39. *The cnoidal waves in Theorem 11.4 are stable in* $H^k_{\text{per}}([0, L])$ *for any integer* $k \geqq 1$. *More prescisely, let* $L > 0$ *be fixed and let* $\{\phi_c\}$ *be any branch of cnoidal waves that are stable in* $H^1_{\text{per}}([0, L])$ *as asserted in Theorem 11.8. Then, for each speed* c *and* $\epsilon > 0$, *there is a* $\delta(\epsilon, c) > 0$ *such that if* $\psi \in H^k_{\text{per}}([0, L])$ *and*

$$d_k(\psi, \phi_c) = \inf_{\tau \in \mathbb{R}} \|\psi - \phi_c(\cdot + \tau)\|_{H^k_{\text{per}}([0,L])} < \delta,$$

then there is a C^{k+1} *function* $r : \mathbb{R} \to \mathbb{R}$ *such that if* u *is the solution of the KdV equation starting at* ψ, *then for all* $t \in \mathbb{R}$,

$$d_k(u(t), \phi_c) \leqq \|u(t) - \phi_c(\cdot - r(t))\|_{H^k_{\text{per}}([0,L])} < \epsilon.$$

Part 6

APPENDICES

von Neumann, John (1903–1957)

Young man, in mathematics you don't understand things. You just get used to them.

Reply to Felix T. Smith who had said, "I'm afraid I don't understand the method of characteristics."— as quoted in *The Dancing Wu Li Masters: An Overview of the New Physics* (1984) by Gary Zukav, footnote on page 208.

If people do not believe that mathematics is simple, it is only because they do not realize how complicated life is.

"Archaeology of computers: Reminiscences, 1945–1947", *Communications of the ACM*, volume 15, issue 7, July 1972, special issue: *twenty-fifth anniversary of the Association for Computing Machinery*, p. 694.

Halmos, Paul Richard (1916-2006)

A good stock of examples, as large as possible, is indispensable for a thorough understanding of any concept, and when I want to learn something new, I make it my first job to build one.

APPENDIX A

Sobolev Spaces and Elliptic Functions

A.1. Introduction

In this appendix we establish the basic tool of the Fourier transform, the space of tempered distributions, the theory of Sobolev spaces $H^s(\mathbb{R}^n)$ of type $L^2(\mathbb{R})$, as well as the periodic Sobolev spaces on $\mathbb{R}$. Also, we group together some facts on the symmetric decreasing rearrangement of a function in $\mathbb{R}$ and on the Jacobian elliptic functions.

A.2. Lebesgue Space $L^p(\Omega)$

Let Ω be an open subset of $\mathbb{R}^n$. We denote by $L^p(\Omega)$ the Banach space of (classes of) measurable functions $f : \Omega \to \mathbb{R}$ (or $\mathbb{C}$) such that $\int_\Omega |f(x)|^p dx < \infty$ if $1 \leqq p < \infty$, or ess $\sup_\Omega |f| < \infty$ if $p = \infty$. We define the norm in $L^p(\Omega)$ for $1 \leqq p < \infty$ by

$$|f|_{L^p} = \Big(\int_\Omega |f(x)|^p dx \Big)^{1/p}$$

and in $L^\infty(\Omega)$ by $|f|_{L^\infty} \equiv |f|_\infty = \text{ess}\sup_\Omega |f|$. The space $L^2(\Omega)$ is a Hilbert space for the scalar product

$$\langle f, g \rangle = \int_\Omega f(x)\overline{g(x)}dx.$$

We denote $|f|_{L^2} = \|f\|$ and $|f|_{L^p} = |f|_p$.

A.3. The Fourier Transform in $L^1(\mathbb{R}^n)$

The Fourier transform of a function $f \in L^1(\mathbb{R}^n)$, denoted by $\widehat{f}$, is defined as

$$\widehat{f}(\xi) = \int_{\mathbb{R}^n} f(x) e^{-i(x\cdot\xi)}\, d\xi \tag{A.1}$$

with $\xi \in \mathbb{R}^n$ and $x \cdot \xi = x_1\xi_1 + ... + x_n\xi_n$ the inner product in $\mathbb{R}^n$.

Note that $L^1(\mathbb{R}^n)$ is the natural space where definition (A.1) makes sense. In fact, $f \to \widehat{f}$ defines a linear transformation from $L^1(\mathbb{R}^n)$ into $L^\infty(\mathbb{R}^n)$ with

$$|\widehat{f}|_\infty \leqq |f|_{L^1}$$

and $\widehat{f}$ continuous. Moreover, we have the following classical result (see Reed and Simon [**241**]).

LEMMA A.1 (Riemann-Lebesgue). *Let $f \in L^1(\mathbb{R}^n)$. Then we have that $\widehat{f}(\xi) \to 0$ as $|\xi| \to \infty$.*

In $L^1(\mathbb{R}^n)$ we have the operation of *convolution*: if $f, g \in L^1(\mathbb{R}^n)$, then

$$f * g(x) = \int_{\mathbb{R}^n} f(x-y)g(y)dy,$$

which is defined for almost everywhere (a.e. henceforth) $x \in \mathbb{R}^n$. More generally, $f * g$ is defined for $f \in L^p(\mathbb{R}^n)$, $1 \leqq p \leqq \infty$, and $g \in L^1(\mathbb{R}^n)$. Moreover the Fourier transform of the convolution of two functions is the product of their Fourier transforms; namely, we have the following results (see Reed and Simon [**242**]).

LEMMA A.2. *We have the following statements for the convolution:*

(1) (*Young's inequality*) *Let $f \in L^p(\mathbb{R}^n)$ and $g \in L^q(\mathbb{R}^n)$, $1 \leqq p, q \leqq \infty$ with $\frac{1}{p} + \frac{1}{q} \geqq 1$. Then $f * g \in L^r(\mathbb{R}^n)$ where $\frac{1}{r} = \frac{1}{p} + \frac{1}{q} - 1$. Moreover,*

$$|f * g|_{L^r} \leqq |f|_{L^p} |g|_{L^q}.$$

(2) *Let $f, g \in L^1(\mathbb{R}^n)$. Then $\widehat{f * g}(\xi) = \widehat{f}(\xi)\widehat{g}(\xi)$.*

A.4. The Fourier Transform in $L^2(\mathbb{R}^n)$

We note that for an arbitrary $f \in L^2(\mathbb{R}^n)$ definition (A.1) of the Fourier transform does not make sense. But, there is a natural extension of this transform on the Hilbert space $L^2(\mathbb{R}^n)$. In fact, initially we have the following wonderful result (see [**242**], [**245**]):

LEMMA A.3. *Let $f \in L^1(\mathbb{R}^n) \cap L^2(\mathbb{R}^n)$. Then $\widehat{f} \in L^2(\mathbb{R}^n)$ and $\|\widehat{f}\| = \|f\|$.*

Therefore, the Fourier transform defines a linear bounded operator from $L^1(\mathbb{R}^n) \cap L^2(\mathbb{R}^n)$ into $L^2(\mathbb{R}^n)$. So, since $L^1(\mathbb{R}^n) \cap L^2(\mathbb{R}^n)$ is a dense subset of $L^2(\mathbb{R}^n)$, there is a unique bounded extension $\mathcal{F}$ defined in $L^2(\mathbb{R}^n)$ (see Appendix B, Lemma B.8). We call $\mathcal{F}$ the Fourier transform in $L^2(\mathbb{R}^n)$ and we shall use the notation $\widehat{f} = \mathcal{F}(f)$ for $f \in L^2(\mathbb{R}^n)$. So, if we denote by $\{h_n\}$ the sequence in $L^1(\mathbb{R}^n) \cap L^2(\mathbb{R}^n)$ defined as $h_k = f$ for $|x| \leqq k$ which vanishes for $|x| > k$, we have that $h_k \to f$ in $L^2(\mathbb{R}^n)$ and therefore $\widehat{h_k} \to \widehat{f}$ in $L^2(\mathbb{R}^n)$. Then,

$$\widehat{f}(\xi) = L^2 - \lim_{k \to \infty} \int_{|x|<k} f(x) e^{-i(x \cdot \xi)} dx.$$

Note that as $\mathcal{F}$ is onto, then from Lemma A.3 it follows immediately that $\mathcal{F}$ is an *isometry* in $L^2(\mathbb{R}^n)$. We denote the inverse of the Fourier transform by $\mathcal{F}^{-1}$ or $f^{\vee}$. So, we have the following classical result (see Rudin [**245**]):

LEMMA A.4 (Plancherel). *The Fourier transform extends uniquely to a unitary map of $L^2(\mathbb{R}^n)$ onto $L^2(\mathbb{R}^n)$.*

A.5. Tempered Distributions

We will use the standard multi-index notation. A multi-index $\alpha = (\alpha_1, ..., \alpha_n)$ is an n-tuple of nonnegative integers. The symbols $|\alpha|$, x^α, and $\mathcal{D}^\alpha$ are defined as follows: $|\alpha| = \sum_{i=1}^n \alpha_i$, $x^\alpha = x_1^{\alpha_1} x_2^{\alpha_2} \cdots x_n^{\alpha_n}$, and

$$\mathcal{D}^\alpha = \frac{\partial^{|\alpha|}}{\partial_{x_1}^{\alpha_1} \partial_{x_2}^{\alpha_2} \cdots \partial_{x_n}^{\alpha_n}}.$$

Define for each $(\alpha, \beta) \in (\mathbb{N})^{2n}$ the *semi-norm*, $\| \cdot \|_{\alpha,\beta}$, by

$$\|f\|_{\alpha,\beta} = \|x^\alpha \mathcal{D}^\beta f\|_\infty.$$

DEFINITION A.5 (Schwartz Space). We define the Schwartz space, $\mathcal{S}(\mathbb{R}^n)$, as

$$\mathcal{S}(\mathbb{R}^n) = \{\phi \in C^\infty(\mathbb{R}^n) : \|\phi\|_{\alpha,\beta} < \infty \text{ for any } (\alpha, \beta) \in (\mathbb{N})^{2n}\}.$$

The topology in $\mathcal{S}(\mathbb{R}^n)$ is that induced by the family of semi-norms:

$$\{\|\cdot\|_{\alpha,\beta}\}_{(\alpha,\beta)\in(\mathbb{N})^{2n}}.$$

So, if $\{\phi_n\} \subseteq \mathcal{S}(\mathbb{R}^n)$, then we will have that $\phi_n \to 0$ as $n \to \infty$, if for any $(\alpha, \beta) \in (\mathbb{N})^{2n}$ one has that $\|\phi_n\|_{\alpha,\beta} \to 0$ as $n \to \infty$.

Next, from the following basic property of the Fourier transform for multi-indices α and β and $\phi \in \mathcal{S}(\mathbb{R}^n) \subseteq L^1(\mathbb{R}^n)$, i.e.,

$$[\mathcal{D}^\alpha(-ix)^\beta \phi(x)]^\wedge(\xi) = (i\xi)^\alpha \mathcal{D}^\beta \widehat{\phi}(\xi),$$

we obtain that the map $\phi \to \widehat{\phi}$ is an isomorphism from $\mathcal{S}(\mathbb{R}^n)$ into itself.

DEFINITION A.6 (Tempered Distributions). We say that a linear functional $\Psi : \mathcal{S}(\mathbb{R}^n) \to \mathbb{C}$ defines a *tempered distribution* if Ψ is continuous.

It is not difficult to see that for every $f \in L^p(\mathbb{R}^n)$ with $1 \leqq p \leqq \infty$, the linear functional

$$\Psi_f(\phi) = \int_{\mathbb{R}^n} f(x)\phi(x)dx, \qquad \phi \in \mathcal{S}(\mathbb{R}^n),$$

defines a tempered distribution (see [**241**]).

DEFINITION A.7 (Fourier Transform in $\mathcal{S}'(\mathbb{R}^n)$). For $\Psi \in \mathcal{S}'(\mathbb{R}^n)$, its Fourier transform $\widehat{\Psi} \in \mathcal{S}'(\mathbb{R}^n)$ is defined as

$$\widehat{\Psi}(\phi) = \Psi(\widehat{\phi}), \qquad \text{for any } \phi \in \mathcal{S}(\mathbb{R}^n).$$

Note that for $f \in L^1(\mathbb{R}^n)$ (or $f \in L^2(\mathbb{R}^n)$) and $\phi \in \mathcal{S}(\mathbb{R}^n)$, we have from Fubini's Theorem that

$$\widehat{\Psi_f}(\phi) = \Psi_f(\widehat{\phi}) = \int_{\mathbb{R}^n} f(x)\widehat{\phi}(x)dx = \int_{\mathbb{R}^n} \widehat{f}(x)\phi(x)dx = \Psi_{\widehat{f}}(\phi).$$

Therefore the definition of the Fourier transform in $\mathcal{S}'(\mathbb{R}^n)$ is consistent with the theory of the Fourier transform in $L^1(\mathbb{R}^n)$ or in $L^2(\mathbb{R}^n)$.

EXAMPLE A.8 (The Hilbert Transform). We have the following:

(1) In $\mathcal{S}'(\mathbb{R})$ we define the principal value of $1/x$, which we denote by p.v.$\frac{1}{x}$, by the expression

$$\text{p.v.}\frac{1}{x}(\phi) = \lim_{\epsilon \downarrow 0} \int_{\epsilon<|x|<1/\epsilon} \frac{\phi(x)}{x}\,dx \qquad \text{for any } \phi \in \mathcal{S}(\mathbb{R}).$$

Since $1/x$ is an odd function, it follows that

$$\text{p.v.}\frac{1}{x}(\phi) = \int_{|x|<1} \frac{\phi(x) - \phi(0)}{x}\,dx + \int_{|x|>1} \frac{\phi(x)}{x}\,dx.$$

So, $|\text{p.v.}\frac{1}{x}(\phi)| \leqq 2|\phi'|_\infty + 2|x\phi|_\infty$. Hence p.v.$\frac{1}{x}(\phi) \in \mathcal{S}'(\mathbb{R})$.

(2) $\widehat{\text{p.v.}\frac{1}{x}}(\xi) = -i\pi\ \text{sgn}(\xi)$.
In fact, by using Fubini's Theorem and the Lebesgue Dominated Convergence Theorem we have that for any $\phi \in \mathcal{S}(\mathbb{R})$

$$\begin{aligned} \widehat{\text{p.v.}\frac{1}{x}}(\phi) &= \text{p.v.}\frac{1}{x}(\widehat{\phi}) = \lim_{\epsilon\downarrow 0}\int_{\epsilon<|x|<1/\epsilon}\frac{1}{x}\Big(\int_{-\infty}^{\infty}\phi(y)e^{-ixy}dx\Big)dx \\ &= \int_{-\infty}^{\infty}\phi(y)\Big(\lim_{\epsilon\downarrow 0}\int_{\epsilon<|x|<1/\epsilon}\frac{e^{-ixy}}{x}dx\Big)dy = -i\pi\int_{-\infty}^{\infty}\text{sgn}(y)\phi(y)dy, \end{aligned} \tag{A.2}$$

where we are using the well-known result

$$\begin{aligned} \lim_{\epsilon\downarrow 0}\int_{\epsilon<|x|<1/\epsilon}\frac{e^{-ixy}}{x}dx &= -2i\int_0^{\infty}\frac{\sin(xy)}{x}dx \\ &= -2i\ \text{sgn}(y)\int_0^{\infty}\frac{\sin x}{x}dx = -i\pi\ \text{sgn}(y). \end{aligned}$$

Therefore, we obtain from (A.2) that $\widehat{\text{p.v.}\frac{1}{x}}(\xi) = -i\pi\ \text{sgn}(\xi)$.

(3) For $\phi \in \mathcal{S}(\mathbb{R})$ we define its *Hilbert transform*, $\mathcal{H}(\phi)$, as the function

$$\mathcal{H}(\phi)(y) = \frac{1}{\pi}\text{p.v.}\ \frac{1}{x} * \phi(y).$$

From (1) and (2) above, we have that $\mathcal{H}(\phi)(y)$ is defined for any $y \in \mathbb{R}$ and $\mathcal{H}(\phi) \in \mathcal{S}'(\mathbb{R})$.

(4) We have that $\widehat{\mathcal{H}(\phi)}(\xi) = -i\text{sgn}(\xi)\widehat{\phi}(\xi)$, for any $\phi \in \mathcal{S}(\mathbb{R})$.
In fact, by using that

$$\mathcal{H}(\phi)(y) = \lim_{\epsilon\to 0}\Big(\frac{1}{\pi}\frac{1}{x}\chi_{\{\epsilon<|x|<1/\epsilon\}} * \phi\Big)(y) \qquad \text{in } \mathcal{S}'(\mathbb{R}),$$

it follows that

$$\lim_{\epsilon\to 0}\Big(\frac{1}{\pi}\frac{1}{x}\chi_{\{\epsilon<|x|<1/\epsilon\}} * \phi\Big)^{\wedge}(\xi) = -i\text{sgn}(\xi)\widehat{\phi}(\xi).$$

Therefore, $\widehat{\mathcal{H}(\phi)}(\xi) = -i\text{sgn}(\xi)\widehat{\phi}(\xi)$ for any $\phi \in \mathcal{S}(\mathbb{R})$. Note that this last relation allows us to extend the Hilbert transform as an isometry in $L^2(\mathbb{R})$: more exactly, we can see that

$$\|\mathcal{H}\phi\| = \|\phi\| \qquad \text{and} \qquad \mathcal{H}(\mathcal{H}(\phi)) = -\phi.$$

(5) Finally, we remember that for any $\psi \in L^p(\mathbb{R})$, with $1 \leqq p < \infty$, the Hilbert transform $\mathcal{H}(\psi)(x)$ exists and is finite a.e. Moreover, the Hilbert transform operator $\mathcal{H} : \psi \to \mathcal{H}\psi$ is bounded from $L^p(\mathbb{R})$ to $L^p(\mathbb{R})$ with $1 < p < \infty$.

A.6. Sobolev Spaces

In this section we will give a brief introduction to classical Sobolev spaces $H^s(\mathbb{R}^n)$ and $H^s(\Omega)$ where Ω is a bounded C^∞ domain in $\mathbb{R}^n$. This will complete the necessary background required by this book. Sobolev spaces measure the differentiability of functions in L^2, and they are a fundamental tool in the study of partial differential equations.

DEFINITION A.9 (Sobolev Space of Type $L^2(\mathbb{R}^n)$). Let $s \in \mathbb{R}$. We define the *Sobolev space* of order s, denoted by $H^s(\mathbb{R}^n)$, as

$$H^s(\mathbb{R}^n) = \{f \in \mathcal{S}'(\mathbb{R}^n) : J^s f(\xi) \equiv [(1+|\xi|^2)^{s/2}\widehat{f}\,]^{\vee}(\xi) \in L^2(\mathbb{R}^n)\},$$

with norm $\|\cdot\|_{s,2}$ defined as

$$||f||^2_{s,2} = ||J^s f||^2 = \int_{\mathbb{R}^n} (1+|\xi|^2)^s |\widehat{f}(\xi)|^2 \, d\xi. \tag{A.3}$$

Note that for every $s \geqq 0$, $H^s(\mathbb{R}^n) \subset L^2(\mathbb{R}^n)$.

DEFINITION A.10 (Sobolev Space of Type $L^2(\Omega)$). Let $s \geqq 0$. The Sobolev space $H^s(\Omega)$ is defined by restriction of elements of $H^s(\mathbb{R}^n)$ to Ω and with norm

$$\|\varphi\|_{s,2,\Omega} = \inf\{\|\psi\|_{s,2} : \psi \text{ coincides with } \varphi \text{ in } \Omega\}. \tag{A.4}$$

If either $W = \mathbb{R}^n$ or $W = \Omega$ and we write $s = m + \sigma$ where m is an integer greater than or equal to zero and $0 < \sigma < 1$, then the following norm, $|||\cdot|||_{s,2}$, i.e.,

$$|||\varphi|||_{s,2} = \left[\|\varphi\|_{s,2} + \sum_{|\alpha|=m} \int_W \int_W \frac{|\mathcal{D}^\alpha \varphi(x) - \mathcal{D}^\alpha \varphi(y)|^2}{|x-y|^{n+2\sigma}} dxdy\right]^{1/2},$$

is an equivalent norm to those defined in (A.3) and (A.4) above.

From the definition of Sobolev spaces we deduce the following properties:

(1) If $s > r$, then $H^s(\mathbb{R}^n) \subset H^r(\mathbb{R}^n)$.
(2) The space $H^s(\mathbb{R}^n)$ is a Hilbert space with respect to the inner product $\langle \cdot, \cdot \rangle_s$ defined by

$$\langle \phi, \psi \rangle_s = \int_{\mathbb{R}^n} J^s \phi(\xi) \overline{J^s \psi(\xi)} \, d\xi.$$

(3) For any $s \in \mathbb{R}^n$, the Schwartz space $\mathcal{S}(\mathbb{R}^n)$ is dense in $H^s(\mathbb{R}^n)$.
(4) If $\psi \in H^s(\mathbb{R}^n)$, then $\frac{\partial \psi}{\partial x_i} \in H^{s-1}(\mathbb{R}^n)$. Therefore, we have that the norms $\|\psi\|_{k,2}$ and $\sum_{|\alpha| \leqq k} \|\partial_x^\alpha \psi\|$ are equivalent in $H^k(\mathbb{R}^n)$.
(5) If either $W = \mathbb{R}^n$ or $W = \Omega$, then $H^s(W)$ is continuously imbedded in $L^p(W)$ if

$$2 \leqq p < \infty \qquad \text{and} \qquad \frac{1}{p} \geqq \frac{1}{2} - \frac{s}{n}.$$

Moreover, $H^s(W)$ is continuously imbedded in $L^\infty(W)$ if $s > \frac{n}{2}$.
(6) The space $H^s(\Omega)$ is compactly imbedded in $L^q(\Omega)$ if $2 \leqq q \leqq \infty$ and $\frac{1}{q} > \frac{1}{2} - \frac{2}{n}$.
(7) If $s > \frac{n}{2} + k$, then $H^s(\mathbb{R}^n)$ is continuously immersed in $C^k_\infty(\mathbb{R}^n)$, the space of functions with k continuous derivatives which vanish at infinity.
(8) If $s > \frac{n}{2}$, then $H^s(\mathbb{R}^n)$ is an algebra with respect to the product of functions. That is, if $\phi, \psi \in H^s(\mathbb{R}^n)$, then $\phi\psi \in H^s(\mathbb{R}^n)$ with

$$\|\phi\psi\|_{s,2} \leqq c_s \|\phi\|_{s,2} \|\psi\|_{s,2}.$$

(9) The space $[H^s(\mathbb{R}^n)]'$, the dual space of $H^s(\mathbb{R}^n)$, is isometrically isomorphic to $H^{-s}(\mathbb{R}^n)$ for all $s \in \mathbb{R}$. The duality is implemented by the pairing

$$(\Psi, \phi)_s = \int_{\mathbb{R}^n} \widehat{\Psi}(\xi) \overline{\widehat{\phi}(\xi)} d\xi, \qquad \Psi \in H^{-s}(\mathbb{R}^n), \text{ for } \phi \in H^s(\mathbb{R}^n).$$

So, if $\Psi, \phi \in L^2(\mathbb{R}^n)$, it follows that $(\Psi, \phi)_s = \langle \Psi, \phi \rangle$.

EXAMPLE A.11. We have the following:

(1) If $n = 1$ and $\psi(x) = \chi_{[-1,1]}(x)$, then we have that $\widehat{\psi}(\xi) = -2\,\sin(\xi)/\xi$. This tells us that $\psi \in H^s(\mathbb{R})$ if $s < 1/2$.
(2) If $n = 1$ and $\phi(x) = \psi * \psi(x)$ with $\psi(x) = \chi_{[-1,1]}(x)$, then

$$\widehat{\phi}(\xi) = 4\,\frac{\sin^2 \xi}{\xi^2}.$$

Thus $\phi \in H^s(\mathbb{R})$ if $s < 3/2$.

A.7. Sobolev Spaces of Periodic Type

To define the Sobolev spaces (of L^2 type) of periodic functions, we need the following notions (for further information see Iorio and Iorio [**139**]). Let $\mathcal{P} = C^\infty_{\text{per}}$ denote the collection of all the functions $f : \mathbb{R} \to \mathbb{C}$ which are C^∞ and periodic with period $2\ell > 0$. The topological dual of $\mathcal{P}$ will be denoted by $\mathcal{P}'$; i.e., $\mathcal{P}'$ is the collection of all continuous linear functionals from $\mathcal{P}$ into $\mathbb{C}$. The set $\mathcal{P}'$ is called the set of all *periodic distributions*. If $\Psi \in \mathcal{P}'$, we denote the value of Ψ in φ by $\Psi(\varphi) = \langle \Psi, \varphi\rangle$. Define the functions $\Theta_k(x) = \exp(ik\pi x/\ell)$, $k \in \mathbb{Z}$, $x \in \mathbb{R}$. The Fourier transform of $\Psi \in \mathcal{P}'$ is the function $\widehat{\Psi} : \mathbb{Z} \to \mathbb{C}$ defined by the formula

$$\widehat{\Psi}(k) = \frac{1}{2l}\langle \Psi, \Theta_{-k}\rangle, \qquad k \in \mathbb{Z}.$$

So if Ψ is a periodic function with period 2ℓ and $\Psi \in L^2([-\ell,\ell])$, we have

$$\widehat{\Psi}(k) = \frac{1}{2l}\int_{-\ell}^{\ell} \Psi(x)\exp(-ik\pi x/\ell)dx.$$

DEFINITION A.12 (Periodic Sobolev Space of Type $L^2(\Omega)$). Let $s \in \mathbb{R}$. The Sobolev space $H^s_{2\ell}(\mathbb{R}) = H^s_{\text{per}}([-\ell,\ell])$ is the set of all $f \in \mathcal{P}'$ such that

$$\|f\|^2_{H^s_{2\ell}} = 2l \sum_{k=-\infty}^{\infty} (1 + |k|^2)^s |\widehat{f}(k)|^2 < \infty.$$

We have that $H^s_{2\ell}(\mathbb{R})$ is a Hilbert space with respect to the inner product

$$\langle f, g\rangle_s = 2l \sum_{k=-\infty}^{\infty} (1 + |k|^2)^s \widehat{f}(k)\overline{\widehat{g}(k)}.$$

In the case when $s = 0$, we obtain a Hilbert space that is isometrically isomorphic to a subspace of $L^2([-\ell,\ell])$, so we have $\langle f, g\rangle_0 = \langle f, g\rangle = \int_{-\ell}^{\ell} f\bar{g}dx$. We will denote $H^0_{2\ell}(\mathbb{R})$ by $L^2_{2\ell}(\mathbb{R})$ with $\|\cdot\|_{L^2_{2\ell}} = |\cdot|_{L^2(-\ell,\ell)}$. Since $H^s_{2\ell}(\mathbb{R}) \subset L^2_{2\ell}(\mathbb{R})$ for every $s \geqq 0$, we have via the Parseval identity that for every $n \in \mathbb{N}$,

$$\|f\|^2_{H^n_{2\ell}} = \sum_{j=0}^{n} \|f^{(j)}\|^2_{L^2_{2\ell}} = \sum_{j=0}^{n} \int_{-\ell}^{\ell} |f^{(j)}(x)|^2\,dx$$

where $f^{(j)}$ represents the j-th derivative of f taken in the sense of $\mathcal{P}'$. Moreover, $[H^s_{2\ell}(\mathbb{R})]'$, the topological dual of $H^s_{2\ell}(\mathbb{R})$, is isometrically isomorphic to $H^{-s}_{2\ell}(\mathbb{R})$ for all $s \in \mathbb{R}$. The duality is implemented by the pairing

$$(f, g)_{-s,s} = 2l \sum_{k=-\infty}^{\infty} \widehat{f}(k)\overline{\widehat{g}(k)}, \qquad f \in H^{-s}_{2\ell}(\mathbb{R}),\ g \in H^s_{2\ell}(\mathbb{R}).$$

So, if $f, g \in L^2_{2\ell}(\mathbb{R})$, it follows that $(f,g)_{-s,s} = \langle f, g\rangle$. We also have Sobolev's Lemma: if $s > \frac{1}{2}$ and

$$C_{\text{per}} = \{f : \mathbb{R} \to \mathbb{C} : f \text{ is continuous and periodic with period } 2\ell\},$$

then $H^s_{2\ell}(\mathbb{R}) \hookrightarrow C_{\text{per}}$. Moreover, if $s \geqq r$, then $H^s_{2\ell}(\mathbb{R})$ is continuously and densely embedded in $H^r_{2\ell}(\mathbb{R})$.

DEFINITION A.13 (The Periodic Hilbert Transform). If ϕ represents a 2ℓ-periodic function of x, the periodic Hilbert transform is defined by the Hilbert kernel as

$$\mathcal{H}\phi(x) = \frac{1}{2\ell}\text{p.v.} \int_{-\ell}^{\ell} \cot\Big[\frac{\pi(x-y)}{2\ell}\Big]\phi(y)dy,$$

where p.v. represents the Cauchy principal value of the integral.

The Fourier transform of $\mathcal{H}\phi$ is given by the sequence $\{\widehat{\mathcal{H}\phi}(k)\}_{k\in\mathbb{Z}}$, where

$$\widehat{\mathcal{H}\phi}(k) = -i\ \text{sgn}(k)\hat{\phi}(k), \quad k \in \mathbb{Z},$$

with $\text{sgn}(0) = 0$ and

$$\text{sgn}(k) = \begin{cases} -1, & k < 0, \\ 1, & k > 0. \end{cases}$$

A.8. The Symmetric Decreasing Rearrangement

In this section, we collect some facts about the symmetric decreasing rearrangement of a function in $\mathbb{R}$. Also, we establish the Riesz Lemma. This material and others in higher dimensions can be found in the papers of Brascamp, Lieb, and Luttinger [**74**] and Riesz [**244**] and in the book of Kawohl [**157**].

Let

$$S = \{f : \mathbb{R} \to [0,\infty] \mid f(x) \leqq f(y) \ \text{ if } \ |x| \geqq |y|\,\}$$

be the set of even decreasing functions. Let f be a nonnegative measurable function on $\mathbb{R}$, $N_y(f) = \{x|\, f(x) \geqq y\}$ and $M_y(f) = m(N_y(f))$, where m denotes the Lebesgue measure. Assume that $M_a(f) < \infty$ for some $a < \infty$. If f^* is another function on $\mathbb{R}$ with the same properties as f and, additionally,

$$\begin{cases} \text{(a)} & f^*(x) = f^*(-x) \qquad \text{for all } x, \\ \text{(b)} & 0 < x < y \Longrightarrow f^*(x) \leqq f^*(y), \\ \text{(c)} & M_y(f^*) = M_y(f) \qquad \text{for all } y > 0, \end{cases}$$

then f^* is called a *symmetric decreasing rearrangement* of f.

Next, we have the following:

(1) If g and h are two symmetric decreasing rearrangements of f, then

$$g(x) = h(x) \qquad \text{a.e.}$$

(2) Construction of a symmetric decreasing rearrangement: Define

$$N^*_y(f) = \begin{cases} \{x \in \mathbb{R} : |x| \leqq \frac{1}{2}m(N_y(f))\} & \text{if } N_y(f) \neq \emptyset, \\ \emptyset & \text{if } N_y(f) = \emptyset, \end{cases}$$

called the symmetric decreasing rearrangement of $N_y(f)$. Then

$$f^*(x) \equiv \sup\,\{y \in \mathbb{R} : x \in N^*_y(f)\} \tag{A.5}$$

is a symmetric decreasing rearrangement of f. Consequently, if f is continuous, then f^* can be found by using (A.5). We also have the fact that $M_a(f) < \infty$ implies that $f^* < \infty$ for all $x \neq 0$.

(3) If $f : \mathbb{R} \to \mathbb{C}$, we define

$$f^* = |f|^*.$$

Clearly $f^* \in S$, and for all y, $m(\{x|f^*(x) \geqq y\}) = m(\{x|\,|f(x)| \geqq y\})$.

In the following section we deal with integrals. By (A.1) above, f^* is unique for our purposes. This implies that

(1) for all $p \in [1, \infty]$, $|f^*|_{L^p} = |f|_{L^p}$,
(2) for all $r > 0$, $(f^r)^* = (f^*)^r$ pointwise,
(3) if $f, g \in L^2(\mathbb{R})$, then

$$\left|\int_{-\infty}^{\infty} f(x)g(x)dx\right| \leqq \int_{-\infty}^{\infty} f^*(x)g^*(x)dx.$$

We finish this section with the main result:

LEMMA A.14 (F. Riesz). *Let $f_1, ..., f_n$ be measurable functions on $\mathbb{R}$ such that they satisfy $m(\{x|f_i(x) \geqq y\}) < \infty$ for all $y > 0$ and all $1 \leqq i \leqq n$. Then*

$$|(f_1 * f_2 * ... * f_n)(0)| \leqq [(f_1^*) * (f_2^*) * ... * (f_n^*)](0)$$

in the sense that if the right-hand side is finite, then the left-hand side exists and the inequality holds.

A.9. The Jacobian Elliptic Functions

In this section, we collect some facts about the theory of Jacobian elliptic functions. We refer the reader to the books Lawden [**178**], Byrd and Friedman [**75**], and Armitage and Eberlein [**40**] as references to this theory.

We start with the definitions of the *normal elliptic integral of the first kind,*

$$\int_0^y \frac{dt}{\sqrt{(1-t^2)(1-k^2t^2)}} = \int_0^\varphi \frac{d\theta}{\sqrt{1-k^2\sin^2\theta}} \equiv F(\varphi, k),$$

where $y = \sin\varphi$, and the *normal elliptic integral of the second kind,*

$$\int_0^y \sqrt{\frac{1-k^2t^2}{1-t^2}}\, dt = \int_0^\varphi \sqrt{1-k^2\sin^2\theta}\, d\theta \equiv E(\varphi, k).$$

In their algebraic forms, these two integrals have the following properties: The first is finite for all real (or complex) values of y, including infinity. The number k is called the *modulus.* This number may take any real or imaginary value. Here we will take $0 < k < 1$. The number k' is called the *complementary modulus* and is related to k by the relation

$$k' = \sqrt{1-k^2}.$$

The variable limit y or φ is the *argument* of the normal elliptic integrals and may be of course either real or complex, but it is usually understood that $0 < y \leqq 1$ (or $-1 \leqq y \leqq 1$) and $0 < \varphi \leqq \pi/2$ (or $-\pi/2 \leqq \varphi \leqq \pi/2$) . When $y = 1$, the integrals above are said to be *complete.* In this case, one writes for $k \in (0, 1)$

$$K \equiv K(k) \equiv F(\pi/2, k) \equiv \int_0^1 \frac{dt}{\sqrt{(1-t^2)(1-k^2t^2)}} = \int_0^{\pi/2} \frac{d\theta}{\sqrt{1-k^2\sin^2\theta}}$$

and

$$E \equiv E(k) \equiv E(\pi/2, k) \equiv \int_0^1 \sqrt{\frac{1-k^2t^2}{1-t^2}}\, dt = \int_0^{\pi/2} \sqrt{1-k^2\sin^2\theta}\, d\theta.$$

Some special properties of K and E are $K(0) = E(0) = \pi/2$, $E(1) = 1$, and $K(1) = +\infty$. For $k \in (0,1)$,

$$\frac{d}{dk}K(k) > 0, \quad \frac{d^2}{dk^2}K(k) > 0, \quad \frac{d}{dk}E(k) < 0, \quad \frac{d^2}{dk^2}E(k) < 0,$$

and $E(k) < K(k)$. We define for $k \in (0,1)$,

$$K'(k) \equiv K(k').$$

So $\frac{d}{dk}K'(k) < 0$.

Using the fact that, for example, the complete elliptic integral K has the following expansion in form series, i.e.,

$$K(k) = \frac{\pi}{2}\sum_{n=0}^{\infty} \frac{(\frac{1}{2})_n(\frac{1}{2})_n}{(1)_n}\frac{k^{2n}}{n!} = \frac{\pi}{2}F(1/2, 1/2; 1; k^2)$$

where $(a)_n$ denotes the shifted factorial defined by $(a)_n = a(a+1)\cdots(a+n-1)$ for $n > 0$, $(a)_0 = 1$ and F is the hypergeometric function, we obtain that K satisfies the following *hypergeometric differential equation*:

$$kk'^2\frac{d^2}{dk^2}K(k) + (1-3k^2)\frac{d}{dk}K(k) - kK(k) = 0. \tag{A.6}$$

Similarly, E satisfies

$$kk'^2\frac{d^2}{dk^2}E(k) + k'^2\frac{d}{dk}E(k) + kE(k) = 0.$$

Moreover, K and E satisfy the differentiable relations

$$\begin{cases} \dfrac{dK}{dk} = \dfrac{E - k'^2K}{kk'^2}, & \dfrac{dE}{dk} = \dfrac{E-K}{k}, \\ \dfrac{d^2E}{dk^2} = -\dfrac{1}{k}\dfrac{dK}{dk} = -\dfrac{E-k'^2K}{k^2k'^2}. \end{cases}$$

So, we can see easily that $k \to K(k)E(k)$ and $k \to K(k) + E(k)$ are strictly increasing functions.

Now we define the *Jacobian Elliptic Functions*. We consider the elliptic integral

$$u(y_1; k) \equiv u = \int_0^{y_1} \frac{dt}{\sqrt{(1-t^2)(1-k^2t^2)}} = \int_0^{\varphi} \frac{d\theta}{\sqrt{1-k^2\sin^2\theta}}$$

which is a strictly increasing function of real variable y_1, so we can define its inverse function by

$$y_1 = \sin\varphi \equiv \mathrm{sn}(u; k),$$

or briefly $y_1 = \mathrm{sn}u$ when it is not necessary to emphasize the modulus. The function $\mathrm{sn}(u)$ is an odd function. Two other basic functions can be defined by

$$\begin{cases} \mathrm{cn}(u;k) = \sqrt{1-y_1^2} = \sqrt{1-\mathrm{sn}^2(u;k)} & \textbf{(cnoidal)}, \\ \mathrm{dn}\,(u;k) = \sqrt{1-k^2y_1^2} = \sqrt{1-k^2\mathrm{sn}^2(u;k)} & \textbf{(dnoidal)}, \end{cases}$$

requiring that $\mathrm{sn}(0,k) = 0$, $\mathrm{cn}(0,k) = 1$, and $\mathrm{dn}(0,k) = 1$. The functions $\mathrm{cn}u$ and $\mathrm{dn}u$ are even functions. The functions $\mathrm{sn}u$, $\mathrm{cn}u$, and $\mathrm{dn}u$ are called *Jacobian elliptic functions* and they are one-valued functions of the argument u. These functions have a real period, namely, $4K, 4K$, and $2K$, respectively. If we take u to be complex, then $\mathrm{sn}u$, $\mathrm{cn}u$, and $\mathrm{dn}u$ are *doubly periodic* and have the periods $(4K, 2iK')$, $(4K, 2K + 2iK')$, and $(2K, 4iK')$, respectively.

Some basic properties of the Jacobian elliptic functions are summarized by the formulas given below:

$$\begin{cases} \mathrm{sn}^2 u + \mathrm{cn}^2 u = 1, \quad k^2\mathrm{sn}^2 u + \mathrm{dn}^2 u = 1, \quad -1 \leqq \mathrm{sn}u \leqq 1, \\ -1 \leqq \mathrm{cn}u \leqq 1, \quad k'^2 \leqq \mathrm{dn}\, u \leqq 1, \quad \mathrm{sn}(u + 4K(k); k) = \mathrm{sn}(u; k), \\ \mathrm{cn}(u + 4K(k); k) = \mathrm{cn}(u; k), \quad \mathrm{dn}(u + 2K(k); k) = \mathrm{dn}(u; k), \\ \mathrm{sn}(u + 2K(k); k) = -\mathrm{sn}(u; k), \quad \mathrm{cn}(u + 2K(k); k) = -\mathrm{cn}(u; k). \end{cases}$$

We also have the very interesting extremal cases:

$$\begin{cases} \mathrm{sn}(u, 0) = \sin u, \quad \mathrm{cn}(u, 0) = \cos u, \quad \mathrm{dn}(u, 0) \equiv 1, \\ \mathrm{sn}(u, 1) = \tanh u, \quad \mathrm{cn}(u, 1) = \mathrm{sech} u, \\ \mathrm{dn}(u, 1) = \mathrm{sech} u. \end{cases}$$

Finally we obtain the following differentiation formulas:

$$\begin{cases} \dfrac{\partial}{\partial u}\mathrm{sn}u = \mathrm{cn}u\ \mathrm{dn}u, \\ \dfrac{\partial}{\partial u}\mathrm{cn}u = -\mathrm{sn}u\ \mathrm{dn}u, \\ \dfrac{\partial}{\partial u}\mathrm{dn}u = -k^2\mathrm{sn}u\ \mathrm{cn}u. \end{cases}$$

APPENDIX B

Operator Theory

B.1. Introduction

In this appendix we establish the basic tools of the theory of closed linear operators in Hilbert spaces which are used in this book. We also give the spectral theory associated with the self-adjoint, closed, unbounded linear operator

$$\mathcal{L} \equiv M + c - \phi^p$$

which is associated with the solitary wave solutions $u(x,t) = \phi(x - ct)$ of the nonlinear evolution equation

$$u_t + u^p u_x - M u_x = 0.$$

Here $x, t \in \mathbb{R}$, $p \in \mathbb{N}$, $p \geqq 1$, and M is a linear operator defined by

$$\widehat{Mu}(\xi) = q(\xi)\widehat{u}(\xi), \quad \text{for } \xi \in \mathbb{R}$$

where the symbol $q(\xi)$ satisfies, for example, the relations established in (B.20) below. In particular, we establish the complete Sturm-Liouville theory on $\mathbb{R}$ associated with the second-order differential operator

$$-\varphi'' + V(\xi)\varphi = 0$$

where $V : \mathbb{R} \to \mathbb{R}$ is continuous and bounded. Also we establish a complete theory of the periodic eigenvalue problem considered on $[0, L]$,

$$\begin{cases} \mathcal{L}_{\text{cn}}\chi = \lambda\chi, \\ \chi(0) = \chi(L), \quad \chi'(0) = \chi'(L). \end{cases}$$

Here the linear operator $\mathcal{L}_{\text{cn}}$ is defined as

$$\mathcal{L}_{\text{cn}} = -\frac{d^2}{dx^2} + c - \phi_c,$$

where the periodic potential ϕ_c is part of the smooth curve $c \in (0, +\infty) \to \phi_c$ determined by Theorem 11.4. Here we include a basic review of Floquet theory.

All this basic information is crucial in the theories established in Chapters 6, 7, and 11.

B.2. Closed Linear Operators: Basic Theory

Let H be a separable Hilbert space with inner product denoted by $\langle\cdot,\cdot\rangle$ and norm $\|\cdot\| = \sqrt{\langle\cdot,\cdot\rangle}$. A linear transformation or operator A is a function $A : H \to H$ with the property

$$A(\alpha u + \beta v) = \alpha A(u) + \beta A(v) \qquad \text{for all } u, v \in H \text{ and all scalars } \alpha, \beta.$$

A slightly more general definition of a linear transformation, which will be used in our study, is obtained if we allow A to be defined only on a linear subspace $D(A)$

in H. Here $D(A)$ is called the *domain* of A. Here we often consider $D(A) = H$, but it is not always possible to define an operator on all of H. For example, if $H = L^2(\mathbb{R})$, the differential operator $A = -\frac{d^2}{dx^2}$ can be defined only on a part of H, since the Sobolev space $H^2(\mathbb{R})$ is the natural domain of A. The set of all images of the operator A,

$$\mathcal{R}(A) = \{Au : u \in D(A)\},$$

is known as the *range* of A. The set

$$\mathcal{N}(A) = \{u \in D(A) : Au = 0\}$$

is called the *null space* or *kernel* of A. The sets $\mathcal{N}$ and $\mathcal{R}$ are always linear subspaces.

An operator A is bounded on its domain $D(A)$ if there exists a constant $m \geqq 0$ such that

$$\|Au\| \leqq m\|u\| \qquad \text{for all } u \in D(A). \tag{B.1}$$

The number

$$\|A\| = \inf\{m : \text{inequality (B.1) is satisfied for all } u \in D(A)\}$$

is called the norm of A. So, to prove that A is unbounded on $D(A)$, one must exhibit a sequence $\{u_n\} \subset D(A)$ with $\|u_n\| \neq 0$ such that

$$\frac{\|Au_n\|}{\|u_n\|} \to \infty.$$

Our interest here is operators of differential type, so we will have that our operators are in general unbounded.

We say that a transformation A is *continuous at the point* u in $D(A)$ if, whenever $\{u_n\} \subset D(A)$ with limit u, then $Au_n \to Au$. An operator A is continuous on its domain if it is continuous at every point of $D(A)$. The following theorems are classical (see Kreyszig [**169**], Stakgold [**257**]).

THEOREM B.1. *If A is continuous at the origin, it is continuous on all of $D(A)$.*

THEOREM B.2. *A transformation is continuous if and only if A is bounded.*

B.2.1. Examples of Linear Operators.

EXAMPLE B.3. Let M be a closed linear subspace in H, and for each $u \in H$ define $v = Pu$ to be the orthogonal projection of u on M. So, P is linear and $\|Pu\| \leqq \|u\|$. Moreover, since $Pu = u$ if and only if $u \in M$, it follows that $\|P\| = 1$, $\mathcal{R}(A) = M$, $\mathcal{N}(A) = M^{\perp}$, and $H = M \oplus M^{\perp}$. Note that $D(P) = H$.

EXAMPLE B.4 (A Differential Operator). Consider the Hilbert space $L^2(\mathbb{R})$ and define the operator $Au \equiv \frac{du}{dx}$ on $L^2(\mathbb{R})$, with domain

$$D(A) = \{u \in L^2(\mathbb{R}) : u \text{ has continuous first derivative in } L^2(\mathbb{R})\}.$$

Since functions in $L^2(\mathbb{R})$ may not be even and continuous, we cannot define the operator for the whole of $L^2(\mathbb{R})$. We note that both $D(A)$ and $\mathcal{R}(A)$ are subsets of $L^2(\mathbb{R})$. Moreover, since $C_0^\infty(\mathbb{R})$ (*the class of infinitely differentiable functions with compact support*) is dense in $L^2(\mathbb{R})$, it follows that $D(A)$ is a dense linear subspace in $L^2(\mathbb{R})$. The operator A is unbounded on $D(A)$. In fact, let $\psi(x) \neq 0$ be any nonnegative continuously differentiable function which vanishes outside the interval $|x| < 1$. For instance, we can take

(B.2) $$\psi(x) = \begin{cases} a \exp\left(\frac{1}{x^2-1}\right), & |x| < 1, \\ 0, & |x| \geqq 1. \end{cases}$$

By multiplying by a suitable constant (the constant a in (B.2)), we may assume that $\int_{-\infty}^{\infty} \psi(x)\, dx = 1$. Set $\psi_n(x) = n\psi(nx)$, $n = 1, 2, \ldots$. Then $\psi_n \in D(A)$ and

$$\|\psi_n\|^2 = n \int_{-\infty}^{\infty} |\psi(y)|^2\, dy = n\|\psi\|.$$

But $\|A\psi_n\|^2 = n^3\|A\psi\|^2$. So, since $A\psi \neq 0$, we obtain that

$$\|A\psi_n\|/\|\psi_n\| = n\|A\psi\|/\|\psi\| \to \infty$$

as $n \to \infty$. This shows that A is not bounded.

EXAMPLE B.5 (General Differential Operators). Consider the differential operator

$$Bu \equiv a_m A^m u + a_{m-1} A^{m-1} u + \cdots + a_0 u, \qquad a_m \neq 0,$$

where $Au = \frac{du}{dx}$. Thus, $A^k u = \frac{d^k u}{dx^k}$ and $A^0 = 1$. We can write B symbolically in the form $B = P(A)$ where $P(z) = a_m z^m + \cdots + a_0$. We take $D(B)$ to be the set of those $u \in L^2(\mathbb{R})$ having continuous derivatives up to order m in $L^2(\mathbb{R})$. The argument used in Example B.4 shows that the operator B is unbounded in $D(B)$.

EXAMPLE B.6 (Multiplication by a Function). Let $q(x)$ be a measurable function defined on $(-\infty, \infty)$. We define the multiplication operator

$$\mathcal{M}_q u \equiv qu$$

on $L^2(\mathbb{R})$, with

$$D(\mathcal{M}_q) = \{u \in L^2(\mathbb{R}) : qu \in L^2(\mathbb{R})\}.$$

The operator $\mathcal{M}_q$ may be bounded or unbounded depending on the function q. In fact, if $q \in L^\infty(\mathbb{R})$, the inequality $\|qu\| \leqq \|q\|_\infty \|u\|$ shows that $D(\mathcal{M}_q) = L^2(\mathbb{R})$ and $\mathcal{M}_q$ is a bounded operator with $\|\mathcal{M}_q\| = \|q\|_\infty$. Now considering

$$q_0(x) = \begin{cases} \frac{1}{x}, & x \neq 0, \\ 1, & x = 0, \end{cases}$$

we have that the operator $\mathcal{M}_{q_0}$ defined by this function is unbounded on $D(\mathcal{M}_{q_0})$. In fact, by considering the characteristic function $u_\epsilon = \chi_{[\epsilon,1]}$ for every $1 > \epsilon > 0$, we have that

$$\|u_\epsilon\|^2 = 1 - \epsilon, \quad \|\mathcal{M}_{q_0} u_\epsilon\|^2 = \frac{1-\epsilon}{\epsilon}.$$

So, $\|\mathcal{M}_{q_0} u_\epsilon\|/\|u_\epsilon\| = \frac{1}{\sqrt{\epsilon}} \to \infty$ as $\epsilon \to 0$. This shows that $\mathcal{M}_{q_0}$ is unbounded. Note that $\mathcal{M}_{q_0}$ has the property that

$$\mathcal{N}(\mathcal{M}_{q_0}) = \{0\}$$

and

$$\overline{\mathcal{R}(\mathcal{M}_{q_0})} = L^2(\mathbb{R}).$$

Then, $\mathcal{M}_{q_0}$ is one-to-one and has an inverse function $(\mathcal{M}_{q_0})^{-1}$ which is defined on $\mathcal{R}(\mathcal{M}_{q_0})$. Since $(\mathcal{M}_{q_0})^{-1} = \mathcal{M}_w$, where $w(x) = x$, it is easy to see that the multiplication operator $\mathcal{M}_w$ with $D(\mathcal{M}_w) \equiv \mathcal{R}(\mathcal{M}_{q_0})$ is also unbounded on $L^2(\mathbb{R})$.

Suppose now that A is a given linear operator, not necessarily bounded, on a domain $D(A)$ in H. When we try to solve the inhomogeneous equation

$$Au = f \tag{B.3}$$

on $D(A)$ for a given function $f \in H$, we face two serious problems. The first one arises when a solution exists but it is not unique. The second, and more serious one, arises when we are able to solve equation (B.3) only for f satisfying specific conditions rather than for every $f \in H$. Naturally, this fault is a consequence of the form, as has been defined on $D(A)$. For instance, if we consider the differential operator defined in Example 2 and suppose that the equation $du/dx = f$ has a solution in $D(A)$, then we should choose f to be continuous rather than choosing any $f \in L^2(\mathbb{R})$ or an f with properties weaker than continuity. A natural solution to this difficulty is to extend the domain A and so to also extend the original operator A such that we can find more functions f. In fact, this is the basic idea of the theory of closed operators.

DEFINITION B.7. The linear operator B is said to be an *extension* of the linear operator A if $D(A) \subset D(B)$ and $Bu = Au$ for each $u \in D(A)$.

Now we have a basic and useful result.

LEMMA B.8. *Let A be a bounded operator defined on a linear subspace $D(A)$ in H. Then A can be extended to the whole of H without affecting continuity and norm.*

PROOF. (1) Suppose that $D(A)$ is closed and $D(A) \neq H$. Define

$$Bu = \begin{cases} Au & \text{if } u \in D(A), \\ 0 & \text{if } u \in D(A)^{\perp}. \end{cases} \tag{B.4}$$

So, using the fact that $H = D(A) \oplus D(A)^{\perp}$, we have that B is a bounded linear extension of A and $\|B\| = \|A\|$.

(2) If $D(A)$ is not closed, we first extend A to $\overline{D(A)}$. To do that, let $\{u_n\} \subset D(A)$ with limit $u \in \overline{D(A)}$. Since $\|Au_n - Au_m\| \leqq \|A\|\|u_n - u_m\|$, it follows that $\{Au_n\}$ is a Cauchy sequence and has a limit f. Define the new operator $\tilde{A}$ over $\overline{D(A)}$ by setting $\tilde{A}u = f$. To see that this definition depends only on u and not on $\{u_n\}$, let $\{v_n\} \subset D(A)$ with limit u. Then $u_n - v_n \to 0$, and by continuity $A(u_n - v_n) \to 0$, so that $\{Au_n\}$ and $\{Av_n\}$ have the same limit f. In this way we extend A to $\overline{D(A)}$ and then by item (1) to H. □

From Lemma B.8 we can extract the following main piece of information: *if A is a bounded operator on its domain $D(A)$ and $\overline{D(A)} = H$, then there exists a unique extension of A to the whole of H with the same norm of A. Therefore, we can always see A as a bounded operator defined on the whole of H.*

DEFINITION B.9 (Closed Operator). Let A be a linear operator on the linear subspace $D(A)$. We say that A is *closed* whenever $\{u_n\} \subset D(A)$ is a sequence satisfying

$$u_n \to u, \quad Au_n \to f \text{ implies } u \in D(A), \quad \text{and} \quad Au = f.$$

Now we establish some basic properties of closed operators (see Reed and Simon [**241**]).

THEOREM B.10 (Closed Graph Theorem). *A closed operator on a closed domain is bounded.*

THEOREM B.11. *Let A be a closed operator. Then its null space is a closed set. If A^{-1} exists, then the operator $A^{-1} : \mathcal{R}(A) \to D(A)$ is closed.*

THEOREM B.12. *Let A be closed, and let A^{-1} exist. Then, $\mathcal{R}(A)$ is closed if and only if A^{-1} is bounded.*

PROOF. Suppose $\mathcal{R}(A)$ is closed. By Theorem B.11, $A^{-1} : \mathcal{R}(A) \to D(A)$ is closed. So, by Theorem B.10 we have that A^{-1} is bounded. Now suppose that $A^{-1} : \mathcal{R}(A) \to D(A)$ is bounded. Then there is an $\eta > 0$ such that

$$\|u\| \leqq \eta \|Au\| \qquad \text{for all } u \in D(A).$$

Let $g \in \overline{\mathcal{R}(A)}$. Then there exists $\{u_n\} \subset D(A)$ such that $Au_n \to g$. Since

$$\|u_n - u_m\| \leqq \eta \|Au_n - Au_m\|,$$

it follows that $\{u_n\}$ is a Cauchy sequence. Therefore, $u_n \to u$. So, since A is closed, we have that $u \in D(A)$ and $Au = g$. Hence $u \in \mathcal{R}(A)$ and so $\mathcal{R}(A)$ is closed. □

As we will see in the following examples, the property of an operator being closed or having a closed extension is a basic one for working with differential operators.

B.2.2. Examples of Closed Operators.

EXAMPLE B.13 (Constructing Closed Operators). Initially we will see that

$$Au = \frac{du}{dx}$$

defined in Example B.4 is not closed on $D(A)$. In fact, define the functions $\psi_n(x) = n\psi(nx)$ where ψ is given by (B.2). For any $u \in L^2(\mathbb{R})$, set

$$J_n u(x) = \psi_n * u(x) = \int_{-\infty}^{\infty} \psi_n(x-y) u(y)\, dy.$$

Then, from Young's inequality (see Appendix A) it follows that $J_n u \in L^2(\mathbb{R})$ with $\|J_n u\| \leqq \|u\|$. Moreover, for any $u \in L^2(\mathbb{R})$, we have

$$\|u - J_n u\| \to 0 \quad \text{as } n \to \infty. \tag{B.5}$$

Now, let u_0 be the function in $L^2(\mathbb{R})$ given by

$$u_0(x) = \begin{cases} 1 + x, & -1 \leqq x \leqq 0, \\ 1 - x, & 0 < x \leqq 1, \\ 0, & |x| > 1. \end{cases}$$

Clearly, u_0 is continuous in $\mathbb{R}$ but its derivative is not. Thus, u_0 is not in $D(A)$. However, $J_n u_0 \in D(A)$ for each n as can be seen from the properties of ψ. Now, u_0' may not be continuous but it is in $L^2(\mathbb{R})$. Moreover,

$$AJ_n u_0(x) = \int_{-1}^{1} \psi_n'(x-y) u_0(y)\, dy = \int_{-1}^{1} \psi_n(x-y) u_0'(y)\, dy = J_n u_0'.$$

Now, from (B.5) it follows that $\|u_0' - AJ_n u_0\| \to 0$ as $n \to \infty$. So, if A were a closed operator, then it would follow that $u_0 \in D(A)$, which we know to be false.

Does A have a closed extension? This question has an affirmative answer. Initially, we will prove that the differential operator A has the following main relation: whenever $u_n\} \subset D(A)$ such that $u_n \to 0$ and $Au_n \to f$ in $L^2(\mathbb{R})$, then $f \equiv 0$. In fact, let $v \in C_0^\infty(\mathbb{R}) \subset D(A)$. Then via integration by parts we have, for each n, that

$$\langle Au_n, v\rangle = -\langle u_n, v'\rangle.$$

So, taking the limit as $n \to \infty$, we get $\langle f, v\rangle = 0$ for all such v. Therefore, since $C_0^\infty(\mathbb{R})$ is dense in $L^2(\mathbb{R})$, we have that $f \equiv 0$.

Now, we construct an extension $\bar{A}$ of A as follows:

An element $u \in H$ is in $D(\bar{A})$ if there is a sequence $\{u_n\} \subset D(A)$ such that $u_n \to u$ in H and Au_n converges in H to some element f. Define $\bar{A}u = f$.

This definition does not depend on the choice of the particular sequence $\{u_n\}$. This follows immediately from the property established above for A. Clearly, $\bar{A}$ is a linear extension of A. Now we will see that $\bar{A}$ is closed. To see this, suppose $\{u_n\} \subset D(\bar{A})$, $u_n \to x$, and $\bar{A}u_n \to y$. Then, for each n there is a sequence $\{w_{n_k}\} \subset D(A)$ such that $w_{n_k} \to u_n$ and $Aw_{n_k} \to \bar{A}u_n$ as $k \to \infty$. In particular, one can find a $z_n \in D(A)$ such that

$$\|z_n - u_n\| < \frac{1}{n} \qquad \text{and} \qquad \|Az_n - \bar{A}u_n\| < \frac{1}{n}.$$

Therefore,

$$\|z_n - x\| \leqq \|z_n - u_n\| + \|u_n - x\| \to 0,$$

and

$$\|Az_n - y\| \leqq \|Az_n - \bar{A}u_n\| + \|\bar{A}u_n - y\| \to 0.$$

This shows that $x \in D(\bar{A})$ and $\bar{A}x = y$. Hence, $\bar{A}$ is closed.

The operator $\bar{A}$ constructed above is called the *closure* of A. It is the "smallest" closed extension of A.

EXAMPLE B.14 (Closed Operators and the Resolvent). The significance of the closed operator $\bar{A}$ constructed in Example B.13 is that its domain can sometimes be characterized concretely and its definition is sufficient to decide which values of λ are such that

$$(\bar{A} - \lambda I)u = f \tag{B.6}$$

has a unique solution $u \in D(\bar{A})$ for each $f \in L^2(\mathbb{R})$. In fact, with regard to the domain of $\bar{A}$ we have, via distributions, that

$$D(\bar{A}) = \{u \in L^2(\mathbb{R}) : u' \in L^2(\mathbb{R})\}$$

where u' represents the distributional derivative of u. Indeed, let $\{u_n\} \subset D(A)$ such that, simultaneously,

$$u_n \to u \qquad \text{and} \qquad \frac{du_n}{dx} \to v \quad \text{as } n \to \infty$$

in $L^2(\mathbb{R})$. Then by definition u belongs to the domain of $D(\bar{A})$ and $\bar{A}u = v$. Now let $\phi \in \mathcal{S}(\mathbb{R})$ (the Schwartz class; see Appendix A). Then, $\langle u_n', \phi\rangle \to \langle v, \phi\rangle$ and $\langle u_n, \phi'\rangle \to \langle u, \phi'\rangle$. So, via integration by parts, we have

$$\langle u_n', \phi\rangle = -\langle u_n, \phi'\rangle \to \langle v, \phi\rangle = -\langle u, \phi'\rangle, \qquad \text{as } n \to \infty.$$

Thus, u has a distributional derivative such that $u' = v \in L^2(\mathbb{R})$. Hence, the closure of A is the generalization of differentiation in the sense of distribution. Note that $D(\bar{A})$ is exactly Sobolev's space $H^1(\mathbb{R})$.

With regard to equation (B.6) we can assume initially that u and f are smooth functions satisfying (B.6). So, $\bar{A}$ reduces to A, and (B.6) becomes the differential equation of first order $u' - \lambda u = f$, which has the solution

$$e^{-\lambda x}u(x) = u(0) + \int_0^x e^{-\lambda y} f(y)\, dy. \tag{B.7}$$

Suppose $\lambda = \alpha + i\beta$. If $\alpha > 0$, it follows from (B.7) that $|u(x)| \to \infty$ as $x \to \infty$ unless

$$u(0) + \int_0^\infty e^{-\lambda y} f(y)\, dy = 0. \tag{B.8}$$

Thus, we have $u \in L^2(\mathbb{R})$ for $\alpha > 0$, when (B.8) holds. Then, in this case we find that given a smooth f, the function

$$u(x) = -\int_x^\infty e^{\lambda(x-y)} f(y)\, dy$$

is continuously differentiable, satisfies (B.6), and belongs to $L^2(\mathbb{R})$, where the last affirmation follows from the inequality

$$\|u\| \leqq \frac{\|f\|}{\alpha}. \tag{B.9}$$

Now, we consider the general case assuming $f \in L^2(\mathbb{R})$. Then, from the density of $C_0^\infty(\mathbb{R})$ in $L^2(\mathbb{R})$, there is a sequence $\{f_n\} \subset C_0^\infty(\mathbb{R})$ converging to f in $L^2(\mathbb{R})$. Hence, for each f_n, there is a solution $u_n \in D(A)$ of $(A - \lambda I)u_n = f_n$. Moreover, by (B.9),

$$\|u_n - u_m\| \leqq \frac{\|f_n - f_m\|}{\alpha} \to 0 \qquad \text{as } m, n \to \infty.$$

Then, $\{u_n\}$ is a Cauchy sequence in $L^2(\mathbb{R})$. Therefore, there exists $u \in L^2(\mathbb{R})$ such that $u_n \to u$. By definition, $u \in D(\bar{A})$ and $(\bar{A} - \lambda I)u = f$. Moreover, u and f satisfy (B.9). Therefore, we have that when $\alpha > 0$, equation (B.6) can be solved for each $f \in L^2(\mathbb{R})$. The same is also true for $\alpha < 0$, in which case

$$u(x) = \int_{-\infty}^x e^{\lambda(x-y)} f(y)\, dy$$

is the corresponding solution.

The following question is about uniqueness. The fact that the solution of $u' - \lambda u = f$ given by (B.7) is unique does not imply that the same is true of (B.6). Let $\alpha \neq 0$ and suppose that for some $f \in L^2(\mathbb{R})$ equation (B.6) has two solutions. Then, their difference u would be a solution of $(\bar{A} - \lambda I)u = 0$. Since $u \in D(\bar{A})$, there is a sequence $\{u_n\} \subset D(A)$ such that $u_n \to u$ while $Au_n \to \lambda u$. Let $v \in D(A)$ and assume that

$$\langle Au_n, v\rangle = -\langle u_n, v'\rangle \tag{B.10}$$

holds. Then we have, in the limit, that $\langle u, v' + \bar{\lambda} v\rangle = 0$ for all $v \in D(A)$. Since $\Re(-\bar{\lambda}) = -\Re\lambda = -\alpha \neq 0$, we know that for each $g \in C_0^\infty(\mathbb{R})$, there exists $v \in D(A)$ such that $v' + \bar{\lambda} v = g$. Hence, $\langle u, g\rangle = 0$. Since $C_0^\infty(\mathbb{R})$ is dense in $L^2(\mathbb{R})$, it follows that $u = 0$.

Now we prove (B.10) for each $v \in D(A)$. Since (B.10) holds whenever $v \in C_0^\infty(\mathbb{R})$ and since for each $v \in D(A)$ there is a sequence $\{\varphi_k\} \subset C_0^\infty(\mathbb{R})$ such that $\varphi_k \to v$ and $\varphi_k' \to v'$ as $k \to \infty$, in $L^2(\mathbb{R})$, it follows that (B.10) is true for each $v \in D(A)$.

DEFINITION B.15 (Resolvent and Spectrum). Let $A : D(A) \to H$ be a closed operator with domain $D(A) \subset H$. The sets

$$\rho(A) = \{\lambda \in \mathbb{C} : A - \lambda I \text{ has a bounded inverse and } \mathcal{R}(A - \lambda I) = H\}$$

and $\sigma(A) = \mathbb{C} - \rho(A)$ are called, respectively, the *resolvent set* of A and the *spectrum set* of A. If $\lambda \in \rho(A)$, $R_\lambda(A) = (A - \lambda I)^{-1}$ is called the *resolvent* of A at λ.

The name "resolvent" is appropriate, since $R_\lambda(A)$ helps to solve the equation

$$(A - \lambda I)x = y.$$

For a point to be in the resolvent set of A, several conditions must be satisfied. These conditions are not all independent. In fact, from Theorem B.10, it follows that if A is a closed operator such that

$$\mathcal{N}(A - \lambda I) = \{0\} \text{ and } \mathcal{R}(A - \lambda I) = H, \text{ then } \lambda \in \rho(A).$$

Moreover, from Theorem B.12, we have that

$$\text{if } \overline{\mathcal{R}(A - \lambda I)} = H \text{ and } (A - \lambda I)^{-1} \text{ exists and is bounded, then } \lambda \in \rho(A).$$

Danger: An operator A can have that $\overline{\mathcal{R}(A)} = H$ and A^{-1} exists but it can be unbounded. (Can you give an example?)

There are many classification schemes for further division of the spectrum. The one used here has been chosen by thinking about the type of operators with which we will work, more precisely, self-adjoint operators. The spectrum $\sigma(A)$ can be partitioned into three disjoint sets as follows:

- **The point spectrum**, $\sigma_p(A)$, is the set of $\lambda \in \mathbb{C}$ such that $(A - \lambda I)u = 0$ has nontrivial solutions; in other words, λ is an *eigenvalue* of A and any corresponding nontrivial solution u is an *eigenvector* of A belonging to λ; the (geometric) multiplicity of λ is the dimension of the kernel $\mathcal{N}(A - \lambda I)$.
- **The continuous spectrum**, $\sigma_c(A)$, is the set of $\lambda \in \mathbb{C}$ such that $R_\lambda(A)$ exists and $\overline{\mathcal{R}(A - \lambda I)} = H$ but $R_\lambda(A) = (A - \lambda I)^{-1}$ is unbounded.
- **The residual spectrum**, $\sigma_r(A)$, is the set of $\lambda \in \mathbb{C}$ such that $R_\lambda(A)$ exists (and may be bounded or not) but $\overline{\mathcal{R}(A - \lambda I)} \neq H$.

So, we have the following disjoint union of the whole complex plane:

$$\mathbb{C} = \rho(A) \cup \sigma(A) = \rho(A) \cup \sigma_p(A) \cup \sigma_c(A) \cup \sigma_r(A).$$

The spectral analysis of closed operators is very important in the study of the stability of travelling wave solutions of nonlinear dispersive evolution equations (see Chapters 6, 7, and 11). For example, the linearization of the equation satisfied by a solitary wave solution of the GKdV equation (see (3.3)) produces the unbounded self-adjoint operator

$$\mathcal{L} \equiv -\frac{d^2}{dx^2} + c - \phi^p$$

on the Hilbert space $L^2(\mathbb{R})$. As we will see below (Theorem B.43 and Theorem B.61), the spectrum of $\mathcal{L}$, which is contained in $\mathbb{R}$, will consist of the interval

$[c, \infty)$ together with a finite number of discrete eigenvalues with finite-dimensional eigenspaces in the interval $(-\infty, c)$.

Now we will see some basic properties of the sets $\rho(A)$, $\sigma(A)$ and the resolvent $R_\lambda(A)$. The resolvent $R_\lambda(A)$ is an operator which depends on the complex parameter λ. This suggests the general setting of an operator function.

By a *vector-valued function* or *operator function* we mean a map

$$\begin{cases} S : \Lambda \to \mathcal{B}(H, H) \\ \quad \lambda \longmapsto S_\lambda \end{cases}$$

where Λ is any subset of the complex plane $\mathbb{C}$ and $\mathcal{B}(H, H)$ is the set of all bounded operators of H in H.

DEFINITION B.16. Let Λ be an open subset of $\mathbb{C}$. An operator function S defined on Λ is said to be locally holomorphic on Λ if for every $x \in H$ and $f \in H'$ (the dual space of H) the function defined by

$$h(\lambda) = f(S_\lambda x)$$

is holomorphic (or analytic) at every $\lambda_0 \in \Lambda$ in the usual sense of complex analysis. The function S is said to be holomorphic or analytic on Λ if S is locally holomorphic on Λ and Λ is a domain (a connected open subset of $\mathbb{C}$).

THEOREM B.17 (First Resolvent Formula and Analyticity of the Resolvent). *Let A be a closed operator. Then we have the following:*

(1) *For $\lambda, \mu \in \rho(A)$, $R_\lambda(A)$ and $R_\mu(A)$ commute and we have the resolvent formula*

$$R_\lambda(A) - R_\mu(A) = (\lambda - \mu) R_\mu(A) R_\lambda(A). \tag{B.11}$$

(2) *The set $\rho(A)$ is an open subset of $\mathbb{C}$.*

(3) *The operator function*

$$\lambda \in \rho(A) \to R_\lambda(A) \in \mathcal{B}(H, H)$$

is an analytic function on each component (maximal connected subset) of $\rho(A)$.

(4) *The set $\sigma(A)$ is a closed subset of $\mathbb{C}$.*

PROOF. Let $x \in H$, and set $u = R_\lambda(A)x$. Thus, $(A - \lambda I)u = x$ and $(A - \mu I)u = x + (\lambda - \mu)u$. Hence,

$$u = R_\mu(A)x + (\lambda - \mu) R_\mu(A) u.$$

Substituting for u, we get (B.11). Note that from (B.11) we have the property that $R_\lambda(A)$ and $R_\mu(A)$ commute. Now, substituting for $R_\lambda(A)$ in the right-hand side of (B.11), we get

$$R_\lambda(A) = R_\mu(A) + (\lambda - \mu) R_\mu^2(A) + (\lambda - \mu) R_\lambda(A) R_\mu(A).$$

Here we have used the notation $[R_\mu(A)]^n = R_\mu^n(A)$. Continuing in this way, we get

$$R_\lambda(A) = (\lambda - \mu)^n R_\lambda(A) R_\mu^n(A) + \sum_{n=1}^{k} (\lambda - \mu)^{n-1} R_\mu^n(A).$$

Now, let $\mu \in \rho(A)$ be fixed. If $|\lambda - \mu| < \frac{1}{\|R_\mu\|}$, it follows that

$$\|(\lambda - \mu)^n R_\lambda(A) R_\mu^n(A)\| \leqq \|R_\lambda(A)\| \Big(|\lambda - \mu|^n \|R_\mu\|^n \Big) \to 0 \qquad \text{as } n \to \infty.$$

Therefore,

$$R_\lambda(A) = \sum_{n=1}^{\infty} (\lambda - \mu)^{n-1} R_\mu^n(A) \tag{B.12}$$

and the series is absolutely convergent in $\mathcal{B}(H, H)$ for every λ in the open disk $B(\mu; \|R_\mu\|^{-1})$. So, the expression (B.12) shows that $R_\lambda(A)$ is analytic on each component (maximal connected subset) of $\rho(A)$. In fact, taking any $x \in H$ and $f \in H'$ and defining h by $h(\lambda) = f(R_\lambda(A)x)$, we obtain from (B.12) the power series representation

$$h(\lambda) = \sum_{n=1}^{\infty} c_n (\lambda - \mu)^{n-1}, \qquad \text{where } c_n = f(R_\mu^n(A)x).$$

This series is absolutely convergent on the disk $\{\lambda : |\lambda - \mu| < \frac{1}{\|R_\mu\|}\}$, and so h is analytic at this disk. This finishes the proof of the theorem. □

When we work with a bounded linear operator, which means $A \in \mathcal{B}(H, H)$, we have some explicit properties of the spectrum. We establish only some of these classical results. To see a proof of them, we recommend that the reader see the references Kato [**153**] and Reed and Simon [**241**].

THEOREM B.18. *Let $A : H \to H$ be a bounded linear operator. Then we have the following:*

1. *The spectrum $\sigma(A)$ of A is a compact set of the complex plane such that $\sigma(A) \neq \emptyset$ and $\sigma(A)$ lies in the disk given by $|\lambda| \leqq \|A\|$.*
2. *The resolvent $\rho(A)$ of A is a nonempty set.*

B.2.3. Examples of Resolvent.

EXAMPLE B.19. When we work with an unbounded operator, Theorem B.18 is false, as the following example shows. Let $AC[0, 1]$ be the set of absolutely continuous functions on $[0, 1]$ whose derivatives are in $L^2([0, 1])$. Define

$$A_j u = i\frac{du}{dx}, \quad j = 1, 2,$$

with domains

$$\begin{cases} D(A_1) = \{u \in L^2([0,1]) : u \in AC[0,1]\}, \\ D(A_2) = \{u \in L^2([0,1]) : u \in AC[0,1] \text{ and } u(0) = 0\}. \end{cases}$$

Both $D(A_1)$ and $D(A_2)$ are dense sets in $L^2([0, 1])$ and both of these operators are closed. But we have

$$\sigma(A_1) = \mathbb{C} \qquad \text{and} \qquad \sigma(A_2) = \emptyset.$$

To see that the spectrum of A_1 is the whole plane, it is enough to observe that

$$(A_1 - \lambda I)e^{-i\lambda x} = 0$$

and $e^{-i\lambda x} \in D(A_1)$ for $\lambda \in \mathbb{C}$. Now, with regard to the operator A_2, we consider the differential equation

$$\begin{cases} \dfrac{d}{dx} f + i\lambda f = ig, \\ f(0) = 0. \end{cases} \tag{B.13}$$

Then, it follows that

$$f(x) = i \int_0^x e^{-i\lambda(x-y)} g(y)dy$$

satisfies, for $g \in L^2([0,1])$, that $f \in D(A_2)$ and it satisfies (B.13) as an equality in $L^2([0,1])$. So, it follows that $A_2 - \lambda I : D(A_2) \to L^2([0,1])$ is a bijection for every $\lambda \in \mathbb{C}$. Therefore, from Theorem B.10 we have that $\lambda \in \rho(A_2)$ for every $\lambda \in \mathbb{C}$.

EXAMPLE B.20 (Spectrum of $A = \frac{d}{dx}$ on $L^2(\mathbb{R})$). It follows from the analysis in the Example B.13 that the half-planes $\Re(\lambda) > 0$ and $\Re(\lambda) < 0$ are in $\rho(\bar{A})$. In fact, since for $\Re(\lambda) \neq 0$ we have that $\bar{A} - \lambda I$ is closed, $(\bar{A} - \lambda I)^{-1}$ exists, and $\mathcal{R}(A - \lambda I) = L^2(\mathbb{R})$, then from Theorem B.10 and Theorem B.11, it follows that $\bar{A} - \lambda I$ has a bounded inverse.

What about the spectrum of the differential operator $\bar{A}u = u'$? We will show that if $\lambda = i\beta$, then $\lambda \in \sigma(\bar{A})$. In fact, let $\psi \in C_0^\infty(\mathbb{R})$ such that $\|\psi\| = 1$. For $\epsilon > 0$, set $\psi_\epsilon(x) = \sqrt{\epsilon} e^{i\beta x} \psi(\epsilon x)$. Then $A\psi_\epsilon = i\beta\psi_\epsilon + \epsilon^{3/2} e^{i\beta x} \psi'(\epsilon x)$, which shows that

$$\|(A - i\beta I)\psi_\epsilon\|^2 = \epsilon^2 \|\psi'\|^2.$$

So, since $\|\psi_\epsilon\|^2 = 1$, we conclude that

$$\frac{\|(A - i\beta I)\psi_\epsilon\|}{\|\psi_\epsilon\|} \to 0 \quad \text{as } \epsilon \to 0,$$

which shows that $i\beta \notin \rho(\bar{A})$, because if $\lambda \in \rho(\bar{A})$, then there will exist a constant $K > 0$ such that

$$\|(\bar{A} - \lambda I)u\| \geqq K\|u\| \qquad \text{for all } u \in D(\bar{A}).$$

Therefore, the differential operator $\bar{A}u = u'$ has $\rho(\bar{A}) = \{\lambda : \Re(\lambda) \neq 0\}$ and $\sigma(\bar{A}) = \{\lambda : \Re(\lambda) = 0\}$.

As will be seen below, via the Spectral Theorem or via the Fourier transform, the spectrum of the closure of the differential operator B defined in Example B.5 depends on the polynomial $P(z) = a_m z^m + \cdots + a_0$; namely, it is possible to show that $\sigma(\bar{B}) = \{P(i\xi) : \xi \in \mathbb{R}\}$. So, we have the basic result

$$\sigma\Big(-\frac{d^2}{dx^2}\Big) = [0, \infty).$$

Now, we consider the notion of adjoint operator. Initially, let A be a bounded operator defined on the whole Hilbert space. With v a fixed element in H, we consider the bounded linear operator $u \to \langle Au, v\rangle$. Then by the Riesz Representation Theorem there exists a unique element $g \in H$ such that

$$\langle Au, v\rangle = \langle u, g\rangle \qquad \text{for all } u \in H.$$

The element g depends on v, and we write $g = A^*v$. A^* is a linear operator defined on the whole of H and is bounded ($\|A^*v\| \leqq \|A\|\|v\|$).

DEFINITION B.21. Let $A : H \to H$ be a bounded linear operator. Then the *adjoint operator* $A^* : H \to H$ is defined by

$$\langle Au, v\rangle = \langle u, A^*v\rangle \qquad \text{for all } u, v \in H.$$

DEFINITION B.22. Let $A : H \to H$ be a bounded linear operator. A is said to be *self-adjoint or Hermitian* if $A = A^*$, and it is *unitary* if A is bijective and $A^* = A^{-1}$.

The following are some basic properties of the adjoint and unitary operators (see [**153**], [**169**], [**241**]): Let $T, U \in \mathcal{B}(H, H)$. Then

(1) $(TU)^* = U^*T^*$, $(T^*)^* = T$,
(2) if T has a bounded inverse, then T^* has a bounded inverse and $(T^*)^{-1} = (T^{-1})^*$,
(3) if U is unitary, U is isometric, and $\|Ux\| = \|x\|$ for all $x \in H$,
(4) $\|T^*\| = \|T\|$.

THEOREM B.23. *Let $T \in \mathcal{B}(H, H)$. Then we have the following:*

(1) $\sigma(T^*) = \{\lambda | \bar{\lambda} \in \sigma(T)\}$ *and* $R_{\bar{\lambda}}(T^*) = [R_\lambda(T)]^*$.
(2) *If* $\lambda \in \sigma_r(T)$, *then* $\bar{\lambda} \in \sigma_p(T^*)$. *If* $\lambda \in \sigma_p(T)$, *then* $\bar{\lambda} \in \sigma_p(T^*) \cup \sigma_r(T^*)$.
(3) *If T is self-adjoint, then T has no residual spectrum.*
(4) *If T is self-adjoint, then* $\sigma(T) \subset \mathbb{R}$.
(5) *If T is self-adjoint, then eigenvectors corresponding to distinct eigenvalues of T are orthogonal.*

PROOF. We are interested in affirmations (2), (3), and (4). The proofs of the other items can be seen in [**153**] and [**169**].

First we give the proof of (4). Let $\lambda, \mu \in \mathbb{R}$. Then, using the fact that $T = T^*$, it follows that

$$\|[T - (\lambda + i\mu)]x\|^2 = \|(T - \lambda)x\|^2 + \mu^2\|x\|^2.$$

This gives $\|[T - (\lambda + i\mu)]x\|^2 \geqq \mu^2\|x\|^2$. So, if $\mu \neq 0$, then $T - (\lambda + i\mu)$ is one-to-one and has a bounded inverse on its range. Then from Theorem B.12, $\mathcal{R}(T - (\lambda + i\mu))$ is closed. If $\mathcal{R}(T - (\lambda + i\mu)) \neq H$, then there is an $f \in H - \{0\}$ such that $\langle [T - (\lambda + i\mu)]x, f\rangle = 0$ for every $x \in D(T)$. So,

$$\langle Tx, f\rangle = \langle x, (\lambda - i\mu)f\rangle \quad \text{where } x \in D(T),$$

and therefore from the definition of adjoint operator and the self-adjoint property of T we get $Tf = (\lambda - i\mu)f$. Therefore $\lambda - i\mu$ would be an eigenvalue of T. Since every eigenvalue of an adjoint operator must be real, it follows that $\mu = 0$, which is a contradiction. Thus if $\mu \neq 0$, $\lambda + i\mu \in \rho(T)$. This proves (4).

Now, we prove (2). Let $\lambda \in \sigma_r(T)$. Then $R_\lambda(T) : \mathcal{R}(T - \lambda) \to D(T - \lambda)$ is well defined and $\mathcal{R}(T_\lambda) \neq H$. Let $f \neq 0$ such that $f \perp \mathcal{R}(T - \lambda)$. Then for each $x \in H$, $\langle (T - \lambda)x, f\rangle = 0$. Hence, $\langle x, T^*f\rangle = \langle Tx, f\rangle = \langle x, \bar{\lambda}f\rangle$. Then, $(T^* - \bar{\lambda})f = 0$ and so $\bar{\lambda} \in \sigma_p(T^*)$.

Now, if a real λ were in the residual spectrum of T, then $\bar{\lambda} = \lambda$ would be in the point spectrum of $T^* = T$, which is impossible since $\sigma_p(T) \cap \sigma_r(T) = \emptyset$. This proves (3) and finishes the proof of the theorem. □

As we will see below, properties (3) and (4) established in Theorem B.23 will also be satisfied by any *unbounded self-adjoint operator defined in a domain dense in H*. Now we have the extension of the definition of adjoint operator to closed operators.

DEFINITION B.24 (Adjoint). Let $A : D(A) \to H$ be a densely defined linear operator. Then the *adjoint operator* $A^* : D(A^*) \to H$ of A is defined as follows:

$$D(A^*) = \{y \in H : \text{there is a } z \in H \text{ satisfying } \langle Ax, y\rangle = \langle x, z\rangle \text{ for all } x \in D(A)\},$$

and for each $y \in D(A^*)$ we define $A^*y = z$. Notice that for z *to be uniquely determined, we need the condition that $D(A)$ is dense in H.*

THEOREM B.25. *Let $A : D(A) \to H$ be a densely defined linear operator. Then we have the following:*

(1) *A^* is a closed operator.*
(2) *If A is closed, then $D(A^*)$ is dense in H. Moreover, $A^{**} = A$.*
(3) *Suppose $\overline{D(A^*)} = H$. Then A is closable and $A \subseteq A^{**}$. Moreover, $A^{**} = \bar{A}$.*
(4) *There are linear operators which have no closed extensions.*

PROOF. (1) Let $\{y_n\} \subset D(A^*)$ such that $y_n \to y_0$ and $A^* y_n \to z_0$. By the definition of A^* we have for every $y \in D(A)$, $\langle Ay, y_n \rangle = \langle y, A^* y_n \rangle$. Since the inner product is continuous, we obtain that $\langle Ay, y_0 \rangle = \langle y, z_0 \rangle$ for every $y \in D(A)$. So, by the definition of A^* we have $y_0 \in D(A^*)$ and $A^* y_0 = z_0$. This finishes the proof of this item.

(2) Suppose that $D(A^*)$ is not dense and that $\psi \in D(A^*)^{\perp}$ with $\psi \neq 0$. Define the *graph* of A as the set

$$\Gamma(A) = \{(u, v) : u \in D(A), v = Au\}.$$

So, $\Gamma(A) \subset H \times H$. Note that $H \times H$ is a Hilbert space with inner product $\langle (u, v), (u_1, v_1) \rangle \equiv \langle u, u_1 \rangle + \langle v, v_1 \rangle$. Therefore A *is a closed operator if and only if* $\Gamma(A)$ *is closed in* $H \times H$. Now define a unitary operator U on $H \times H$ by

$$U(x, y) = (y, -x).$$

So, $U^2 = -I$ and $U[E^{\perp}] = (U[E])^{\perp}$. Moreover, it is easy to see that $\Gamma(A^*) = (U[\Gamma(A)])^{\perp}$. Therefore, $(\psi, 0) \in [\Gamma(A^*)]^{\perp}$ and so $U([\Gamma(A^*)]^{\perp})$ is not the graph of a single-valued operator. Finally, since A is closed, it follows that

$$U([\Gamma(A^*)]^{\perp}) = (U[\Gamma(A^*)])^{\perp} = \overline{\Gamma(A)} = \Gamma(A),$$

which is a contradiction. Finally, from the relation $\overline{\Gamma(A)} = (U[\Gamma(A^*)])^{\perp}$ we obtain that $\Gamma(A) = \Gamma(A^{**})$ and so $A = A^{**}$.

(3) Initially, we have that $\langle Ax, y \rangle = \langle x, A^* y \rangle$ for all $x \in D(A)$ and $y \in D(A^*)$. Then, taking conjugates, we have

$$\langle A^* y, x \rangle = \langle y, Ax \rangle \qquad \text{for all } y \in D(A^*) \text{ and } x \in D(A).$$

So, since $D(A^*)$ is dense in H, we have that A^{**} is well defined. By definition, $x \in D(A^{**})$ and $A^{**} x = Ax$. Therefore, $A \subseteq A^{**}$ and so A is closable.

Now we show that $(\bar{A})^* = A^*$. In fact, since $A \subseteq \bar{A}$, then $(\bar{A})^* \subseteq A^*$ and so $D((\bar{A})^*) \subseteq D(A^*)$. Now we need to show that $D(A^*) \subseteq D((\bar{A})^*)$. Let $y \in D(A^*)$. If $x \in D(\bar{A})$, then there is a sequence $\{u_n\} \subseteq D(A)$, such that $u_n \to x$ and $Au_n \to \bar{A}x$. Therefore, the equality $\langle Au_n, y \rangle = \langle u_n, A^* y \rangle$ implies that

$$\langle \bar{A}x, y \rangle = \langle x, A^* y \rangle \qquad \text{for all } x \in D(\bar{A})$$

and so $y \in D(A^*)$. This shows that $(\bar{A})^* = A^*$.

Finally we show that $A^{**} = \bar{A}$. Since A is closable and $A \subseteq A^{**}$, then $\bar{A}$ exists. Moreover, $\overline{D((\bar{A})^*)} = \overline{D(A^*)} = H$. Therefore,

$$(\bar{A})^* = A^*; \text{ then} (\bar{A})^{**} = A^{**}; \text{ then } \bar{A} = A^{**}.$$

(4) Let $f \in L^{\infty}(\mathbb{R})$, but $f \notin L^2(\mathbb{R})$. Let

$$D(A) = \Big\{ \psi \in L^2(\mathbb{R}) : \int_{-\infty}^{\infty} |\psi(x)||f(x)| dx < \infty \Big\}.$$

So, $C_0^\infty(\mathbb{R}) \subset D(A)$. Therefore, $D(A)$ is dense in $L^2(\mathbb{R})$. Let $\psi_0 \in L^2(\mathbb{R})$ and define $A\psi = \langle \psi, f\rangle \psi_0$ for $\psi \in D(A)$. Suppose that $\varphi \in D(A^*)$. Then

$$\langle \psi, A^*\varphi\rangle = \langle A\psi, \varphi\rangle = \langle \psi, f\rangle\langle \psi_0, \varphi\rangle = \langle \psi, \langle \varphi, \psi_0\rangle f\rangle$$

for all $\psi \in D(A)$. Thus, $A^*\varphi = \langle \varphi, \psi_0\rangle f$. Since $f \notin L^2(\mathbb{R})$, $\langle \varphi, \psi_0\rangle = 0$. So, $D(A^*) \subset [\psi_0]^\perp$ and therefore $D(A^*)$ is not dense. Now suppose that A has a closed extension. Then, if we consider the closure of A, $\bar{A}$, we have that $D(A) \subset D(\bar{A})$ and $A \subset \bar{A}$. Since $D(\bar{A})$ is dense, it is easy to see that $(\bar{A})^* \subset A^*$. So,

$$D((\bar{A})^*) \subset D(A^*) \subset [\psi_0]^\perp,$$

which is impossible by part (2). □

DEFINITION B.26 (Self-Adjoint). Let $A : D(A) \to H$ be a densely defined linear operator. Then A is called a *self-adjoint operator* if $A = A^*$, that is, if and only if $D(A) = D(A^*)$ and

$$\langle Au, v\rangle = \langle u, Av\rangle, \qquad \text{for all } u, v \in D(A).$$

Now we have the following basic characterization of the spectrum of an unbounded self-adjoint operator.

THEOREM B.27 (Spectrum of Self-Adjoint Operators). *Let $A : D(A) \to H$ be a self-adjoint operator ($D(A)$ is dense in H). Then we have the following:*

(1) $\sigma(A) \subset \mathbb{R}$.
(2) *A has no residual spectrum. So,* $\sigma(A) = \sigma_p(A) \cup \sigma_c(A)$.

PROOF. The proof follows along the same lines as the proof of parts (3) and (4) in Theorem B.23. □

REMARK B.28. An operator A with dense domain in H is *symmetric* on its domain if

$$\langle Au, v\rangle = \langle u, Av\rangle \qquad \text{for all } u, v \in D(A).$$

Every self-adjoint operator is symmetric and every *bounded* symmetric operator is self-adjoint, but there are unbounded symmetric operators which are **not** self-adjoint. In fact, define

$$A_0 u = \frac{du}{dx}$$

with domain in $L^2([0,1])$ given by

$$D(A_0) = \{u : u \text{ is absolutely continuous, } u' \in L^2([0,1]), u(0) = 0, u(1) = 0\}.$$

Then, A_0 is closed, $D(A_0)$ is dense in $L^2([0,1])$, and

$$\begin{cases} A_0^* v = -\dfrac{dv}{dx}, \\ D(A_0^*) = \{v \in L^2([0,1]) : v \text{ is absolutely continuous and } v' \in L^2([0,1])\}. \end{cases}$$

Now, define

$$Au = -i\frac{du}{dx}, \qquad D(A) = D(A_0).$$

Then A is symmetric but

$$\begin{cases} A^* v = -i\dfrac{dv}{dx}, \\ D(A^*) = \{v \in L^2([0,1]) : v \text{ is absolutely continuous and } v' \in L^2([0,1])\}. \end{cases}$$

REMARK B.29. Let A be a linear operator on a domain $D(A)$ dense in H. Then the relation

$$[\mathcal{R}(A)]^{\perp} = \mathcal{N}(A^*)$$

characterizes the range of A (see [**153**]). If $\mathcal{R}(A)$ is closed, then

$$\mathcal{R}(A) = [\mathcal{N}(A^*)]^{\perp}.$$

The following theorem is a very useful tool for finding the spectrum of a self-adjoint operator.

THEOREM B.30. *Let $A : D(A) \to H$ be a self-adjoint operator ($D(A)$ is dense in H). Then a real number λ is in $\sigma(A)$ if and only if there is a sequence $\{u_n\} \subset D(A)$ such that*

$$\|u_n\| = 1 \qquad \textit{and} \qquad \|(A - \lambda I)u_n\| \to 0 \ \textit{ as } \ n \to \infty. \tag{B.14}$$

PROOF. Suppose (B.14) holds. If $\lambda \in \rho(A)$, we have that $R_\lambda(A)(A - \lambda I)u_n = u_n \to 0$ by the second part of (B.14). This contradicts the first part of (B.14). Conversely, suppose (B.14) does not hold. Then there is a constant C such that

$$\|u\| \leqq C\|(A - \lambda I)u\|, \qquad u \in D(A). \tag{B.15}$$

In fact, if (B.15) does not hold, there is a sequence $\{v_n\} \subset D(A)$ such that

$$\|(A - \lambda I)v_n\|^{-1}\|v_n\| \to \infty \qquad \text{as } n \to \infty.$$

Set $v_n/\|v_n\|$. Then u_n satisfies (B.14). Thus (B.15) holds.

Inequality (B.15) *tells us two basic things.* The first is that $A - \lambda I$ is a one-to-one function on $D(A)$. The second big piece of information from (B.15) is that $\mathcal{R}(A - \lambda I)$ is closed. In fact, if $\{f_n\} \subset \mathcal{R}(A - \lambda I)$ and $f_n \to f$ in H, let u_n be the unique solution of $(A - \lambda I)u_n = f_n$. By (B.15), u_n is a Cauchy sequence. Thus there exists $u \in H$ such that $u_n \to u$. Therefore if $v \in D(A)$, we have

$$\langle u, (A - \lambda I)v\rangle = \lim_{n\to\infty} \langle u_n, (A - \lambda I)v\rangle = \lim_{n\to\infty} \langle f_n, v\rangle = \langle f, v\rangle.$$

Hence $u \in D(A)$ and $(A - \lambda I)u = f$. Once we know this, we can show that $\mathcal{R}(A - \lambda I) = H$ and this in turn implies that $\lambda \in \rho(A)$. Let $f \in H$ and $w \in \mathcal{R}(A - \lambda I)$. Let v be the unique element of $D(A)$ such that $(A - \lambda I)v = w$. Set $F(w) = \langle v, f\rangle$. This defines a linear functional on $\mathcal{R}(A - \lambda I)$, which is a Hilbert space. F is also bounded. In fact,

$$|F(w)| \leqq \|v\|\|f\| \leqq C\|f\|\|w\|.$$

By the Riesz Representation Theorem, there is a $u \in \mathcal{R}(A - \lambda I)$ such that $F(w) = \langle w, u\rangle$ for all $w \in \mathcal{R}(A - \lambda I)$. This gives

$$\langle u, (A - \lambda I)v\rangle = \langle f, v\rangle, \qquad v \in D(A).$$

Since A is self-adjoint, this implies that $u \in D(A)$ and $(A - \lambda I)u = f$. Hence $f \in \mathcal{R}(A - \lambda I)$, and the proof is complete. □

As a consequence of the proof of Theorem B.30 we have the following basic result.

COROLLARY B.31. If $A : D(A) \to H$ is a self-adjoint operator and

$$\|u\| \leqq C\|Au\|, \qquad u \in D(A),$$

then A is one-to-one and $\mathcal{R}(A)$ is closed.

B.2.4. Examples of Explicit Spectrum.

EXAMPLE B.32 (The Key Point: Multiplication Operator). Let $q(x)$ be a real-valued mensurable function defined on $(-\infty, \infty)$. We define the multiplication operator as

$$\begin{cases} \mathcal{M}_q : D(\mathcal{M}_q) \to L^2(\mathbb{R}), \\ \mathcal{M}_q u = qu \end{cases}$$

with $D(\mathcal{M}_q) = \{u \in L^2(\mathbb{R}) : qu \in L^2(\mathbb{R})\}$. Since $C_0^\infty(\mathbb{R}) \subset D(\mathcal{M}_q)$ and $C_0^\infty(\mathbb{R})$ is dense in $L^2(\mathbb{R})$, then $D(\mathcal{M}_q)$ is dense in $L^2(\mathbb{R})$. Now, we have the following properties of the operator $\mathcal{M}_q$ which will be crucial in our future work.

(1) $\mathcal{M}_q$ *is self-adjoint.* In fact, initially $\mathcal{M}_q$ is symmetric:

$$\langle \mathcal{M}_q u, v\rangle = \int_{-\infty}^{\infty} q(x)u(x)\overline{v(x)}dx = \int_{-\infty}^{\infty} u(x)\overline{q(x)v(x)}dx = \langle u, \mathcal{M}_q v\rangle,$$

for all $u, v \in \mathcal{M}_q$. Hence, $\mathcal{M}_q \subset \mathcal{M}_q^*$. Now we show that $D(\mathcal{M}_q^*) \subset D(\mathcal{M}_q)$. In fact, let $f \in D(\mathcal{M}_q^*)$ and define

$$\chi_N(x) = \begin{cases} 1 & \text{if } |q(x)| \leqq N, \\ 0 & \text{otherwise.} \end{cases}$$

Then $\chi_N(x)|\mathcal{M}_q^* f(x)|^2 \leqq \chi_{N+1}(x)|\mathcal{M}_q^* f(x)|^2$ for every $x \in \mathbb{R}$ and $N \in \mathbb{R}$. So, using the Monotone Convergence Theorem and $\chi_N \psi \in D(\mathcal{M}_q)$ for every $\psi \in L^2(\mathbb{R})$, we have

$$\begin{aligned}
\|\mathcal{M}_q^* f\| &= \lim_{N\to\infty} \|\chi_N \mathcal{M}_q^* f\| = \lim_{N\to\infty} \Big(\sup_{\|\psi\|=1} |\langle \psi, \chi_N \mathcal{M}_q^* f\rangle| \Big) \\
&= \lim_{N\to\infty} \Big(\sup_{\|\psi\|=1} |\langle \mathcal{M}_q(\chi_N\psi), f\rangle| \Big) = \lim_{N\to\infty} \Big(\sup_{\|\psi\|=1} |\langle \psi, \chi_N q f\rangle| \Big) \\
&= \lim_{N\to\infty} \|\chi_N q f\| = \|qf\|.
\end{aligned}$$

Then $qf \in L^2(\mathbb{R})$. Hence $f \in \mathcal{M}_q$ and so $\mathcal{M}_q$ is self-adjoint. Since $\mathcal{M}_q$ is self-adjoint, it follows from Theorem B.25 that $\mathcal{M}_q$ is closed. Moreover, by Theorem B.27, $\sigma(\mathcal{M}_q) \subset \mathbb{R}$.

(2) Suppose that for every $r \in \mathbb{R}$ the set $I_r = \{x \in \mathbb{R} : q(x) = r\}$ has Lebesgue measure zero. Then, $\mathcal{M}_q$ *has no eigenvalues.* So, $\sigma(\mathcal{M}_q) = \sigma_c(\mathcal{M}_q)$. In fact, for any λ, let $f \in D(\mathcal{M}_q)$ be such that $\mathcal{M}_q f = \lambda f$. Then

$$\begin{aligned}
0 = \|(\mathcal{M}_q - \lambda)f\|^2 &= \int_{-\infty}^{\infty} |q(x) - \lambda|^2 |f(x)|^2 \, dx \\
&= \int_{\mathbb{R} - I_\lambda} |q(x) - \lambda|^2 |f(x)|^2 \, dx.
\end{aligned}$$

Since $|q(x) - \lambda| > 0$ for all $x \in \mathbb{R} - I_\lambda$, we have that $f(x) = 0$ almost everywhere in $\mathbb{R} - I_\lambda$. So, $f \equiv 0$ and λ is not an eigenvalue of $\mathcal{M}_q$.

(3) Suppose q is continuous. Then, $\sigma(\mathcal{M}_q) = \overline{q(\mathbb{R})}$. In fact, we consider $\lambda \notin \overline{q(\mathbb{R})}$ and define $M = \operatorname{dis}(\lambda, q(\mathbb{R})) > 0$. Then, for $f \in D(\mathcal{M}_q)$

$$\|(\mathcal{M}_q - \lambda)f\|^2 = \int_{-\infty}^{\infty} |q(x) - \lambda|^2 |f(x)|^2 \, dx \geqq K^2 \|f\|^2.$$

Therefore, $\mathcal{M}_q - \lambda$ is one-to-one and has a bounded inverse on its range. By Theorem B.12, $\mathcal{R}(\mathcal{M}_q - \lambda)$ is closed. If $\mathcal{R}(\mathcal{M}_q - \lambda) \neq H$, then $\lambda \in \sigma_r(\mathcal{M}_q)$, which is impossible by Theorem B.27. So, $\mathcal{R}(\mathcal{M}_q - \lambda) = H$ and by the Closed Graph Theorem (Theorem B.10) we have that $\lambda \in \rho(\mathcal{M}_q)$. Therefore, $\sigma(\mathcal{M}_q) \subset \overline{q(\mathbb{R})}$. To obtain the other relation, it is sufficient to show $q(\mathbb{R}) \subset \sigma(\mathcal{M}_q)$. Let $\lambda = q(r)$. Then, by continuity, for every $\epsilon > 0$ the set $B_\epsilon = \{x : |\lambda - q(x)| < \epsilon\}$ has Lebesgue measure not zero, that is, $m(B_\epsilon) > 0$. We consider the functions $f_\epsilon = \frac{1}{\sqrt{m(B_\epsilon)}}\chi_{B_\epsilon}$. Then $\|f_\epsilon\| = 1$ and

$$\begin{aligned}\|(\mathcal{M}_q - \lambda)f_\epsilon\|^2 &= \int_{-\infty}^{\infty} |q(x) - \lambda|^2 |f_\epsilon(x)|^2 dx \\ &= \frac{1}{m(B_\epsilon)} \int_{B_\epsilon} |q(x) - \lambda|^2 dx \leqq \epsilon^2.\end{aligned}$$

Suppose that $\lambda \in \rho(\mathcal{M}_q)$. Then λ is not an eigenvalue of $\mathcal{M}_q$ and so $(\mathcal{M}_q - \lambda)f_\epsilon \neq 0$ for every $\epsilon > 0$. Define the functions

$$g_\epsilon = \frac{1}{\|(\mathcal{M}_q - \lambda)f_\epsilon\|}[(\mathcal{M}_q - \lambda)f_\epsilon].$$

Here g_ϵ belongs to the domain of $R_\lambda(\mathcal{M}_q) = (\mathcal{M}_q - \lambda)^{-1}$ and has norm 1. Thus

$$\|R_\lambda(\mathcal{M}_q)g_\epsilon\| = \frac{1}{\|(\mathcal{M}_q - \lambda)f_\epsilon\|}\|f_\epsilon\| \geqq \frac{1}{\epsilon}.$$

This shows that the resolvent $R_\lambda(\mathcal{M}_q)$ is an unbounded operator, which is a contradiction. So, $q(\mathbb{R}) \subset \sigma(\mathcal{M}_g)$.

(4) We consider q which is real-valued (or complex-valued) and measurable. We say that λ is in the *essential range* of q (er(q) henceforth) if and only if

$$m[q^{-1}(B(\lambda;\epsilon))] > 0$$

for all $\epsilon > 0$, where $B(\lambda;\epsilon) = \{y \in \mathbb{C} : |y - \lambda| < \epsilon\}$. It follows immediately from the definition of essential range that $\mathrm{er}(q) \subseteq \overline{q(\mathbb{R})}$. In general, we have that $\mathrm{er}(q) \neq \overline{q(\mathbb{R})}$. In fact, we consider the measurable function $q(x) = 1$ for $x \in [0, 1)$, $q(1) = 2$, and $q(x) = 0$ otherwise. Then, $2 \in \overline{q(\mathbb{R})}$ and $q^{-1}(B(2; \frac{1}{2})) = \{1\}$. So, $2 \notin \mathrm{er}(q)$.

Now, we will prove that $\sigma(\mathcal{M}_q)$ is the essential range of q. First, we will see that $0 \in \rho(\mathcal{M}_q)$ if and only if there exists $\delta > 0$ such that $|q(x)| \geqq \delta$, a.e. In fact, if $0 \in \rho(\mathcal{M}_q)$, then there is an $S \in \mathcal{B}(L^2(\mathbb{R}), L^2(\mathbb{R}))$ such that $S\mathcal{M}_q\psi = \psi$ for all $\psi \in D(\mathcal{M}_q)$. So,

$$\|\psi\| \leqq \|S\|\|\mathcal{M}_q\psi\|.$$

Now, let

$$G = \{x : |q(x)| < \|S\|^{-1}\}$$

and consider E which is measurable with $m(E)$ positive and finite. Then, $\chi_{G\cap E} \in D(\mathcal{M}_q)$ and so we get

$$\int_{G\cap E} (1 - \|S\|^2|q|^2)dx = 0.$$

Then from the definition of G, we need to have $m(G \cap E) = 0$. Since the measure space $(\mathbb{R}, m)$ is σ-finite, it follows that $m(G) = 0$. Now, if

$|q(x)| \geqq \delta$ a.e., then $|q^{-1}(x)| \leqq \delta^{-1}$ a.e. and so $S = \mathcal{M}_{q^{-1}}$ is a bounded inverse to $\mathcal{M}_q$.

Therefore we have that $\lambda \in \rho(\mathcal{M}_q)$ if and only if $0 \in \rho(\mathcal{M}_q - \lambda)$ if and only if there is a $\delta > 0$ such that $|q(x) - \lambda| \geqq \delta$ a.e. if and only if there is a $\delta > 0$ such that

$$m(\{x : |q(x) - \lambda| < \delta\}) = m[q^{-1}(B(\lambda; \delta))] = 0,$$

which is equivalent to saying that λ does not belong to the essential range of q.

REMARK B.33. We can show that $\mathcal{M}_q$ is a closed operator without using the fact that it is self-adjoint. More precisely, let (Ω, μ) be a measure space and let $q : \Omega \to \mathbb{C}$ be a measurable function. Then the multiplication operator $\mathcal{M}_q$ with domain

$$D(\mathcal{M}_q) = \{\psi \in L^2(\Omega, \mu) : q\psi \in L^2(\Omega, \mu)\}$$

is densely defined and closed in $L^2(\Omega, \mu)$. In fact, let $\psi \in L^2(\Omega, \mu)$. Define for every $n \in \mathbb{N}$

$$\psi_n(x) = \begin{cases} \psi(x) & \text{if } |q(x)| \leqq n, \\ 0 & \text{otherwise.} \end{cases}$$

Then, from the relations $|q\psi_n| \leqq n|\psi|$, $|\psi_n| \leqq |\psi|$, and $\psi_n(x) \to \psi(x)$, it follows that $\psi_n \in D(\mathcal{M}_q)$ and $\psi_n \to \psi$ in $L^2(\Omega, \mu)$ (the last affirmation is obtained via the Dominated Convergence Theorem). Now, let $\{\psi_n\} \subset D(\mathcal{M}_q)$ such that $\psi_n \to \psi$ and $\mathcal{M}_q\psi_n \to \phi$ in $L^2(\Omega, \mu)$. Then, taking a subsequence if it is necessary, we can assume $\psi_n \to \psi$ and $q\psi_n \to \phi$ a.e. So, $q\psi_n \to q\psi$ a.e. and therefore $q\psi = \phi$ a.e. Hence, $\psi \in D(\mathcal{M}_q)$ and $\mathcal{M}_q\psi = \phi$.

REMARK B.34. Note that item (2) above is still true if we consider $q : \mathbb{R} \to \mathbb{C}$, but in this case $\mathcal{M}_q$ is not self-adjoint.

Unless $q \in L^\infty(\mathbb{R})$, the operator $\mathcal{M}_q$ will be unbounded, and therefore we have found a lot of unbounded self-adjoint operators. The main *Spectral Theorem to Self-Adjoint Operators* below says that we have found all of them.

DEFINITION B.35. Let $A : D(A) \subset H \to H$, $B : D(B) \subset H_1 \to H_1$ be linear operators. Then A and B are *unitarily equivalent* if and only if there is a unitary operator $U : H \to H_1$ such that

(1) $U[D(A)] = D(B)$,
(2) $U^{-1}BUv = Av$, for all $v \in D(A)$.

THEOREM B.36 (Spectral Theorem: Multiplication Operator Form). *Let A be a self-adjoint operator on a separable Hilbert space H (densely defined with domain $D(A)$). Then there is a separable, σ-finite measure space (Ω, μ), a unitary operator $U : H \to L^2(\Omega, \mu)$, and a real-valued function q on Ω which is finite a.e., such that A is unitarily equivalent to the multiplication operator $\mathcal{M}_q$.*

PROOF. See [**241**], [**267**]. □

COROLLARY B.37. Let A be a self-adjoint operator with domain $D(A)$, unitarily equivalent to $\mathcal{M}_q$. Then $\sigma(A) = \sigma(\mathcal{M}_q)$.

PROOF. Let U be the unitary operator obtained via Theorem B.36. Then the relations for $\lambda \in \mathbb{C}$,

$$\mathcal{M}_q - \lambda = U(A-\lambda)U^{-1} \quad \text{and} \quad A-\lambda = U^{-1}(\mathcal{M}_q - \lambda)U,$$

show the corollary. □

B.3. Pseudo-Differential Operators and Their Spectrum

Now we establish a more accurate spectral theory associated with differential operators that appear in the study of stability of solitary wave solutions. Initially, we have two results about the stability of the closedness and self-adjointness of operators (see Kato [**153**]).

DEFINITION B.38. Let T and A be operators with the same domain space H. We say that A is *relatively bounded with respect to* T (or simply T*-bounded*) if

$$\text{(B.16)} \qquad \begin{cases} D(T) \subseteq D(A) & \text{and} \\ \|Au\| \leqq a\|u\| + b\|Tu\|, & u \in D(T), \end{cases}$$

where $a, b \geqq 0$. The number $b_0 = \inf\{b : b \text{ satisfies } (B.16)\}$ is called the *relative bound* of A with respect to T or simply the T*-bound*.

It follows immediately from (B.16) that a bounded operator $A : H \to H$ is T*-bounded* for any T with $D(T) \subseteq H$ and with T-bound equal to zero.

THEOREM B.39 (Stability of Closedness). *Let T and A be operators defined on H, and let A be T-bounded with T-bound smaller than 1. Then $S = T + A$ is closed if and only if T is closed. In particular $T + A$ is closed if A is bounded and T is closed.*

PROOF. Using inequality (B.16) with $b < 1$, we have that

$$\text{(B.17)} \qquad -a\|u\| + (1-b)\|Tu\| \leqq \|Su\| \leqq a\|u\| + (1+b)\|Tu\|$$

with $u \in D(T)$. So, if S is closed, then it follows from the second inequality in (B.17) that if $\{u_n\} \subseteq D(T) = D(S)$ and we have that $u_n \to u$ and $Tu_n \to f$, then $\{Su_n\}$ is Cauchy and therefore $u \in D(S)$ with $Su_n \to Su$. Now, the first inequality in (B.17) shows that $Tu_n \to Tu$. So, $Tu = f$ and therefore T is closed. Similarly, S is closed if T is. □

We note that is easy to see that the inequality in (B.16) is equivalent to the condition

$$\text{(B.18)} \qquad \|Au\|^2 \leqq a_1^2\|u\|^2 + b_1^2\|Tu\|^2, \qquad u \in D(T),$$

where $a_1, b_1 \geqq 0$. Thus the T-bound of A may be defined as $b_0 = \inf\{b_1 : b_1 \text{ satisfies } (B.18)\}$.

Before giving our stability theorem for self-adjointness, we establish a basic criterion for self-adjointness (see [**242**]).

THEOREM B.40. *Let T be a symmetric operator on H with dense domain $D(T)$. Then the following three statements are equivalent:*

(1) *T is self-adjoint.*
(2) *T is closed and $\mathcal{N}(T^* \pm iI) = \{0\}$.*
(3) *$\mathcal{R}(T \pm iI) = H$.*

PROOF. It follows immediately from Theorems B.25 and B.27 that (1) implies (2). Suppose that (2) holds; then the relation $\overline{\mathcal{R}(A)} = \mathcal{N}(A^*)^\perp$ shows that $\mathcal{R}(T-iI)$ is dense in H. Now, we see that $\mathcal{R}(T-iI)$ is closed. Initially, since T is symmetric, it follows that for all $\varphi \in D(T)$

$$\|(T-i)\varphi\|^2 = \|T\varphi\|^2 + \|\varphi\|^2.$$

Thus if $\psi_0 \in H$ and $\{\varphi_n\} \subseteq D(T)$ with $(T-iI)\varphi_n \to \psi_0$, then $\varphi_n \to \varphi_0$ and $T\varphi_n \to \psi_0 + i\varphi_0$. Since T is closed, $\varphi_0 \in D(T)$ and $T\varphi_0 = \psi_0 + i\varphi_0$. Hence, $\psi_0 \in \mathcal{R}(T-iI)$. Similarly, $\mathcal{R}(T+iI) = H$.

Finally, we will show that (3) implies (1). Since T is symmetric, then T^* is an extension of T and so $D(T) \subseteq D(T^*)$. Hence we will show that $D(T^*) \subseteq D(T)$. Let $\varphi \in D(T^*)$. Since $\mathcal{R}(T-iI) = H$, then there is $\psi \in D(T)$ such that $(T-iI)\psi = (T^*-iI)\varphi$. So, $(T^*-i)(\varphi-\psi) = 0$. Since $\mathcal{R}(T+iI) = H$, then

$$\mathcal{N}(T^*-iI) = \mathcal{R}(T+iI)^\perp = \{0\},$$

so $\varphi = \psi \in D(T)$. This proves that T is self-adjoint. □

THEOREM B.41 (Stability of Self-Adjointness). *Let T be self-adjoint. If A is symmetric and T-bounded with T-bound smaller than 1, then $T+A$ is also self-adjoint. In particular $T+A$ is self-adjoint if A is bounded and symmetric with $D(T) \subseteq D(A)$.*

PROOF. It is immediate that $D(T+A) = D(T)$ and $T+A$ is symmetric. We may assume without loss of generality that (B.18) holds with constants a_1, b_1 such that $a_1 > 0$, $0 < b_1 < 1$. Since T is self-adjoint, we have from (B.18) that

$$\|Au\|^2 \leqq a_1^2\|u\|^2 + b_1^2\|Tu\|^2 = \|(b_1T \mp ia_1I)u\|^2, \qquad u \in D(T).$$

Let $c_1 = a_1/b_1$ and $(T \mp ic_1I)u = v$. Then considering $R_{\pm ic_1} = R_{\pm ic_1}(T \mp ic_1I)$,

$$\|AR_{\pm ic_1}v\| \leqq b_1\|(T \mp ic_1I)R_{\pm ic_1}v\| = b_1\|v\| \qquad \text{for all } v \in H.$$

So, we have that the operators $B_\pm = -AR_{\pm ic_1} \in \mathcal{B}(H,H)$ and they satisfy $\|B_\pm\| \leqq b_1 < 1$. Then by the Neumann series we have that $(I-B_\pm)^{-1}$ exists and belongs to $\mathcal{B}(H,H)$. Since

$$T + A \mp ic_1I = (I - B_\pm)(T \mp ic_1)$$

and $T \mp ic_iI$ has range H (since $ic_1\mp \in \rho(T)$), then $\mathcal{R}(T+A\mp ic_1I) = H$. So, by Theorem B.40, $T+A$ is self-adjoint. □

B.3.1. Examples of the Spectrum of Pseudo-Differential Operators.

EXAMPLE B.42. We consider the linear operator $\mathcal{A} : L^2(\mathbb{R}) \to L^2(\mathbb{R})$ defined as a Fourier multiplier operator by

$$\widehat{\mathcal{A}f}(\xi) = [q(\xi) + c]\widehat{f}(\xi) \qquad \text{for all } \xi \in \mathbb{R}, \tag{B.19}$$

where $c \in \mathbb{R}$. We assume that $q : \mathbb{R} \to \mathbb{R}$ is a continuous, even function and satisfies

$$\begin{cases} a_1|\xi|^\mu \leqq q(\xi) \leqq a_2|\xi|^\mu & \text{for } |\xi| \geqq \xi_0, \\ q(\xi) \geqq b & \text{for all } \xi \in \mathbb{R} \end{cases} \tag{B.20}$$

where ξ_0 and b are finite real numbers, $\mu \geqq 1$, and $a_1, a_2 > 0$. Then we have the following results:

(1) *Spectrum of* $\mathcal{A} = [c + \inf_{\xi\in\mathbb{R}} q(\xi), +\infty)$.
It follows from (B.19) and (B.20) that a natural domain of $\mathcal{A}$ can be chosen as $D(\mathcal{A}) = H^\mu(\mathbb{R})$. So, via the Fourier transform, this in turn implies that $\mathcal{A}$ is a self-adjoint, closed operator on $L^2(\mathbb{R})$. Now, we consider the space with weight

$$L^2_{(1+\xi^2)^\mu d\xi}(\mathbb{R}) = \{f : \mathbb{R} \to \mathbb{C} : \int_{-\infty}^{\infty} (1+\xi^2)^\mu |f(\xi)|^2 d\xi < \infty\}. \tag{B.21}$$

Then, defining the multiplication operator $\mathcal{M}_{q+C}$ on $L^2(\mathbb{R})$, it follows that $D(\mathcal{M}_{q+C}) = L^2_{(1+\xi^2)^\mu d\xi}(\mathbb{R})$. Since the Fourier transform $\mathcal{F} : L^2(\mathbb{R}) \to L^2(\mathbb{R})$ is a unitary operator and $\mathcal{F}[H^\mu(\mathbb{R})] = L^2_{(1+\xi^2)^\mu d\xi}(\mathbb{R})$, it follows from (B.19) that $\mathcal{A}$ and $\mathcal{M}_{q+c}$ are *unitarily equivalent.* So, from the Spectral Theorem (Corollary B.37) it follows that $\sigma(\mathcal{A}) = \sigma(\mathcal{M}_{q+c})$. Therefore, we have from Example B.32 that

$$\sigma(\mathcal{A}) = [c + \inf_{\xi\in\mathbb{R}} q(\xi), +\infty). \tag{B.22}$$

(2) *Linear operators associated to the GKdV, GBO, and ILW equations.*
Some important operators that appear in the study of solitary wave solutions are defined by the functions

$$q_1(\xi) = \xi^2, \quad q_2(\xi) = |\xi|, \quad q_3(\xi) = \xi\coth(\xi H) - \frac{1}{H},$$

for each $H \in (0,\infty)$, namely, the linear operators associated with the GKdV, GBO, and ILW equations, respectively (see (2.5), (2.7), and (2.8)). So, if we consider the operators

$$\mathcal{A}_1 = -\frac{d^2}{dx^2} + c, \quad \mathcal{A}_2 = \mathcal{H}\partial_x + c, \quad \text{and} \quad \mathcal{A}_3 f = \mathcal{F}^{-1}\mathcal{M}_{q_3+c}\mathcal{F}f, \tag{B.23}$$

we obtain from (B.22) and Example B.32 that

$$\sigma(\mathcal{A}_i) = \sigma_c(\mathcal{A}_i) = [c, \infty), \qquad i = 1, 2, 3. \tag{B.24}$$

B.4. Spectrum of Linear Operators Associated to Solitary Waves

Now, we establish the main theorem of this appendix, which is about the spectrum of linear operators associated to some solitary wave solutions of nonlinear wave equations. In fact, we consider the following general model for one-dimensional long-wavelength nonlinear waves:

$$u_t + f(u)_x - Mu_x = 0, \tag{B.25}$$

where $u = u(x,t)$, $f : \mathbb{R} \to \mathbb{R}$ is a smooth function with $f(0) = 0$, $f'(0) = 0$, and M is a differential or pseudo-differential operator with symbol q, more precisely,

$$\widehat{Mu}(\xi) = q(\xi)\widehat{u}(\xi)$$

for all $\xi \in \mathbb{R}$. It is assumed that the symbol q satisfies the conditions established in (B.20) and that $q'(x) \neq 0$ a.e. on $\mathbb{R}$. The travelling wave solutions $u(x,t) = \phi(x-ct)$ of equation (B.25) which vanish at infinity are called solitary waves. Substituting this form of u into (B.25) and integrating once, with zero boundary conditions imposed at infinity, we obtain the differential or pseudo-differential equation for ϕ,

$$(M + c)\phi - f(\phi) = 0. \tag{B.26}$$

In general it is assumed that the constant c, called the wave speed of the solution ϕ, satisfies $c > -b$ where b is the constant appearing in (B.20). So, $M + c$ represents a positive operator. We note that if $\phi \in H^{\mu}(\mathbb{R})$ is a solution of (B.25), then a bootstrap argument shows that $\phi \in H^{s}(\mathbb{R})$ for all $s \in \mathbb{R}$. Hence, ϕ is infinitely differentiable with all derivatives in $L^2(\mathbb{R})$. So, taking one derivative in (B.26), we obtain that ϕ' satisfies

$$(M + c)\phi' - f'(\phi)\phi' = 0.$$

As is known from the theories about stability of solitary wave solutions for equations of the form (B.26) (see Chapters 6 and 7), spectral information for the linear operator $\mathcal{L} : L^2(\mathbb{R}) \to L^2(\mathbb{R})$ defined by

$$\mathcal{L}u \equiv (M + c)u - f'(\phi)u \tag{B.27}$$

is a crucial ingredient. Here we consider $f'(\phi)$ as being the multiplication operator $\mathcal{M}_{f'(\phi)}$. Since $f'(\phi(\xi)) \to 0$ as $|\xi| \to \infty$, it follows that $\mathcal{M}_{f'(\phi)}$ is a bounded operator on $L^2(\mathbb{R})$. Therefore, $D(\mathcal{L}) = D(M+c) = H^{\mu}(\mathbb{R})$ and it is self-adjoint (see Theorem B.41). Note that $\mathcal{L}$ has the eigenvalue zero with eigenfunction ϕ'.

The objective of this section is to show the next basic theorem.

THEOREM B.43. *Suppose* $\inf_{\xi\in\mathbb{R}} q(\xi) = 0$. *The operator* $\mathcal{L}$ *defined in* (B.27) *is a closed, unbounded, self-adjoint operator on* $L^2(\mathbb{R})$ *whose spectrum consists of the essential part* $[c, \infty)$ *together with a finite number of discrete eigenvalues (with finite-dimensional eigenspaces) in the interval* $(-\infty, c)$.

Note that condition $\inf_{\xi\in\mathbb{R}} q(\xi) = 0$ implies that the linear operator M is positive. This means $\langle M\psi, \psi\rangle \geqq 0$ for all $\psi \in D(M) = H^{\mu}(\mathbb{R})$.

Initially, we start by showing that $\mathcal{L}$ *has the same essential spectrum as* $M + c$. To show this, we need some preliminaries. We start by doing another partition of the spectrum.

DEFINITION B.44 (Essential Spectrum). Let $A : D(A) \subset H \to H$ be a linear operator.

(1) We say $\lambda \in \sigma(A)$ is *nonessential* if and only if the following conditions are satisfied:

$$\begin{cases} \lambda \text{ is an isolated eigenvalue of } \sigma(A) \text{ and} \\ \text{the space } \mathcal{N}(A - \lambda I) \text{ is of finite dimension.} \end{cases}$$

We call the set of all λ the *discrete spectrum* and we denote it by $\sigma_{\text{disc}}(A)$.

(2) The complement in $\sigma(A)$ of the nonessential points is called the *essential spectrum* of A. We denote this set by $\sigma_{\text{ess}}(A)$. So, $\sigma(A) = \sigma_{\text{disc}}(A) \cup \sigma_{\text{ess}}(A)$.

REMARK B.45. We note that there is considerable divergence in the literature concerning the definition of $\sigma_{\text{ess}}(A)$ (see Kato [**153**], Reed and Simon [**241**], [**242**], Schechter [**250**]). But all of these definitions coincide when A is self-adjoint.

Now we have a classical characterization of $\sigma_{\text{ess}}(A)$ for self-adjoint operators (see [**242**]).

THEOREM B.46 (Weyl's Criterion). *Let* A *be a self-adjoint operator with domain* $D(A)$. *For* $\lambda \in \mathbb{R}$, *the following conditions are equivalent:*

(1) $\lambda \in \sigma_{\text{ess}}(A)$.

(2) *There exists a sequence* $\{\psi_n\} \subset D(A)$ *orthogonal in* H *with* $\|\psi_n\| = 1$, *such that*

$$\lim_{n\to\infty} \|(A - \lambda I)\psi_n\| = 0.$$

EXAMPLE B.47. We consider the operator $\mathcal{A}$ defined in (B.19) with $q'(x) \neq 0$ a.e. in $\mathbb{R}$. Then, from (B.22) it follows that

$$\sigma(\mathcal{A}) = [c + \inf_{\xi\in\mathbb{R}} q(\xi), +\infty).$$

Moreover, it follows from Example B.42 that $\mathcal{A}$ has no eigenvalues and so

$$\sigma(\mathcal{A}) = \sigma_c(\mathcal{A}) = \sigma_{\text{ess}}(\mathcal{A}).$$

Now we will see that the essential spectrum is preserved when we perturb the operator of interest by a special class of operators.

THEOREM B.48 (Invariance of Essential Spectrum). *Let* A *and* B *be self-adjoint operators on* H *such that*

(1) $D(A) = D(B)$ *and*
(2) $A - B$ *is a compact operator from* $D(A)$ *to* H. *Here* $D(A)$ *is considered with the graph norm generated by* A, $\|u\|_A^2 = \|Au\|^2 + \|u\|^2$. *(Note that* $D(A)$ *with the graph norm* $\|\cdot\|_A$ *is a Banach space since* A *is closed).*

Then, $\sigma_{\text{ess}}(A) = \sigma_{\text{ess}}(B)$.

PROOF. Initially, we consider the operator identity $i : D(A) \to D(B)$ where $D(A)$ and $D(B)$ have the graph norms $\|\cdot\|_A$ and $\|\cdot\|_B$, respectively. Since i is a closed operator on the closed domain $D(A)$, we have from Theorem B.10 that i is bounded. So, there is a positive constant η such that $\|u\|_B \leqq \eta\|u\|_A$. Similarly we obtain a positive constant η_0 such that $\|u\|_A \leqq \eta_0\|u\|_B$. So, $D(A)$ and $D(B)$ have equivalent norms. In particular,

$$A - B : D(B) \to H$$

is compact. Hence, by symmetry, we only need to show that $\sigma_{\text{ess}}(A) \subset \sigma_{\text{ess}}(B)$. Let $\lambda \in \sigma_{\text{ess}}(A)$. Then, by Theorem B.46, there is a sequence $\{\psi_n\} \subset D(A)$, orthogonal in H, and $\|\psi_n\| = 1$ such that $(A - \lambda I)\psi_n \to 0$ in H. Since

$$\|\psi_n\|_A^2 = \|\psi_n\|^2 + \|A\psi_n\|^2 = 1 + \|\lambda\psi_n + (A - \lambda)\psi_n\|^2 \to 1 + \lambda^2,$$

it follows that $\{\psi_n\}$ is bounded in $(D(A), \|\cdot\|_A)$. So, since $A - B$ is compact, we have that there is a subsequence $\{\psi_{n_k}\}$ such that $\{(A - B)\psi_{n_k}\}$ is convergent in H.

Now, for $\varphi \in H$ we consider the Fourier coefficients of φ with respect to $\{\psi_{n_k}\}$: $\alpha_k = \langle\varphi, \psi_{n_k}\rangle$. Then

$$\begin{aligned}\|\varphi - \sum_{k=1}^{m} \alpha_k\psi_{n_k}\|^2 &= \|\varphi\|^2 + \sum_{k=1}^{m} |\langle\varphi, \psi_{n_k}\rangle - \alpha_k|^2 - \sum_{k=1}^{m} |\langle\varphi, \psi_{n_k}\rangle|^2 \\ &= \|\varphi\|^2 - \sum_{k=1}^{m} |\langle\varphi, \psi_{n_k}\rangle|^2.\end{aligned}$$

Hence, $\sum_{k=1}^{m} |\langle\varphi, \psi_{n_k}\rangle|^2 \leqq \|\varphi\|^2$. Letting $m \to \infty$, we have

$$\sum_{k=1}^{\infty} |\langle\varphi, \psi_{n_k}\rangle|^2 \leqq \|\varphi\|^2 \quad \text{(Bessel's inequality)}.$$

Therefore, $\langle \psi_{n_k}, \varphi \rangle \to 0$ as $k \to \infty$. By what we have just shown, for every $\varphi \in D(A)$

$$\langle (A-B)\psi_{n_k}, \varphi \rangle = \langle \psi_{n_k}, (A-B)\varphi \rangle \to 0, \qquad k \to \infty.$$

Let $\psi = \lim_{k\to\infty} (A-B)\psi_{n_k}$. Then, $\langle \psi, \varphi \rangle = 0$. Since $D(A)$ is dense in H, it follows that $\psi = 0$. Therefore,

$$(B-\lambda)\psi_{n_k} = (A-\lambda)\psi_{n_k} - (A-B)\psi_{n_k} \to 0 \quad \text{as } k \to \infty.$$

So, $\lambda \in \sigma_{\text{ess}}(B)$. □

In the next theorem we have that q satisfies (B.20), $q'(x) \neq 0$ a.e., and $\inf_{\xi\in\mathbb{R}} q(\xi) = 0$.

THEOREM B.49 (Essential Spectrum of $\mathcal{L}$). *We consider the linear operator defined in* (B.27)*:*

$$\mathcal{L}u = (M+c)u - f'(\phi)u.$$

Then, $\sigma_{\text{ess}}(\mathcal{L}) = [c, \infty)$.

PROOF. Initially we note that $D(\mathcal{L}) = D(M+c)$. From Example B.42 and Theorem B.48 we only need to show that $\mathcal{L} - (M+c)$ is compact, in other words, $f'(\phi)$ is relatively compact with respect to $M+c$. So, suppose that $\{\psi_n\}$ and $\{(M+c)\psi_n\}$ are bounded sequences in $L^2(\mathbb{R})$. We are required to show that the sequence $\{f'(\phi)\psi_n\}$ has a convergent subsequence in $L^2(\mathbb{R})$. The assumptions on the sequence $\{\psi_n\}$ and the symbol q defining M (see (B.20)) imply that $\{\psi_n\}$ is bounded in $H^\mu(\mathbb{R})$, so there is a function $\psi \in H^\mu(\mathbb{R})$ such that (passing to some subsequence if necessary) $\psi_n \rightharpoonup \psi$ in $H^\mu(\mathbb{R})$ ("$\rightharpoonup$" means weak convergence). Since the inclusion of $H^\mu(\Omega)$ into $L^2(\Omega)$ is compact on bounded domains Ω, one may conclude (passing to subsequences if necessary and using a Cantor diagonalization argument) that $\{\psi_n\}$ converges to ψ in the $L^2(\Omega)$-norm on any bounded interval Ω in $\mathbb{R}$. Let φ be a smooth function with $\varphi(x) = 1$ if $|x| \leqq 1$ and $\varphi(x) = 0$ if $|x| \geqq 2$, and define $\varphi_R(x) = \varphi(x/R)$ for $x \in \mathbb{R}$. Then for each fixed R, we can see that

$$\|\varphi_R f'(\phi)(\psi_n - \psi)\| \leqq D\|\psi_n - \psi\|_{L^2(-2R,2R)},$$

where D is a constant positive depending on $\|f'(\phi)\|_\infty$. Therefore,

$$\|\varphi_R f'(\phi)(\psi_n - \psi)\| \to 0 \quad \text{as } n \to \infty.$$

Now,

$$\begin{aligned}&\|(1-\varphi_R)f'(\phi)(\psi_n - \psi)\|^2 \\ &\qquad \leqq D\|\psi_n - \psi\|^2\Big(\sup_{|x|\leqq 2R} |1-\psi_R(x)|^2 + \sup_{|x|\geqq 2R} |f'(\phi(x))|^2\Big).\end{aligned}$$

Since $\{\psi_n\}$ is bounded in $H^\mu(\mathbb{R})$ and $\psi \in H^\mu(\mathbb{R})$,

$$\|(1-\varphi_R)f'(\phi)(\psi_n - \psi)\| \to 0$$

uniformly in n as $R \to \infty$. Hence, writing

$$\|f'(\phi)(\psi_n - \psi)\| \leqq \|\varphi_R f'(\phi)(\psi_n - \psi)\| + \|(1-\varphi_R)f'(\phi)(\psi_n - \psi)\|,$$

one sees that $\|f'(\phi)(\psi_n - \psi)\|$ can be made arbitrarily small by first choosing R sufficiently large in the second term on the right-hand side of the last expression and then choosing n sufficiently large in the first term. This proves that

$$f'(\phi)\psi_n \to f'(\phi)\psi,$$

as desired. □

Now, to finish the proof of Theorem B.43, we need two results. The first is a strong result about the property of a semi-Fredholm operator to be stable under small perturbations. The second result is a characterization of the essential spectrum of an operator. The proof of these results can be seen in Kato ([**153**], Theorem IV.5.31) and Schechter [**250**].

DEFINITION B.50 (Fredholm and Semi-Fredholm Operators). Let $A : D(A) \subset H \to H$ be a closed linear operator.

(1) We say A is *Fredholm* if $\mathcal{R}(A)$ is closed and both numbers $\dim(\mathcal{R}(A)^{\perp})$ and $\dim(\mathcal{N}(A))$ are finite.
(2) We say A is *semi-Fredholm* if $\mathcal{R}(A)$ is closed and at least one of $\dim(\mathcal{N}(A))$ and $\dim(\mathcal{R}(A)^{\perp})$ is finite.

THEOREM B.51 (Stability of Semi-Fredholm Operators). *Let A be a closed semi-Fredholm operator in H. We consider B, an A-bounded operator from H to H. Then $A+\eta B$ is semi-Fredholm, and* $\dim(\mathcal{N}(A+\eta B))$ *and* $\dim(\mathcal{R}(A+\eta B)^{\perp})$ *are constants for sufficiently small* $|\eta| > 0$.

THEOREM B.52. *Let A be a self-adjoint operator in H. Then $\lambda \in \sigma_{\text{ess}}(A)$ if and only if either $\mathcal{R}(A-\lambda I)$ is not closed or $\mathcal{R}(A-\lambda I)$ is closed but* $\dim(\mathcal{N}(A-\lambda I)) = \dim(\mathcal{R}(A-\lambda I)^{\perp}) = \infty$.

Now we prove Theorem B.43.

PROOF OF THEOREM B.43. Initially we show that $\mathcal{L}$ is bounded below. This means that there exists $\beta \in \mathbb{R}$ such that

$$\langle \mathcal{L}\psi, \psi\rangle \geqq \beta \langle \psi, \psi\rangle$$

for all $\psi \in D(\mathcal{L})$. In fact, since the map $\xi \to f'(\phi(\xi))$ is a continuous function and

$$\lim_{|\xi|\to\infty} [c - f'(\phi(\xi))] = c, \tag{B.28}$$

it follows that

$$\beta \equiv \inf\{c - f'(\phi(\xi)) : \xi \in \mathbb{R}\} < \infty.$$

Therefore, for all $\psi \in D(\mathcal{L}) = H^{\mu}(\mathbb{R})$,

$$\langle \mathcal{L}\psi, \psi\rangle = \langle M\psi, \psi\rangle + \int_{-\infty}^{\infty} [c - f'(\phi(\xi))] |\psi(\xi)|^2 \, d\xi \geqq \beta \int_{-\infty}^{\infty} |\psi(\xi)|^2 \, d\xi. \tag{B.29}$$

So, relation (B.29) shows that the spectrum of $\mathcal{L}$ is bounded below. In fact, initially we have from Theorem B.49 that $\sigma_{\text{ess}}(\mathcal{L}) = [c, \infty)$. Now, let $\lambda \in \sigma(\mathcal{L}) - \sigma_{\text{ess}}(\mathcal{L})$ and $\mathcal{L}\psi = \lambda\psi$. Then from (B.29) it follows that $\lambda \geqq \beta$. Note that since $\mathcal{L}\phi' = 0$, it follows that $\beta \leqq 0$. So, since $c > 0$, it follows that $\sigma(\mathcal{L}) \cap (-\infty, \beta) = \emptyset$.

Now, since $\sigma_{\text{ess}}(\mathcal{L}) = [c, \infty)$, we only need to show that $\sigma_{\text{disc}}(\mathcal{L})$ is finite. In fact, suppose that there is an infinite enumerable set $\{\lambda_n\} \subseteq \sigma_{\text{disc}}(\mathcal{L}) \subseteq [\beta, c)$. Then there is a subsequence $\{\lambda_{n_k}\}$ such that $\lambda_{n_k} \to \lambda$ as $k \to \infty$. Moreover, $\lambda \in [\beta, c]$. Next, suppose that $\lambda \in [\beta, c)$. Then since $\sigma(\mathcal{L})$ is closed, we have that $\lambda \in \sigma(\mathcal{L})$ and therefore $\lambda \in \sigma_{\text{disc}}(\mathcal{L})$, which is a contradiction. Suppose now that $\lambda = c$, and we write

$$\mathcal{L} - cI = \mathcal{L} - \lambda_n I + (\lambda_n - c)I \equiv \mathcal{J} + (\lambda_n - c)I.$$

Then, we have initially from Theorems B.39–B.41 that $\mathcal{J} \equiv \mathcal{L} - \lambda_n I$ is a closed and self-adjoint operator. Moreover, since

$$\dim(\mathcal{R}(\mathcal{J})^{\perp}) = \dim(\mathcal{N}(\mathcal{J})) < \infty$$

and $\mathcal{R}(\mathcal{J})$ is closed (Theorem B.52), we obtain that $\mathcal{J}$ is semi-Fredholm (indeed it is Fredholm). Therefore by choosing n sufficiently large and applying Theorem B.51, we have that $\mathcal{L} - cI$ is semi-Fredholm. Then, since $\mathcal{L} - cI$ is self-adjoint, we have that

$$\dim(\mathcal{N}(\mathcal{L} - cI)) = \dim(\mathcal{R}(\mathcal{L} - cI)^{\perp}) < \infty,$$

and so from Theorem B.52 we obtain that $c \notin \sigma_{\text{ess}}(\mathcal{L})$, which is a contradiction. Therefore, $\sigma_{\text{disc}}(\mathcal{L})$ must be finite. This finishes the proof of the theorem. □

We finish this section with a basic result of the theory of self-adjoint operators which is used in Chapter 6. Here we give a nonstandard proof of it.

THEOREM B.53. *Let A be a self-adjoint operator in H. If $\lambda \in \sigma(A)$ and λ is an isolated point, then λ is an eigenvalue of A.*

PROOF. By the Spectral Theorem for self-adjoint operators (Theorem B.36) we can reduce the problem so that we are working with a multiplication operator $\mathcal{M}_g$, where g is a mensurable real-valued function defined on a measure space (Ω, μ) which is separable and σ-finite. So, we suppose that $\lambda \in \sigma(\mathcal{M}_g)$ and that there exists an $\epsilon_0 > 0$ such that

$$(\lambda - \epsilon_0, \lambda + \epsilon_0) \cap \sigma(\mathcal{M}_g) = \{\lambda\}.$$

Now, by Example B.32 we know that $\sigma(\mathcal{M}_g) =$ *the essential range of* g, namely,

$$\sigma(\mathcal{M}_g) = \{\beta \in \mathbb{R} \ : \ \mu[g^{-1}(B(\beta;\epsilon))] > 0, \text{ for all } \epsilon > 0\}$$

where $B(\beta;\epsilon) = \{y \in \mathbb{C} : |y - \beta| < \epsilon\}$. In particular we have $\mu[g^{-1}(B(\lambda;\epsilon_0))] > 0$. Next, we define $A_{\epsilon_0} = \{x \in \Omega : |g(x) - \lambda| < \epsilon_0\}$. Then we can write

$$A_{\epsilon_0} = \{x \in A_{\epsilon_0} : g(x) = \lambda\} \cup \{x \in A_{\epsilon_0} : g(x) \neq \lambda\} \equiv B_1 \cup B_2.$$

So, if we prove that $\mu(B_1) > 0$, then we finish the proof. In fact, let $D \subseteq \Omega$ such that $0 < \mu(B_1 \cap D) < \infty$ and define $f = \chi_{B_1 \cap D}$. Then the equalities

$$(g - \lambda)f = \begin{cases} 0, & x \in B_1 \cap D, \\ 0, & x \notin B_1 \cap D, \end{cases}$$

and $\|fg\|^2 = \lambda^2 \mu(B_1 \cap D) < \infty$ will imply that λ is an eigenvalue for $\mathcal{M}_g$.

Suppose now that $\mu(B_1) = 0$. Initially, we have that $\lambda \notin \sigma_p(\mathcal{M}_g)$. Indeed, suppose that f is an eigenfunction for λ. Then it follows that for all

$$x \in \widetilde{B_2} = \{y \in \Omega : g(y) \neq \lambda\}$$

we have that $f(x) = 0$. So, since $\Omega - \widetilde{B_2} = B_1$, it follows immediately that $f \equiv 0$. Therefore, we obtain from Theorem B.27 that $\lambda \in \sigma_c(\mathcal{M}_g)$ and so

$$\overline{\mathcal{R}(\mathcal{M}_g - \lambda I)} = L^2(\Omega) \qquad \text{and} \qquad (\mathcal{M}_g - \lambda I)^{-1} \text{ exists and is unbounded.}$$

Next, since $\mu(B_2) > 0$, then there exits $y \in B_2$ such that $|g(y) - \lambda| < \epsilon_0$ and $g(y) \neq \lambda$. Then $g(y) \in \rho(\mathcal{M}_g)$ (hence $\mathcal{M}_g - g(y)I$ is Fredholm). So, from the relation

$$\mathcal{M}_g - \lambda I = [\mathcal{M}_g - g(y)I] + [g(y) - \lambda]I,$$

we obtain from Theorem B.51 by choosing $\epsilon_0 < \eta$ that $\mathcal{M}_g - \lambda I$ is semi-Fredholm and therefore $\mathcal{R}(\mathcal{M}_g - \lambda I)$ is closed. Then, $\mathcal{M}_g - \lambda I$ is one-to-one and onto. Then by the Closed Graph Theorem we obtain that $\lambda \in \rho(\mathcal{M}_g)$, which is a contradiction. This finishes the proof of the theorem. □

B.5. Sturm-Liouville Theory

In this section we establish the basic results of the Sturm-Liouville theory associated to differential equations of the form

$$\mathcal{J}\varphi \equiv -\varphi'' + V(\xi)\varphi = 0, \tag{B.30}$$

where the potential $V(\xi)$ is assumed to be real-valued and continuous. These results will give us more accurate information about the spectrum of the linear operator

$$\mathcal{L}_0\psi \equiv -\psi'' + [c - f'(\phi)]\psi, \tag{B.31}$$

which is associated to the solitary wave solutions, ϕ, for the GKdV equations (2.5) with $f(u) = u^{p+1}/(p+1)$.

Initially we have the classical *Sturm Oscillation Theorem*.

THEOREM B.54. *Let φ_1 and φ_2 be two nontrivial solutions of the differential equations*

$$-\varphi_1'' + V_1(\xi)\varphi_1 = 0, \qquad -\varphi_2'' + V_2(\xi)\varphi_2 = 0. \tag{B.32}$$

(1) *If $V_1(\xi) \geqq V_2(\xi)$ for $\xi \in [a,b]$ and $\varphi_1(a) = \varphi_1(b) = 0$, then there is a $\zeta \in [a,b]$ such that $\varphi_2(\zeta) = 0$. In other words, between two contiguous zeros of φ_1 there is a zero of φ_2.*
(2) *If $V_1(\xi) > V_2(\xi)$ for $\xi \in W \subset [a,b]$, with W having a positive Lebesgue measure, then a point ζ such that $\varphi_2(\zeta) = 0$ can be found in (a,b) between two zeros of φ_1.*

PROOF. Suppose that φ_1 has no zeros in (a,b) and $\varphi_1 > 0$ in (a,b). Then, $\varphi_1'(a) > 0$ and $\varphi_1'(b) < 0$ (if, for example, $\varphi_1'(a) = 0$, then it follows from the uniqueness theorem for ordinary differential equations that $\varphi_1 \equiv 0$). Suppose that φ_2 has no zeros in $[a,b]$ and $\varphi_2(\xi) > 0$ for $\xi \in [a,b]$. From (B.32) it follows that

$$0 = \int_a^b (\varphi_1\varphi_2'' - \varphi_1''\varphi_2)\, d\xi + \int_a^b (V_1 - V_2)\varphi_1\varphi_2\, d\xi.$$

Then by hypothesis and assumption

$$\begin{cases} 0 \geqq \displaystyle\int_a^b (\varphi_1\varphi_2'' - \varphi_1''\varphi_2)\, d\xi = \int_a^b \frac{d}{dx}(\varphi_1\varphi_2' - \varphi_1'\varphi_2)\, d\xi \\ = \varphi_1\varphi_2' - \varphi_1'\varphi_2\Big|_a^b = \varphi_1'(a)\varphi_2(a) - \varphi_1'(b)\varphi_2(b) > 0, \end{cases}$$

which is a contradiction. This proves the first part of the theorem.

Now suppose that $\varphi_1(a) = \varphi_1(b) = 0$, $\varphi_1(\xi) > 0$, $\varphi_2(\xi) > 0$ for $\xi \in (a,b)$, and $V_1 > V_2$ in W. Then, as above, we find that $0 > \varphi_1'(a)\varphi_2(a) - \varphi_1'(b)\varphi_2(b)$. But, since $\varphi_1(a) > 0$, $\varphi_2(a) \geqq 0$, $\varphi_1'(b) < 0$, and $\varphi_2(b) \geqq 0$, it follows that $\varphi_1'(a)\varphi_2(a) - \varphi_1'(b)\varphi_2(b) \geqq 0$, which is a contradiction. This finishes the proof of the theorem. □

COROLLARY B.55. If $V(\xi) \geqq 0$ for $\xi \in [a,b]$ in (B.30), then any nontrivial solution has at most one zero in $[a,b]$.

PROOF. Using Theorem B.54, we compare equation (B.30) with $-\varphi'' = 0$, which has solution $\varphi \equiv 1$ without zeros. □

Now we consider the potential V in (B.30) satisfying the condition

$$\lim_{|\xi|\to\infty} V(\xi) = c > 0. \tag{B.33}$$

THEOREM B.56. *If φ is a nontrivial solution of* (B.30), *where V satisfies* (B.33), *then the set of zeros of φ is finite (probably empty).*

PROOF. Initially from the uniqueness theorem for differentials equations, all the zeros of φ are isolated. Now, choosing R such that $V(\xi) \geqq 0$ for all $|\xi| \geqq R$, it follows from Corollary B.55 that φ has at most one zero in each interval $(-\infty, -R)$ and (R, ∞). This finishes the proof. □

COROLLARY B.57. Let $\lambda < c$ be an eigenvalue of $\mathfrak{J}$ with eigenfunction φ. Then the set of zeros of φ is finite (possibly empty).

PROOF. We consider $V_1(\xi) \equiv V(\xi) - \lambda$. Then we have

$$-\varphi'' + V_1(\xi)\varphi = 0.$$

Since $V_1(\xi) \to c-\lambda$ and $c-\lambda > 0$, then by Theorem B.56, the proof is complete. □

Now we have a main result about the relation between the zeros of solutions of equation (B.30) with V satisfying (B.33).

THEOREM B.58. *Let $\varphi_1, \varphi_2 \in L^2(\mathbb{R})$ be eigenfunctions of $\mathfrak{J}$ with eigenvalues $\lambda_1, \lambda_2 < c$ and with number of zeros n_1 and n_2, respectively. If $\lambda_2 > \lambda_1$, then $n_2 > n_1$.*

PROOF. Suppose $n_1 > 0$. Let $\alpha_1 < \alpha_2 < \cdots < \alpha_{n_1}$ be all the zeros of φ_1. By Sturm's Oscillation Theorem, there is a zero of φ_2 in each interval $[\alpha_i, \alpha_{i+1}]$, $i = 1, ..., n_1 - 1$. Now, we will show that at least one zero can be found in each interval $(-\infty, \alpha_1)$ and $(\alpha_{n_1}, +\infty)$. In other words we need to extend Sturm's Oscillation Theorem to a half-line.

We consider the interval $(\beta, +\infty)$, with $\beta = \alpha_{n_1}$, and suppose that $\varphi_2(\xi) > 0$ for $\xi \in (\beta, +\infty)$ with $\varphi_1(\beta) = 0$. We can also assume, without loss of generality, that $\varphi_1(\xi) > 0$ for $\xi \in (\beta, +\infty)$. Now, using the fact that φ_1, φ_2 are eigenfunctions, it follows that

$$\text{(B.34)} \qquad \begin{cases} 0 = \displaystyle\int_\beta^N (\varphi_1\varphi_2'' - \varphi_1''\varphi_2)\, d\xi + \int_\beta^N (\lambda_2 - \lambda_1)\varphi_1\varphi_2\, d\xi \\ \quad = \varphi_1\varphi_2' - \varphi_1'\varphi_2\Big|_\beta^N + \displaystyle\int_\beta^N (\lambda_2 - \lambda_1)\varphi_1\varphi_2\, d\xi. \end{cases}$$

Moreover, if we take $\alpha \equiv \varphi_1(\beta)\varphi_2'(\beta) - \varphi_1'(\beta)\varphi_2(\beta)$ and

$$\epsilon \equiv \int_\beta^{\beta+1} (\lambda_2 - \lambda_1)\varphi_1\varphi_2\, d\xi,$$

then from our assumptions it follows that $\alpha \leqq 0$ and $\epsilon > 0$. Moreover, from (B.34) we have for $N \geqq \beta + 1$ that

$$\text{(B.35)} \quad \varphi_1(N)\varphi_2'(N) - \varphi_1'(N)\varphi_2(N) = \alpha - \epsilon - \int_{\beta+1}^N (\lambda_2 - \lambda_1)\varphi_1\varphi_2\, d\xi \leqq -\epsilon < 0.$$

Since $\varphi_1, \varphi_2 \in L^2(\mathbb{R})$ and $-\varphi_i'' + V(\xi)\varphi_i = \lambda_i\varphi_i$, we have that $\varphi_i(\xi), \varphi_i'(\xi) \to 0$ as $\xi \to +\infty$. Therefore from (B.35) we obtain that, taking $N \to \infty$, $0 \leqq -\epsilon < 0$, which is impossible.

Finally, the proof of the case $n_1 = 0$ follows immediately from the property of orthogonality of the eigenfunctions φ_1, φ_2, since the two functions cannot preserve their sign. This finishes the proof of the theorem. □

Next, using the Sturm-Liouville theory, we can obtain a more accurate description of the spectrum associated to the linear operator (B.31). Initially, we have the property of preservation of sign of the eigenfunction associated to the lowest eigenvalue of $\mathcal{L}_0$.

THEOREM B.59. *Consider the operator $\mathcal{L}_0$ defined in* (B.31), *and let λ_0 be the least eigenvalue associated to $\mathcal{L}_0$. Then λ_0 is a simple eigenvalue such that any corresponding eigenfunction ψ satisfies either $\psi(\xi) > 0$ a.e. or $\psi(\xi) < 0$ a.e.*

SKETCH OF THE PROOF. Several proofs of this fact can be found in the literature (see [**243**]). Here we use an approach introduced in Albert, Bona, and Henry [**11**]. Define $R_c = (-\partial_x^2 + c)^{-1}$. Then R_c is a Fourier multiplier operator with symbol $q(\xi) = 1/(c+\xi^2)$. Since $q \in L^2(\mathbb{R})$, there is a function $K_c \in L^2(\mathbb{R})$ satisfying

$$\widehat{K_c}(k) = \frac{1}{c+k^2}.$$

It may be easily verified that $K_c(\xi) = e^{-\sqrt{c}|\xi|}/(2\sqrt{c})$. Hence $K_c(\xi) > 0$ for all $\xi \in \mathbb{R}$, and so from Proposition 2 in [**11**] it follows that λ_0 is a simple eigenvalue such that any corresponding eigenfunction ψ satisfies either $\psi(\xi) > 0$ a.e. or $\psi(\xi) < 0$ a.e. This proves the theorem. □

REMARK B.60. We note that if we consider the general linear operator defined in (B.27),

$$\mathcal{L}u = (M+c)u - f'(\phi) \cdot u$$

where M has a symbol $q(\xi)$, we can obtain a result similar to that of Theorem B.59 for the least eigenvalue of $\mathcal{L}$. From Proposition 2 in [**11**] it is sufficient to see that $Q_c \in L^2(\mathbb{R})$ satisfying

$$\widehat{Q_c}(k) = \frac{1}{c+q(k)}$$

is strictly positive.

Now we establish an exact description of the spectrum of $\mathcal{L}_0$ defined in (B.31). This information is crucial in the stability theory of solitary wave solutions for many evolution equations (see Chapters 6 and 7). We will consider $f : \mathbb{R} \to \mathbb{R}$, a smooth function satisfying $f(0) = f'(0) = 0$.

THEOREM B.61. *Suppose that $\phi \in L^2(\mathbb{R})$ satisfies the differential equation*

$$-\phi'' + c\phi - f(\phi) = 0, \tag{B.36}$$

with $c > 0$ and ϕ' having exactly one unique zero. Then the differential operator

$$\mathcal{L}_0\psi = -\psi'' + [c - f'(\phi)]\psi$$

defined in $L^2(\mathbb{R})$ has exactly one simple negative eigenvalue λ_0; the eigenvalue 0 is simple with associated eigenfunction ϕ'; and there exists $\delta > 0$ such that every $\lambda \in \sigma(\mathcal{L}_0) - \{\lambda_0, 0\}$ satisfies $\lambda > \delta$.

PROOF. From (B.36) it follows easily that $\phi \in H^s(\mathbb{R})$ for all $s \in \mathbb{R}$ and $\mathcal{L}_0\phi' = 0$. Moreover, from the theory of ordinary differential equations we obtain that zero is a simple eigenvalue of $\mathcal{L}_0$ (the Wronskian of two solutions in $L^2(\mathbb{R})$ of equation $\mathcal{L}_0\psi = 0$ is zero).

Now, since $\mathcal{L}_0\phi' = 0$, it follows from Theorem B.43 and Theorem B.59 that the least eigenvalue of $\mathcal{L}_0$, λ_0, satisfies $\lambda_0 < 0 < c$, is simple, and has a positive associated eigenfunction.

Next, we will see that there are no eigenvalues of $\mathcal{L}_0$ in $(\lambda_0, 0)$. Suppose that there exists $\lambda \in (\lambda_0, 0)$ such that $\mathcal{L}_0\psi_1 = \lambda\psi_1$ with $\psi_1 \in D(\mathcal{L}_0) - \{0\}$. Then, since $[c - f'(\phi(\xi))] \to c$ as $|\xi| \to \infty$, it follows from Theorem B.58 that ψ_1 must have at least one zero. Applying Theorem B.58 again but this time to the eigenvalues λ and 0, we must have that ϕ' has at least two zeros, which is a contradiction. Finally, the existence of a positive number δ follows from Theorem B.43. □

B.6. Floquet Theory

In this section we develop a complete study based on the Floquet theory [**206**] about the periodic eigenvalue problem considered on $[0, L]$:

$$\text{(B.37)} \qquad \begin{cases} \mathcal{L}_{\text{cn}}\chi = \lambda\chi, \\ \chi(0) = \chi(L), \ \ \chi'(0) = \chi'(L), \end{cases}$$

where the linear operator $\mathcal{L}_{\text{cn}}$ is defined as

$$\mathcal{L}_{\text{cn}} = -\frac{d^2}{dx^2} + c - \phi_c.$$

Here we consider the periodic potential ϕ_c to be part of the smooth curve $c \in (0, +\infty) \to \phi_c$ determined by Theorem 11.4. So, ϕ_c has the specific cnoidal wave form

$$\phi_c(\xi) \equiv \phi_c(\xi; \beta_1, \beta_2, \beta_3) = \beta_2 + (\beta_3 - \beta_2)\,\text{cn}^2\Big[\sqrt{\frac{\beta_3 - \beta_1}{12}}\,\xi; k\Big]$$

and satisfies the nonlinear equation

$$\phi_c'' + \frac{1}{2}\phi_c - c\phi_c = A_{\phi_c},$$

where $\beta_1, \beta_2, \beta_3$ are real constants satisfying $\beta_1 < \beta_2 < \beta_3$, $\beta_1 + \beta_2 + \beta_3 = 3c$,

$$k^2 = \frac{\beta_3 - \beta_2}{\beta_3 - \beta_1}, \quad \text{and} \quad A_{\phi_c} = -\frac{1}{6}\sum_{i<j}\beta_i\beta_j.$$

So we shall show the following theorem.

THEOREM B.62. *Let ϕ_c be the cnoidal wave given in Theorem 11.4 for $c \in (0, \infty)$. Let*

$$\lambda_0 \leqq \lambda_1 \leqq \lambda_2 \leqq \lambda_3 \leqq \lambda_4 \leqq \cdots$$

denote the eigenvalues of problem (B.37)*. Then $\lambda_0 < \lambda_1 = 0 < \lambda_2$ are all simple while, for $j \geq 3$, the λ_j are double eigenvalues. The λ_j only accumulate at $+\infty$.*

Theorem B.62 is a consequence of Floquet's theory together with some particular facts about the *Lamé equation.* For the reader's convenience, we quickly outline the basic results that are needed in the proof of this theorem.

We introduce the so-called semi-periodic eigenvalue problem:

$$\text{(B.38)} \qquad \begin{cases} \mathcal{L}_{\text{cn}}\xi = \mu\xi, \\ \xi(0) = -\xi(L), \ \ \xi'(0) = -\xi'(L). \end{cases}$$

It follows directly from the spectral theory of compact symmetric operators that both (B.37) and (B.38) have a countable infinite number of real eigenvalues,

$$\lambda_0 \leqq \lambda_1 \leqq \lambda_2 \leqq \lambda_3 \leqq \cdots$$

and

$$\mu_0 \leqq \mu_1 \leqq \mu_2 \leqq \mu_3 \leqq \cdots,$$

that accumulate only at $+\infty$. As the operators are of second order, the eigenvalues are either simple or double, and double eigenvalues are displayed twice in the listings above. The eigenfunctions associated with λ_n and μ_n are denoted χ_n and ξ_n, respectively, for $n = 0, 1, 2, \ldots$. The boundary conditions on χ_n in (B.37) imply that χ_n can be extended to all of $\mathbb{R}$ as a continuously differentiable function which is periodic of period L. Similarly, the boundary conditions on ξ_n allow it to be extended as a periodic solution of period $2L$ of

$$\mathcal{L}_{\mathrm{cn}} f = \gamma f \tag{B.39}$$

with $\gamma = \mu_n$ (define $\xi_n(L + x) = \xi_n(L - x)$ for $0 \leq x \leq L$). Indeed, the theory goes on to assert that the only periodic solutions of (B.39) of period L correspond to $\gamma = \lambda_j$ for some j while the only periodic solutions of period $2L$ are either those associated with $\gamma = \lambda_j$, but viewed on $[0, 2L]$, or those corresponding to $\gamma = \mu_j$, but extended as just indicated, for some $j = 0, 1, \ldots$. We remind the reader that

(B.40)
(1) χ_0 has no zeros in $[0, L]$,
(2) χ_{2n+1} and χ_{2n+2} have exactly $2n + 2$ zeros in $[0, L)$,
(3) ξ_{2n} and ξ_{2n+1} have exactly $2n + 1$ zeros in $[0, L)$.

Sturm's Oscillation Theorem then implies that the sequences above are intertwined, *viz.*

$$\lambda_0 < \mu_0 \leqq \mu_1 < \lambda_1 \leqq \lambda_2 < \mu_2 \leqq \mu_3 < \lambda_3 \leqq \cdots.$$

For a given value γ, if all solutions of (B.39) are bounded, then γ is called a *stable* value, whereas if there is an unbounded solution, γ is called *unstable.* The open intervals

$$(\lambda_0, \mu_0), (\mu_1, \lambda_1), (\lambda_2, \mu_2), (\mu_3, \lambda_3), \ldots$$

are called *intervals of stability.* The endpoints of these intervals are generally unstable. This is always so for $\gamma = \lambda_0$, as λ_0 is always simple. The intervals

$$(-\infty, \lambda_0), (\mu_0, \mu_1), (\lambda_1, \lambda_2), (\mu_2, \mu_3), \ldots$$

and so on are called *intervals of instability.* Of course, at a double eigenvalue, the interval is empty and is omitted from this discussion. The absence of an instability interval means there is a value of γ for which all solutions of (B.39) are either periodic of period L or periodic with basic period $2L$. In other words, the coexistence of solutions of (B.39) with period L or period $2L$ occurs for that value of γ.

We note that for *Mathieu's equation,*

$$f'' + (\lambda + \alpha \cos 2x) f = 0,$$

no interval of instability can ever disappear unless $\alpha = 0$ (see Eastham [**101**]).

Next we give the proof of Theorem B.62.

PROOF OF THEOREM B.62. We certainly know that $\lambda_0 < \lambda_1 \leqq \lambda_2$. Since $\mathcal{L}_{\mathrm{cn}} \phi_c' = 0$ and ϕ_c' has 2 zeros in $[0, L)$, it follows that 0 is either λ_1 or λ_2. We will show that $0 = \lambda_1 < \lambda_2$. First, define the transformation

$$T_\eta \chi(x) \equiv \chi(\eta x) \qquad \text{for } \eta^2 = \frac{12}{\beta_3 - \beta_1}.$$

Then, using the explicit form of ϕ_c, the problem (B.37) is equivalent to the eigenvalue problem

$$\begin{cases} \dfrac{d^2}{dx^2}\Lambda + [\rho - 12k^2\mathrm{sn}^2(x)]\Lambda = 0, \\ \Lambda(0) = \Lambda(2K), \quad \Lambda'(0) = \Lambda'(2K), \end{cases} \tag{B.41}$$

for $\Lambda \equiv T_\eta \chi$, where

$$\rho = -\frac{12(c - \beta_3 - \lambda)}{\beta_3 - \beta_1}. \tag{B.42}$$

The second-order differential equation in (B.41) is the *Jacobian form of Lamé's equation.* Floquet's theory ([**206**], Theorem 7.8) informs us that this equation has exactly 4 intervals of instability, namely

$$(-\infty, \rho_0), \quad (\mu_0', \mu_1'), \quad (\rho_1, \rho_2), \quad (\mu_2', \mu_3'),$$

where the μ_i' are the eigenvalues associated to the semi-periodic problem determined by (B.41). Therefore, the first three eigenvalues ρ_0, ρ_1, ρ_2 are simple and the remainder of the eigenvalues for (B.41) are double. So, $\rho_3 = \rho_4$, $\rho_5 = \rho_6, \ldots$.

The first three eigenvalues, ρ_0, ρ_1, ρ_2, and their corresponding eigenfunctions, $\Lambda_0, \Lambda_1, \Lambda_2$, are known explicitly. Since $\rho_1 = 4 + 4k^2$ is an eigenvalue of (B.41) with eigenfunction

$$\Lambda_1(x) \equiv \mathrm{cn}(x)\mathrm{sn}(x)\mathrm{dn}(x) = CT_\eta \phi_c'(x),$$

it follows from (B.42) that $\lambda = 0$ is a simple eigenvalue of problem (B.37) with eigenfunction ϕ_c'. The functions Λ_0 and Λ_2 (called *Lamé polynomials*) defined by

$$\begin{cases} \Lambda_0(x) = \mathrm{dn}(x)\big[1 - (1 + 2k^2 - \sqrt{1 - k^2 + 4k^4}\,)\mathrm{sn}^2(x)\big] \quad \text{and} \\ \Lambda_2(x) = \mathrm{dn}(x)\big[1 - (1 + 2k^2 + \sqrt{1 - k^2 + 4k^4}\,)\mathrm{sn}^2(x)\big] \end{cases}$$

are the eigenfunctions associated to the other two eigenvalues, ρ_0 and ρ_2. These eigenvalues must satisfy the equation

$$\rho = k^2 + \frac{5k^2}{1 + \frac{9}{4}k^2 - \frac{1}{4}\rho}$$

(see Ince [**134**]). This equation has two roots:

$$\rho_0 = 2 + 5k^2 - 2\sqrt{1 - k^2 + 4k^4}$$

and

$$\rho_2 = 2 + 5k^2 + 2\sqrt{1 - k^2 + 4k^4}.$$

Since Λ_0 has no zeros in $[0, 2K]$ and Λ_2 has exactly 2 zeros in $[0, 2K)$, it must be the case that Λ_0 is the eigenfunction associated to ρ_0, the first eigenvalue of (B.41). On the other hand, since $\rho_0 < \rho_1$ for every $k \in (0, 1)$, we obtain from (B.42) and the relation

$$-\beta_1(1 + k^2) = (2 - k^2)\alpha_3 - 3c$$

the inequality

$$12\lambda_0 = 3\frac{\beta_3 - c}{k^2 + 1}\rho_0 + 12(c - \beta_3) < 0.$$

As a consequence, the first eigenvalue λ_0 of $\mathcal{L}_{\text{cn}}$ is negative and has the eigenfunction $\chi_0(x) = \Lambda_0(\frac{1}{\eta}x)$. It is also true that $\rho_1 < \rho_2$ for every $k \in (0,1)$, so it follows from (B.42) that

$$12\lambda_2 = 3\frac{\beta_3 - c}{k^2+1}\rho_2 + 12(c - \beta_3) > 0.$$

So λ_2 is the third eigenvalue of $\mathcal{L}_{\text{cn}}$ with eigenfunction $\chi_2(x) = \Lambda_2(\frac{1}{\eta}x)$.

It is straightforward to ascertain that the first two eigenvalues of Lamé's equation in the semi-periodic case are

$$\mu_0' = 5 + 2k^2 - 2\sqrt{4 - k^2 + k^4} \quad \text{and} \quad \mu_1' = 5 + 5k^2 - 2\sqrt{4 - 7k^2 + 4k^4}.$$

The associated eigenfunctions are

$$\begin{cases} \xi_0(x) = \text{cn}(x)\Big[1 - (2 + k^2 - \sqrt{4 - k^2 + k^4}\,)\text{sn}^2(x)\Big], \\ \xi_1(x) = 3\text{sn}(x) - \Big(2 + 2k^2 - \sqrt{4 - 7k^2 + 4k^4}\,\Big)\text{sn}^3(x), \end{cases}$$

respectively, both of which have exactly one zero in $[0, 2K)$. Since $\mu_0' < \mu_1' < 4k^2+4$, we conclude from the relation

$$\mu_i' = -\frac{12(c - \beta_3 - \mu_i)}{\beta_3 - \beta_1}, \qquad \text{for } i \geqq 0, \tag{B.43}$$

that the first three instability intervals associated to $\mathcal{L}_{\text{cn}}$ are

$$(-\infty, \lambda_0), \ (\mu_0, \mu_1), \ (\lambda_1, \lambda_2).$$

The third and fourth eigenvalues are

$$\mu_2' = 5 + 2k^2 + 2\sqrt{4 - k^2 + k^4} \quad \text{and} \quad \mu_3' = 5 + 5k^2 + 2\sqrt{4 - 7k^2 + 4k^4}$$

with associated eigenfunctions

$$\begin{cases} \xi_2(x) = \text{cn}(x)\Big[1 - (2 + k^2 + \sqrt{4 - k^2 + k^4}\,)\text{sn}^2(x)\Big], \\ \xi_3(x) = 3\text{sn}(x) - \Big(2 + 2k^2 + \sqrt{4 - 7k^2 + 4k^4}\,\Big)\text{sn}^3(x), \end{cases}$$

respectively, both of which have exactly three zeros in $[0, 2K)$. Finally, it follows from (B.43) that the last instability interval of $\mathcal{L}_{\text{cn}}$ is (μ_2, μ_3). This concludes the proof. □

Bibliography

1. Abdelouhab, L., Bona, J.L., Felland, M., and Saut, J.-C., *Nonlocal models for nonlinear dispersive waves*, Physica D. 40 (1989), 360–392.
2. Ablowitz, M.J. and Clarkson, P.A., *Solitons, Nonlinear Evolution Equations and Inverse Scattering*, Cambridge University Press, 1991.
3. Ablowitz, M.J. and Segur, H., *Solitons and Inverse Scattering*, SIAM Publication, 1981.
4. Adams, R.A., *Sobolev Spaces*, Academic Press, New York, 1975.
5. Alarcon, E., Angulo, J., and Montenegro, F., *Stability and instability of solitary waves for a nonlinear dispersive system*, Nonlinear Anal. TMA. 36 (1999), 1015–1035.
6. Albert, J.P., *Positivity properties and stability of solitary-wave solutions of model equations for long waves*, Comm. PDE 17 (1992), 1–22.
7. Albert, J.P., *Positivity properties and uniqueness of solitary wave solutions of the intermediate long-wave equation*, Evolution equations (Baton Rouge, LA, 1992), 11–20, Lecture Notes in Pure and Appl. Math., 168, Dekker, New York, 1995.
8. Albert, J.P., *Concentration compactness and the stability of solitary-wave solutions to nonlocal equations*, Contemp. Math. 221 (1999), 1–29.
9. Albert, J.P. and Angulo, J., *Existence and stability of ground-state solutions of a Schrödinger-KdV system*, Proc. Royal Soc. Edinburgh, Sect. A 133, 5 (2003), 987–1029.
10. Albert, J.P. and Bona, J.L., *Total positivity and the stability of internal waves in fluids of finite depth*, IMA J. Applied Math. 46 (1991), 1–19.
11. Albert, J.P., Bona, J.L., and Henry, D., *Sufficient conditions for stability of solitary-wave solutions of model equations for long waves*, Physica D 24 (1987), 343–366.
12. Albert, J.P., Bona, J.L., and Nguyen, N., *Stability of solitary waves in higher-order Sobolev spaces*, Commun. Math. Sci. 2 (2004), 35–52.
13. Albert, J.P., Bona, J.L., and Restrepo, J., *Solitary-wave solutions of the Benjamin equation*, SIAM J. Appl. Math. 59 (1999), 2139–2161.
14. Albert, J.P., Bona, J.L., and Saut, J.-C., *Model equations for waves in stratified fluids*, Proc. Royal Soc. Edinburgh, Sect. A 453 (1997), 1233–1260.
15. Albert, J.P. and Linares, F., *Stability of solitary-wave solutions to long-wave equations with general dispersion*, Fifth Workshop on Partial Differential Equations (Rio de Janeiro, 1997). Mat. Contemp. 15 (1998), 1–19.
16. Albert, J.P. and Toland, J.F., *On the exact solutions of the intermediate long-wave equation*, Diff. Integral Eq. 7 (1994), 601–612.
17. Alexander, J.C., Pego, R.L., and Sachs, R.L., *On the transverse instability of solitary waves in the Kadomtsev-Petviashvili equation*, Phys. Lett. A 226 (1997), 187–192.
18. Amick, C.J. and Toland, J.F., *Uniqueness of Benjamin's solitary-wave solution of the Benjamin-Ono equation*, IMA J. Appl. Math. 46 (1991), 21–28.
19. Amick, C.J. and Toland, J.F., *On periodic water-waves and their convergence to solitary waves in the long-wave limit*, Philos. Trans. Roy. Soc. London Ser. A 303, 1481 (1981), 633–669.
20. Angulo, J., *On the Cauchy problem for a Boussinesq-type system*, Adv. Differential. Equations. 4 (1999), 457–492.
21. Angulo, J., *Existence and stability of solitary-wave solutions of the Benjamin equation*, J. Differential Equations 152 (1999), 136–159.
22. Angulo, J., *Stability of solitary waves solutions for equations of short and long dispersive waves*, Electronic Journal of Differential Equations, v. 2006, 72 (2006), 1–18.
23. Angulo, J., *On the instability of solitary waves solutions of the generalized Benjamin equation*, Adv. Differential Equations 8 (2003), 55–82.

24. Angulo, J., *On the instability of solitary-wave solutions for fifth-order water wave models*, Electron. J. Differential Equations (electronic) 6 (2003), 1–18.
25. Angulo, J., *Non-linear stability of periodic travelling wave solutions to the Schrödinger and Modificada Korteweg-de Vries*, J. of Differential Equations 235, No. 1 (2007), 1–18.
26. Angulo, J., *Stability of dnoidal waves to Hirota-Satsuma system*, Differential and Integral, Equations 18, No. 6 (2005), 611–645.
27. Angulo, J., *Stability of cnoidal waves for the Hirota-Satsuma system*, Matemática Contemporanea 27 (2005), 189–223.
28. Angulo, J., Bona, J.L., and Scialom, M., *Stability of cnoidal waves*, Advances in Differential Equations 11, No. 12 (2006), 1321-1374.
29. Angulo, J., Bona, J.L., Linares, F., and Scialom, M., *On the structure of singularities in solutions of the nonlinear Schrödinger equation for the critical case,* $p = 4/n$, Nonlinear theory of generalized functions (M. Oberguggenberger *et al.*, eds.), Chapman & Hall/CRC, Boca Raton, FL (1999), 3–22.
30. Angulo, J., Bona, J.L., Linares, F., and Scialom, M., *Scaling, stability and singularities for nonlinear, dispersive wave equations: the critical case*, Nonlinearity 15 (2002), 759–786.
31. Angulo, J. and Linares, F., *Global existence of solutions of a nonlinear dispersive model*, J. Math. Anal. Appl. 195 (1995), 797–808.
32. Angulo, J. and Linares, F., *Periodic pulses of coupled nonlinear Schrödinger equations in optics*, Indiana University Mathematics Journal 56, no. 2 (2007), 847–877.
33. Angulo, J., Lopes, O., and Neves, A. *Instability of travelling waves for weakly coupled KdV systems*, Nonlinear Anal. TMA. 69 (2008), 1870–1887.
34. Angulo, J. and Montenegro, F., *Existence and evenness of solitary-wave solutions for an equation of short and long dispersive waves*, Nonlinearity 13 (2000), 1595–1611.
35. Angulo, J. and Montenegro, F., *Orbital stability of solitary wave solutions for an interaction equation of short and long dispersive waves*, J. Differential Equations 174 (2001), 181–199.
36. Angulo, J. and Moura, R. P., *Ill-posedness and the nonexistence of standing-wave solutions for the nonlocal nonlinear Schrödinger equation*, Differential and Integral Equation 20 (2007), 1107–1130.
37. Angulo, J. and Natali, F., *Positivity properties of the Fourier transform and the stability of periodic travelling-wave solutions*, SIAM J. Math. Anal., v. 40, 3 (2008), 1123-1151.
38. Angulo, J. and Natali, F., *Stability and instability of periodic travelling-wave solutions for the critical Korteweg-de Vries and non-linear Schrödinger equations*, Physica D 238 (2009), 603-621.
39. Angulo, J. and Quintero, R., *Existence and orbital stability of cnoidal waves for a1D Bousssinesq equation*, International Journal of Mathematics and Mathematical Sciences, 2007, ID 52020 (2007).
40. Armitage, J. V. and Eberlein, W. F., *Elliptic Functions*, London Mathematical Society, Student Texts, 67, Springer-Verlag, Cambridge, U.P., Cambridge, 2006.
41. Benjamin, T.B., *Internal waves of permanent form in fluids of great depth*, J. Fluid Mech. 29 (1967), 559–592.
42. Benjamin, T.B., *The stability of solitary waves*, Proc. Royal Soc. London. A 338 (1972), 153–183.
43. Benjamin, T.B., *A new kind of solitary wave*, J. Fluid Mech. 245 (1992), 401–411.
44. Benjamin, T.B., *Solitary and periodic waves of a new kind*, Philos. Trans. Roy. Soc. London Ser. A 354 (1996), 1775–1806.
45. Benjamin, T.B., *Lectures on nonlinear wave motion*, Nonlinear Wave Motion, A. C. Newell, ed., AMS, Providence, R. I., 15 (1974), 3–47.
46. Benjamin, T.B., Bona, J.L., and Bose, D.K., *Solitary-wave solutions of nonlinear problems*, Phil. Trans. Royal Soc. London Ser. A 331 (1990), 195–244.
47. Benjamin, T.B., Bona, J.L., and Mahony, J.J., *Model equations for long waves in nonlinear dispersive systems*, Phil. Trans. Royal Soc. A272 (1972), 47–78.
48. Benjamin, T.B. and Feir, J.E., *The disintegration of wavetrains on deep water, Part 1, Theory*, J. Fluid Mech. 27 (1967), 417–430.
49. Bennet, D., Bona, J.L., Brown, R., Stansfield, S., and Stroughair, J., *The stability of internal waves*, Math. Proc. Cambridge Philos. Soc. 94 (1983), 351–379.
50. Benney, D.J., *A general theory for interactions between short and long waves*, Stud. Appl. Math. 56 (1977), 81–94.

51. Berestycki, H. and Cazenave, T., *Instabilité des états stationnaires dans les équations de Schrödinger et de Klein-Gordon non linéaires*, C. Rend. Acad. Sci. Paris. 293 (1981), 489–492.
52. Berestycki, H., Gallouet, T., and Kavian, O., *Équations de champs scalaires Euclidiens non-linéaires dans le plan*, C. Rend. Acad. Sc. Paris. 297 (1983), 307–310.
53. Berestycki, H. and Lions, P.-L., *Existence of stationary states in nonlinear scalar field equations*, Bifurcation Phenomena in Mathematical Physics and Related Topics (C. Bardos and D. Bessis, eds.), K. Reidel Publ. Co., Dordrecht, Boston, London (1976), 269–292.
54. Berestycki, H. and Lions, P.-L., *Nonlinear scalar field equations I*, Arch. Rat. Mech. Anal. 82 (1983), 313–346.
55. Berestycki, H., Lions, P.-L., and Peletier, L. A., *An ODE approach to the existence of positive solutions for semilinear problems in* $\mathbb{R}^n$, Indiana Univ. Math. J. 30 (1981), 141–157.
56. Bergh, J. and Löfström, J., *Interpolations Spaces*, Springer, New York, 1976.
57. Billingham, J. and King, A.C., *Wave Motion*, Cambridge Texts in Applied Mathematics, 2000.
58. Bona, J.L., *On the stability theory of solitary waves*, Proc. Roy. Soc. London 344 (1975), 363–374 .
59. Bona, J.L., *Nonlinear Waves Phenomena. 51 Seminário Brasileiro de Análise*, Universidade Federal de Santa Catarina, Florianópolis, SC, 2000.
60. Bona, J.L., Dougalis, V.A., Karakashian, O.A., and McKinney, W.R., *Numerical simulation of singular solutions of the generalized Korteweg de Vries equation*, Mathematical problems in the theory of water waves (F. Diaz, J.-M. Ghidaglia and J.-C. Saut, eds.), Amer. Math. Soc., Providence, RI (1996), 17–29 .
61. Bona, J.L. and Li, Y.A., *Decay and analyticity of solitary waves*, J. Math. Pures Appliq. 76 (1997), 377–430 .
62. Bona, J.L. and Sachs, R., *Global existence of smooth solutions and stability of solitary waves for a generalized Boussinesq equation*, Comm. Math. Phys. 118 (1988), 15–29 .
63. Bona, J.L. and Soyeur, A., *On the stability of solitary-wave solutions of model equations for long waves*, J. Nonlinear Sci. 4 (1994), 449–470.
64. Bona, J.L., Souganidis, P.E., and Strauss, W.A., *Stability and instability of solitary waves of Korteweg-de Vries type*, Proc. Royal Soc. London Ser. A 411 (1987), 395–412.
65. Bourgain, J., *Fourier transform restriction phenomena for certain lattice subsets and applications to nonlinear evolution equations I, II*, Geometric and Functional Anal. 3 (1993), 107–156, 209–262.
66. Bourgain, J., *On the Cauchy problem for the Kadomtsev-Petviashvili equation*, Geometric and Functional Anal. 3 (1993), 315–341.
67. Bourgain, J., *Aspects of long time behaviour of solutions of nonlinear Hamiltonian evolution equations*, Geometric and Functional Anal. 5 (1995), 105–140.
68. Bourgain, J., *Periodic Korteweg-de Vries equation with measures as initial data*, Selecta Math. (N.S.) 3 (1997), 115–159.
69. Bourgain, J., *Global solutions of nonlinear Schrödinger equations*, American Mathematical Society Colloquium Publications, AMS, Providence, RI, 46, 1999.
70. Boussinesq, J., *Théorie de l'intumescence liquide appelée onde solitaire ou de translation se propageant dans un canal re rectangular*, Comptes Rendus 72 (1871), 755–759.
71. Boussinesq, J., *Théorie des ondes et des remous qui se propagent le long d'un canal rectangular horizontal, en communiquant au liquide contenu dans ce canal des vitesses sensiblement pareilles de la surface au fond*, J. Math. Pures Appl. (2) 17 (1872), 55–108.
72. Boyd, J. P., *Cnoidal waves as exact sums of repeated solitary waves: New series for elliptic functions*, SIAM J. Appl. Math. 44 (1984), 952–955.
73. Boyd, J. P., *The special modular transformation for polycnoidal waves of the Korteweg-de Vries equation*, J. Math. Phys. 25 (1984), 3402–3414.
74. Brascamp, H.J., Lieb, E.H., and Luttinger, J.M., *A general rearrangement inequality for multiple integral*, J. Funct. Anal. 17 (1974), 227–237.
75. Byrd, P. F. and Friedman, M. D., *Handbook of Elliptic Integrals for Engineers and Scientists*, 2nd edition, Springer-Verlag, New York, 1971.
76. Calderon, A.P., *Commutators of singular integral operators*, Proc. Nat. Acad. Sci. U.S.A. 53 (1965), 1092–1099.

77. Cazenave, T., *Stable solutions of the logarithmic Schrödinger equation*, Nonlinear Anal. T.M.A. 7 (1983), 1127–1140.
78. Cazenave, T., *Stability and instability of stationary states in nonlinear Schrödinger equations, in contributions to nonlinear partial differential equations I*, Research Notes in Mathematics, 89, Pitman, London, 1983.
79. Cazenave, T., *An Introduction to Nonlinear Schrödinger Equations*, Textos de Métodos Matemáticos, Universidade Federal do Rio de Janeiro, 26, third edition, 1989.
80. Cazenave, T., *Blow Up and Scattering in the Nonlinear Schrödinger Equation*, Textos de Métodos Matemáticos, Universidade Federal do Rio de Janeiro, 30, second edition, 1996.
81. Cazenave, T. and Esteban, M.J., *On the Stability of stationary states for nonlinear Schrödinger equations with an external magnetic field*, Mat. Apl. Comp. 7 (1988), 155–168.
82. Cazenave, T. and Lions, P.-L., *Orbital stability of standing waves for some nonlinear Schrödinger equations*, Comm. Math. Phys. 85 (1982), 549–561.
83. Chen, L., *Orbital stability of solitary waves of the Schrödinger-KDV equation*, J. Partial Diff. Eqs. 12 (1999), 11–25.
84. Chen, H. and Bona, J.L., *Existence and asymptotic properties of solitary-wave solutions of Benjamin-type equations*, Adv. Diff. Eq. 3 (1998), 51–84.
85. Coddington, E.A. and Levinson, N., *Theory of Ordinary Differential Equations*, McGraw-Hill Book Co., 1955.
86. Coifman, R. and Meyer, Y., *Au delá des opérateurs pseudo-différentiels*, Asterisque, no. 57, Société Mathématique de France, Paris (1978).
87. Colin, T. and Weinstein, M., *On the ground states of vector nonlinear Schrödinger equations*, Ann. Inst. H. Poincaré Phys. Théor. 65 (1996), 57–79.
88. Colliander, J., Keel, M., Staffilani, G., Takaoka, H., and Tao, T., *Sharp global well-posedness for periodic and non-periodic KdV and mKdV on* $\mathbb{R}$ *and* $\mathbb{T}$, JAMS 16 (2003), 705–749.
89. Colliander, J., Kenig, C., and Staffilani, G., *On solutions for the Kadomtsev-Petviashvili I equation. Dedicated to the memory of I. G. Petrovskii on the occasion of his 100th anniversary*, Mosc. Math. J. 1 (2001), 491–520.
90. Comech, A., Cuccagna, S., and Pelinovsky, D., *Nonlinear instability of a critical traveling wave in the generalized Korteweg-de Vries equation*, SIAM J. Math. Anal. 39 (2007), 1–33.
91. Constantin, A. and Strauss, W., *Stability of peakons*, Comm. Pure Appl. Math. 53 (2000), 603–610.
92. Constantin, A. and Strauss, W., *Stability of the Camassa-Holm solitons*, J. Nonlinear Sci. 12 (2002), 415–422.
93. Craig, W. and Groves, M.D., *Hamiltonian long-wave approximations to the water-wave problem*, Wave Motion 19 (1994), 367–389.
94. de Bouard, A. and Saut, J.-C., *Remarks on the stability of generalized KP solitary waves. Mathematical problems in the theory of water waves (Luminy, 1995)*, Contemp. Math., 200, Amer. Math. Soc., Providence, RI (1996), 75–84.
95. de Bouard, A. and Saut, J.-C., *Solitary waves of generalized Kadomtsev Petviashvili equations*, Ann. Inst. H. Poincaré Anal. Non Linéaire 14 (1997), 211–236.
96. de Bouard, A. and Saut, J.-C., *Symmetries and decay of the generalized Kadomtsev-Petviashvili solitary waves*, SIAM J. Math. Anal. 28 (1997), 1064–1085.
97. Debnath, L., *Nonlinear Partial Differential Equations, for Scientist and Engineers*, Birkhäuser, Boston, 1997.
98. Dodd, R.K., Eilbeck, J.C., Gibbon, J.D., and Morris, H.C., *Solitons and Nonlinear Wave equations*, Academic Press, London, 1982.
99. Drazin, P.G., *Solitons*, Cambridge University Press, Cambridge, 1983.
100. Dutta, M. and Debnath, L., *Elements of the Theory of Elliptic and Associated Functions with Applications*, World Press Pub. Ltd., Calcutta, 1965.
101. Eastham, M.S.P., *The Spectral Theory of Periodic Differential Equations*, Scottish Academic Press, London, UK, 1973.
102. Erdélyi, A., *On Lamé functions*, Philos. Mag. (7) 31 (1941), 123–130.
103. Erdélyi, A., Magnus, W., Oberhettinger, F., and Tricomi, F., *Tables of Integral Transform*, vol. 2, McGraw-Hill, New York, 1954.
104. Falkovitch, G. and Turitsyn, S., *Stability of magneto elastic solitons and self-focusing of sound in antiferromagnet*, Soviet Phys. JEETP 62 (1985), 146–152.

105. Fang, Y.-F. and Grillakis, M., *Existence and uniqueness for Boussinesq type equations on a circle*, Comm. PDE 21 (1996), 1253–1277.
106. Fermi, E., Pasta, J., and Ulam, S., *Studies of nonlinear problems, I (1955), Los Alamos Report, LA, 1940*, Lectures in Applied Mathemathics (ed., A.C. Newell), A.M.S. 15 (1974), 143–156.
107. Fibich, G. and Wang, X.-P., *Stability of solitary waves for nonlinear Schrödinger equations with inhomogeneous nonlinearities*, Phys. D. 175 (2003), 96–108.
108. Fokas, A.S. and Ablowitz, M.J., *The inverse scaterring transform for the Benjamin-Ono equation— a pivot to multidimensional equations*, Stud. Appl. Math. 68 (1983), 1–10.
109. Fonseca, G. and Linares, F., *Benjamin-Ono equation with unbounded data*, J. Math. Anal. Appl. 247 (2000), 426–447.
110. Fonseca, G., Linares, F., and Ponce, G., *Global existence for the critical generalized KdV equation*, Proc. Amer. Math. Soc. (electronic) 131 (2003), 1847–1855.
111. Fukuizumi, R. and Ohta, M., *Stability of standing waves for nonlinear Schrödinger equations with potentials*, Differential Integral Equations 16 (2003), 111–128.
112. Funakoshi, M. and Oikawa, M., *The resonant interaction between a long internal gravity wave and a surface gravity wave packet*, J. Phys. Soc. Japan 52 (1983), 1982–1995 .
113. Gallay, T. and Hărăguş, M., *Stability of small periodic waves for the nonlinear Schrödinger equation*, J. Differential Equations 234 (2007), 544–581.
114. Gallay, T. and Hărăguş, M., *Orbital stability of periodic waves for the nonlinear Schrödinger equation*, J. Dyn. Diff. Eqns. 19 (2007), 825-865.
115. Gardner, R. A., *Spectral analysis of long wavelength periodic waves and applications*, J. Für Die Reine und Angewandte Mathematik 491 (1997), 149-181.
116. Gardner, C.S., Greene, J.M., Kruskal, M.D., and Miura, R.M., *Korteweg de Vries equation and generalizations, VI, Methods for exact solution*, Comm. Pure Appl. Math. 27 (1974), 97–133.
117. Gardner, C.S. and Morikawa, G.K., *Similarity in the asymptotic behavior of collision-free hydromagnetic waves and water waves*, Courant Inst. Math. Sci. Rep. NYO-9082 (1960), 1–30.
118. Gidas, B., Ni, W.M., and Nirenberg, L., *Symmetry of positive solutions of nonlinear elliptic equations in* $\mathbb{R}^n$, Advances in Math. Suppl. Studies 7A (1981), 369–402.
119. Ginibre, J. and Velo, G., *Smoothing properties and existence of solutions for the generalized Benjamin-Ono equation*, J. Differential Equations 93 (1991), 150–212.
120. Glassey, R.T., *On the blowing up of solutions to the Cauchy problem for the nonlinear Schrödinger equation*, J. Math. Phys. 18 (1977), 1794–1797.
121. Gonçalves Ribeiro, J.M., *Instability of symmetric stationary states for some nonlinear Schrödinger equations with an external magnetic field*, Ann. Inst. Henri Poincaré, Physique Théorique 54 (1991), 403–433.
122. Grillakis, M., *Linearized instability for nonlinear Schrödinger and Klein-Gordon equations*, Comm. Pure Appl. Math. 41 (1988), 747–774.
123. Grillakis, M., Shatah, J., and Strauss, W., *Stability theory of solitary waves in the presence of symmetry I*, J. Funct. Anal. 74 (1987), 160–197.
124. Grillakis, M., Shatah, J., and Strauss, W., *Stability theory of solitary waves in the presence of symmetry II*, J. Funct. Anal. 94 (1990), 308–348.
125. Grünrock, A., *A bilinear Airy-estimate with application to gKdV-3*, Diff. Int. Eq. 18 (2005), 1333–1339.
126. Hardy, G., Littlewood, J.E., and Polya, G., *Inequalities*, Cambridge, Cambridge University Press, 1934.
127. Hartree, D.,*The wave mechanics of an atom with a non-Coulomb central field, Part I. Theory and methods*, Proc. Camb. Philos. Soc. 24 (1968), 89–132.
128. Hasegawa, A.H., *Optical Solitons in Fibers*, 2nd edition, Springer-Verlag, 1990.
129. Henry, D., *Geometric Theory of Semilinear Parabolic Equations*, Lectures Notes in Mathematics, Springer-Verlag, 840, 1981.
130. Henry, D., Perez, B., and Wreszinski, W.F., *Stability theory for solitary-wave solutions of scalar field equations*, Comm. Math. Phys. 85 (1982), 351–361.
131. Hirota, R., *Exact N-solutions of the wave equation of long wave in shallow water and in nonlinear lattices*, J. Math. Phys. 14 (1973), 810–814.

132. Hirota, R., *Exact solutions of the Korteweg-de Vries equation for multiple collisions of solitons*, Phys. Rev. Lett. 27 (1971), 1192–1194.
133. Ichikawa, Y.H., *Topics on solitons in plasma*, Physica Scripta 20 (1979), 296–305.
134. Ince, E.L., *The periodic Lamé functions*, Proc. Roy. Soc. Edin. 60 (1940), 47–63.
135. Ionescu, A., Kenig, C., and Tataru, D., *Global well-posedness of the KP-I initial value problem in the energy space*, Inventiones Mathematicae 173 (2008), 265–304.
136. Iorio, R.J. Jr., *On the Cauchy problem for the Benjamin-Ono equation*, Comm. PDE 11 (1986), 1031–1081.
137. Iorio, R.J. Jr., *KdV, BO, and friends in weighted Sobolev spaces*, Lectures Notes in Mathematics, Springer-Verlag, 1450 (1990), 105–121.
138. Iorio, R.J. Jr., *Unique continuation principle for the Benjamin-Ono equation*, Differential and Integral Equations 11 (2003), 1281–1291.
139. Iorio, R.J. Jr. and Iorio de Magalhães, V., *Fourier Analysis and Partial Differential Equations*, Cambridge Stud. in Advan. Math. 70, 2001.
140. Iorio, R.J. Jr. and Leite Nunes, W.V., *Introdução às Equações de Evolução Não Lineares*, 18° Colóquio Brasileiro de Matemática, 1991.
141. Iorio, R.J. Jr. and Leite Nunes, W.V., *On equations of KP-type*, Proc. Royal Soc. Edinburgh, Sect. A 128 (1998), 725–743 .
142. Iorio, R.J. Jr., Linares, F., and Scialom, M., *Living under the Sobolev barrier*, Proceedings of the Third Summer School on Differential Geometry, Partial Differential Equations and Numerical Analysis (Spanish) (Bogota, 1995), Colec. Mem. Acad. Colombiana Cienc. Exact. Fís. Natur., Bogotá 7 (1996), 61–73 .
143. Iorio, R.J. Jr., Linares, F., and Scialom, M., *KdV, and BO equations with bore-like data*, Differential Integral Equations 11 (1998), 895–915.
144. Jeffrey, A. and Engelbrecht, J., *Nonlinear Waves in Solids*, Springer-Verlag, New York, 1994.
145. Jeffrey, A. and Kakutani, T., *Weak nonlinear dispersive waves: A discussion centered around the KdV equation*, SIAM Rev. 14 (1972), 582–643.
146. Joseph, R., *Solitary wave in a finite depth fluid*, J. Phys. A 10 (1977), L225–L227.
147. Kadomtsev, B.B. and Petviashvili, V.I., *On stability of solitary waves in weakly dispersive media*, Dokl. Akad. Nauk. SSSR. 192 (1970), 753–756.
148. Kappeler, T. and Topalov, P., *Global well-posedness of KdV in* $H^{-1}(\mathbb{T},\mathbb{R})$, Duke Math. J. 135 (2006), 327–360.
149. Karlin, S., *Total Positivity*, Stanford University Press, 1968.
150. Kato, T., *Quasilinear equations of evolution with applications to partial differential equations*, Lecture Notes in Math., Springer-Verlag, 448 (1975), 25–70.
151. Kato, T., *On the Korteweg-de Vries equation*, Manuscripta 28 (1979), 89–99.
152. Kato, T., *On the Cauchy problem for the (generalized) Korteweg-de Vries equation*, Advances in Mathematics Supplementary Studies (M.G. Crandall, ed.), Academic Press, 8 (1983), 93–128.
153. Kato, T., *Perturbation Theory for Linear Operators*, 2nd edition, Springer, Berlin, 1984.
154. Kato, T., *Abstract evolution equations, linear and quasilinear, revisited*, Lecture Notes in Math., Springer-Verlag, 1540 (1992), 103–125.
155. Kawahara, T., *Oscillatory solitary waves in dispersive media*, J. Phys. Soc. Japan 33 (1972), 260-264.
156. Kawahara, T., Sugimoto, N., and Kakutani, T., *Nonlinear interaction between short and long capillary-gravity waves*, J. Phys. Soc. Japan 39 (1975), 1379-1386.
157. Kawohl, B., *Rearragements and convexity of level sets in PDE*, Lecture Notes in Mathematics, Springer-Verlag, 1150, 1985.
158. Kelley, P.L., *Self-focusing of optical beams*, Phys. Rev. Lett. 15 (1965), 1005–1008.
159. Kenig, C.E., and Koenig, K., *On the local well-posedness of the Benjamin-Ono and modified Benjamin-Ono equations*, Math. Res. Lett. 10 (2003), 879–895.
160. Kenig, C.E., Ponce, G., and Vega, L., *Well-posedness and scattering results for the generalized Korteweg-de Vries equation via the contraction principle*, Comm. Pure Appl. Math. 46 (1993), 527–620.
161. Kenig, C.E., Ponce, G., and Vega, L., *On the generalized Benjamin-Ono equation*, Trans. Amer. Math. Soc. 342 (1994), 155–172.
162. Kenig, C.E., Ponce, G., and Vega, L., *A bilinear estimate with applications to the KdV equation*, J. Amer. Math. Soc. 9 (1996), 573–603.

163. Kenig, C.E., Ponce, G., and Vega, L., *On the ill-posedness of some canonical dispersive equations*, Duke Math. J. 106 (2001), 617–633.

164. Kichenassamy, S., *Existence of solitary waves for water-wave models*, Nonlinearity 10 (1997), 133–151.

165. Kichenassamy, S. and Olver, P., *Existence and nonexistence of solitary wave solutions to higher-order model evolution equations*, SIAM J. Math. Anal. 23 (1992), 1141-1166.

166. Koch, H. and Tzvetkov, N., *On the local well-posedness of the Benjamin-Ono equation in* $H^s(\mathbb{R})$, IMRN 2003 (2003), 1449-1464 .

167. Kodama, Y., Satsuma, J., and Ablowitz, M., *Nonlinear intermediate long-wave equation: analysis and method of solution*, Phys. Rev. Lett. 46 (1981), 687–690.

168. Korteweg, D.J. and de Vries, G., *On the change of form of long waves advancing in a rectangular channel, and on a new type of long stationary waves*, Phil. Mag. (5) 39 (1895), 422–443.

169. Kreyszig, E., *Introductory Functional Analysis with Applications*, John Wiley & Sons, New York, 1978.

170. Kruskal, M.D. and Zabusky, N.J., *Progress on the Fermi-Pasta-Ulam nonlinear string problem*, Princeton Plasma Physics Laboratory Annual Rep. MATT-Q-21, Princeton, N.J. (1963), 301–308.

171. Kubota, T., Ko, D., and Dobbs, L., *Weakly nonlinear internal gravity waves in stratified fluids of finite depth*, J. Hydrodynamics 12 (1978), 157–165.

172. Kuznetsov, E., *Soliton stability in equations of the KdV type*, Phys. Lett. A 101, 7 (1984), 314–316.

173. Kuznetsov, E., Rubenchik, A., and Zakharov, V., *Soliton stability in plasma and hydrodynamics*, Phys. 142 (1986), 103–165.

174. Kwong, M.K., *Uniqueness of positive solutions of* $\Delta u - u + u^p = 0$ *in* $\mathbb{R}^n$, Arch. Rat. Mech. Anal. 105 (1989), 243–266.

175. Laedke, E.W., Blaha, L., and Spatschek, K.H., *Collapsing states of generalized Korteweg-de Vries equations*, Physica D 40 (1989), 249–264.

176. Laedke, E.W., Blaha, L., Spatschek, K.H., and Kuznetsov, E.A., *On stability of collapse in the critical case*, J. Math. Phys. 33 (3) (1992), 967–973.

177. Lamb, G.L., *Elements of Soliton Theory*, John Wiley, New York, 1980.

178. Lawden, D. F., *Elliptic Functions and Applications*, Applied Mathematical Sciences, V. 80, Springer-Verlag, New York, 1989.

179. Lax, P.D., *Integrals of nonlinear equations of evolution and solitary waves*, Comm. Pure Appl. Math. 21 (1968), 467–490.

180. Lax, P.D., *Periodic solutions of the Korteweg-de Vries equation*, Comm. Pure Appl. Math. 28 (1975), 141–188.

181. Leibovich, S., *Weakly non-linear waves in rotating fluids*, J. Fluid Mech. 42 (1970), 803–822.

182. Levandosky, S.P., *Stability and instability of fourth-order solitary waves*, J. Dynam. Differential Equations, no. 1, 10 (1998), 151–188.

183. Levandosky, S.P., *A stability analysis of fifth-order water wave models*, Physica D 125 (1999), 222–240.

184. Levandosky, S.P., *Stability of solitary waves of a fifth-order water wave model*, Physica D 227 (2007), 162–172.

185. Li, Y.A. and Bona, J.L., *Analyticity of solitary-wave solutions of model equations for long waves*, SIAM J. Math. Anal. 27 (1996), 725–737.

186. Lieb, E., *On the lowest eigenvalue of the Laplacian for the intersection of two domains*, Invent. Math. 74 (1983), 441–448.

187. Lighthill, M.J., *Waves in Fluids*, Cambridge University Press, Cambridge, 1978.

188. Lin, C., *Orbital stability of solitary waves of the nonlinear Schrödinger KDV equation*, J. Partial Diff. Eqs. 12 (1999), 11–25.

189. Linares, F. and Scialom, M., *On generalized Benjamin type equations*, Discrete and Continuous Dynamical Systems 12, 1 (2005), 161-174.

190. Liu, Y., *Blow up and instability of solitary-wave solutions to a generalized Kadomtsev-Petviashvili equation*, Trans. Amer. Math. Soc. 353 (2001), 191–208.

191. Liu, A., Kubota, T., and Ko, D., *Resonant transfer of energy between nonlinear waves in neighboring pycnoclines*, Stud. Appl. Math. 63 (1980), 25–45.

192. Liu, Y. and Wang, X.-P., *Nonlinear stability of solitary waves of a generalized Kadomtsev-Petviashvili equation*, Comm. Math. Phys. 183 (1997), 253–266.
193. Lions, P.-L., *The concentration-compactness principle in the calculus of variations. The locally compact case, part 1*, Ann. Inst. H. Poincaré, Anal. Non Linéare. 1 (1984), 109–145.
194. Lions, P.-L., *The concentration-compactness principle in the calculus of variations. The locally compact case, part 2*, Ann. Inst. H. Poincaré, Anal. Non Linéare. 4 (1984), 223–283.
195. Lopes, O., *A constrained minimization problem with integrals on the entire space*, Bol. Soc. Brasil. Mat. (N.S.), no. 1, 25 (1994), 77–92.
196. Lopes, O., *Radial symmetry of minimizers for some translation and rotation invariant functionals*, J. Differential Equations 124 (1996), 378–388.
197. Lopes, O., *Variational systems defined by improper integrals*, International Conference on Differential Equations (L. Magalhaes *et al.*, eds.) World Scientific, Singapore (1998), 137–151.
198. Lopes, O., *Nonlocal variational problems arising in long wave propagation*, ESAIM Control Optim. Calc. Var. 5 (2000), 501–528.
199. Lopes, O. *A linearized instability result for solitary waves*, Discrete and Continuous Dynamical Systems. Series A 8 (2002), 115–119.
200. Lopes, O., *Stability of peakons for the generalized Camassa-Holm equation*, Electron. J. Differential Equations (electronic) 5 (2003), 1–12.
201. Lopes, O., *Stability of solitary waves of some coupled systems*, Nonlinearity 19 (2006), 95-113.
202. Lopes, O. and Neves, A., *Orbital stability of double solitons for the Benjamin-Ono equation*, Comm. Math. Phys. 262 (2006), 557–591.
203. Luenberger, D., *Optimization by Vector Space Methods*, Wiley and Sons, New York, 1969.
204. Maddocks, J. and Sachs, R., *On the stability of KdV multi-solitons*, Comm. Pure Appl. Math. 46 (1993), 867–901.
205. Magnus, W. and Oberhettinger, E., *Formulas and Theorems for the Special Functions of Mathematical Physics*, Springer, NY, 1986.
206. Magnus, W. and Winkler, S., *Hill's Equation*, Interscience, Tracts in Pure and Appl. Math., Wiley, NY, 20, 1976.
207. Makhankov, V., *On stationary solutions of the Schrödinger equation with a self-consistent potential satisfying Boussinesq's equation*, Phys. Lett. A. 50 (1974), 42–44.
208. Martel, Y., *Asymptotic N soliton-like solutions of the subcritical and critical generalized Korteweg-de Vries equations*, Amer. J. Math. 127 (2005), 1103–1140.
209. Martel, Y., *Linear problems related to asymptotic stability of solitons of the generalized KdV equations*, SIAM J. Math. Anal. 38 (2006), 759–781.
210. Martel, Y. and Merle, F., *A Liouville theorem for the critical generalized Korteweg-de Vries equation*, J. Math. Pures Appl. (9) 79 (2000), 339–425.
211. Martel, Y. and Merle, F., *Asymptotic stability of solitons for subcritical generalized KdV equations*, Arch. Ration. Mech. Anal. 157 (2001), 219–254.
212. Martel, Y. and Merle, F., *Instability of solitons for the critical generalized Korteweg-de Vries equation*, Geom. Funct. Anal. 11 (2001), 74–123.
213. Martel, Y. and Merle, F., *Correction: "Asymptotic stability of solitons for the subcritical generalized KdV equations" [Arch. Ration. Mech. Anal. 157 (2001), no. 3, 219–254]*, Arch. Ration. Mech. Anal. 162 (2002), 191.
214. Martel, Y. and Merle, F., *Blow up in finite time and dynamics of blow up solutions for the L^2-critical generalized KdV equation*, J. Amer. Math. Soc. (electronic) 15 (2002), 617–664.
215. Martel, Y. and Merle, F., *Stability of blow-up profile and lower bounds for blow-up rate for the critical generalized KdV equation*, Ann. of Math. (2) 155 (2002), 235–280.
216. Martel, Y. and Merle, F., *Asymptotic stability of solitons of the subcritical gKdV equations revisited*, Nonlinearity 18 (2005), 55–80.
217. Martel, Y. and Merle, F., *Refined asymptotics around solitons for the gKdV equations*, Discrete and Continuous Dynamical Systems. Series A 20, 2 (2008), 177-218.
218. Matsuno, Y., *Exact multi-soliton solution of the Benjamin-Ono equation*, J. Phys. A: Math. Gen. 12 (1979), 619–621.
219. Maz'ya, V. and Shaposhnikova, T., *Jacques Hadamard, A Universal Mathematician*, History of Mathematics, AMS and London Mathematical Society, Vol. 14 (1998), 196.
220. Merle, F., *Existence of blow-up solutions in the energy space for the critical generalized KdV equation*, J. Amer. Math. Soc. 14 (2001), 555–578.

221. Merle, F. and Vega, L., *L^2 stability of solitons for KdV equation*, IMRN 13 (2003), 735–753.
222. Miller, J. and Weinstein, M. I., *Asymptotic stability of solitary waves for the regularized long-wave equation*, Comm. Pure Appl. Math. 49 (1996), 399–441.
223. Miura, R.M., *The Korteweg-de Vries equation: A survey of results*, SIAM Rev. 18 (1976), 412–459.
224. Molinet, L., *Global well-posedness in L^2 for the periodic Benjamin-Ono equation*, American Journal of Mathematics 130, 3 (2008), 635-683.
225. Molinet, L. and Ribau, T., *Well-posedness results for the generalized Benjamin-Ono equation with small initial data*, J. Math. Pures Appl. 83 (2004), 277–311.
226. Molinet, L., Saut, J. C., and Tzvetkov, N., *Global well-posedness for the KP-I equation*, Math. Ann. 324 (2002), 255–275.
227. Nariboli, G.A., *Nonlinear longitudinal dispersive waves in elastic rods*, Iowa State Univ. Engineering Res. Inst Preprint 442 (1969).
228. Nawa, H., *Limiting profiles of blow-up solutions of the nonlinear Schrödinger equation with critical power nonlinearity*, Proc. Japan Acad. Ser. A 73 (1997), 171–175.
229. Nishikawa, K., Hojo, H., Mima, K., and Ikesi, H., *Coupled nonlinear electron-plasma and ion-acoustic wave*, Phys. Rev. Lett. 33 (1974), 148–151.
230. Ogawa, T. and Tsutsumi, Y., *Blow-up of H^1-solutions for the one-dimensional nonlinear Schrödinger equation with critical power nonlinearity*, Proc. Amer. Math. Soc. 111 (1991), 487-496.
231. Ohta, M., *Stability and instability of standing waves for one dimensional nonlinear Schrödinger equations with double power nonlinearity*, Kodai Math. J. 18 (1995), 68–74.
232. Ohta, M., *Stability of solitary waves for coupled nonlinear Schrödinger equations*, Nonlinear Anal. 26 (1996), 933–939.
233. Olver, P.J., *Hamiltonian and non-Hamiltonian models for water waves*, Lecture Notes in Physics, Springer, Berlin, 195 (1984), 273–290.
234. Ono, H., *Algebraic solitary waves in stratified fluids*, J. Phys. Soc. Japan 39 (1975), 1082-1091.
235. Pego, R. L. and Weinstein, M.I., *Eigenvalues, and instabilities of solitary waves*, Philos. Trans. Roy. Soc. London Ser. A 340 (1992), 47–94.
236. Pego, R.L. and Weinstein, M.I., *Asymptotic stability of solitary waves*, Comm. Math. Phys. 164 (1994), 305–349.
237. Pelinovsky, D.E. and Grimshaw, R.H.J., *Nonlocal models for envelope waves in a stratified fluid*, Stud. Appl. Math. 97 (1996), 369–391.
238. Phillips, P., *Wave interactions: The evolution of an idea*, J. Fluid Mech. 106 (1981), 215–227.
239. Ponce, G., *On the global well-posedness of the Benjamin-Ono equation*, Diff. Integral Equations 4 (1991), 527–542.
240. Rayleigh, Lord, *On waves*, Phil. Mag. 1 (1876), 257–279.
241. Reed, S. and Simon, B., *Methods of Modern Mathematical Physics: Functional Analysis*, Academic Press, Vol. I, 1980.
242. Reed, S. and Simon, B., *Methods of Modern Mathematical Physics: Fourier Analysis, Self-Adjointness*, Academic Press, Vol. II, 1975.
243. Reed, S. and Simon, B., *Methods of Modern Mathematical Physics: Analysis of Operator*, Academic Press, Vol. IV, 1978.
244. Riesz, F., *Sur une inégalité intégrale*, J. LMS 4 (1930), 162-168.
245. Rudin, W., *Real and Complex Analysis*, McGraw-Hill International Editions: Mathematics Series, 1987.
246. Russell, J.S., *Report on waves*, Rep. 14th Meet. Brit. Assoc. Adv. Sci., York (1844), 311-390.
247. Sandstede, B. and Scheel, A., *On the stability of periodic travelling waves with large spatial period*, J. Differential Equations 172 (2001), 134–188.
248. Saut, J.-C., *Recent results on the generalized Kadomtsev-Petviashvili equations*, Acta Appl. Math. 39 (1995), 477–487.
249. Saut, J.-C. and Tzvetkov, N., *On periodic KP-I type equations*, Comm. Math. Phys. 221 (2001), 451–476.
250. Schechter, M., *Principles of Functional Analysis*, second edition, AMS, 2002.
251. Shatah, J., *Stable standing waves of nonlinear Klein-Gordon equations*, Comm. Math. Phys. 91 (1983), 313–327.

252. Shatah, J., *Unstable ground state of nonlinear Klein-Gordon equations*, Trans. Amer. Math. Soc. 290 (1985), 701–710.
253. Shatah, J. and Strauss, W., *Instability of nonlinear bound states*, Comm. Math. Phys. 100 (1985), 173–190.
254. Shatah, J. and Strauss, W., *Spectral condition for instability*, Nonlinear PDE's, dynamics and continuum physics (South Hadley, MA, 1998), Contemp. Math. Amer. Math. Soc., Providence, RI, 255 (2000), 189–198.
255. Souganidis, P.E. and Strauss, W.A., *Instability of a class of dispersive solitary waves*, Proc. Royal Soc. Edinburgh, Sect. A 114 (1980), 195–212.
256. Staffilani, G., *On solutions for periodic generalized KdV equations*, IMRN 18 (1997), 899–917.
257. Stakgold, I., *Green's Functions and Boundary Value Problems*, second edition, Wiley-Interscience, New York, 1998.
258. Stein, L., *Singular Integrals and Differentiability of Functions*, Princeton University Press, Princeton, NJ, 1970.
259. Stein, L., *Harmonic Analysis: Real-Variable Methods, Orthogonality, and Oscillatory Integrals*, Princeton University Press, Princeton, NJ, 1993.
260. Stoker, J.I., *Water Waves*, Interscience, New York, 1957.
261. Strauss, W.A., *Existence of solitary waves in higher dimensions*, Comm. Math. Phys. 55 (1977), 149–162.
262. Strauss, W.A., *Nonlinear Wave Equations*, CBMS Regional Conference Series in Mathematics. Published for the Conference Board of the Mathematical Sciences, Washington, DC, by the American Mathematical Society, Providence, RI, 73, 1989.
263. Strauss, W.A., *Partial Differential Equations. An Introduction*, John Wiley & Sons, Inc., New York, 1992.
264. Taniuti, W. and Washini, H., *Propagation of ion-acoustic solitary waves of small amplitude*, Phys. Rev. Lett. 17 (1966), 996–998.
265. Tao, T., *Scattering for the quartic generalized Korteweg-de Vries*, J. Diff. Eq. 232 (2007), 623–651.
266. Tao, T., *Nonlinear dispersive equations: Local and global analysis*, AMS-CBMS, 2007.
267. Thayer, J., *Operadores Auto-Adjuntos e Equações Diferenciais Parciais*, Projeto Euclides, IMPA, 1998.
268. Toda, M., *Wave propagation in anharmonic lattices*, J. Phys. Soc. Japan. 23 (1967), 501–506.
269. Tom, M., *Smoothing properties of some weak solutions of the Benjamin Ono equation*, Diff. Integral Eq. 3 (1990), 683–694.
270. Tsutsumi, Y., *The Cauchy problem for the Korteweg-de Vries equation with measures as initial data*, SIAM J. Math. Anal. 20 (1989), 582–588.
271. Wang, X. P., Ablowitz, M. J., and Segur, H., *Wave collapse and instability of solitary waves of a generalized Kadomtsev-Petviashvili equation*, Phys. D 78 (1994), 241–265.
272. Weinstein, M.I., *Nonlinear Schrödinger equation and sharp interpolation estimates*, Comm. Math. Phys. 87 (1983), 567–576.
273. Weinstein, M.I., *On the structure and formation of singularities in solutions to nonlinear dispersive evolution equations*, Comm. PDE 11 (1986), 545–565.
274. Weinstein, M.I., *Liapunov stability of ground states of nonlinear dispersive evolution equations*, Comm. Pure Appl. Math. 39 (1986), 51–68.
275. Weinstein, M.I., *Solitary waves of nonlinear dispersive evolution equations with critical power nonlinearities*, J. Diff. Eq. 69 (1987), 192–203.
276. Weinstein, M.I., *Existence and dynamic stability of solitary wave solutions of equations arising in long wave propagation*, Comm. PDE 12 (1987), 1133–1173.
277. Whitham, G.B., *Linear and Nonlinear Waves*, John Wiley, New York, 1974.
278. Yajima, N. and Satsuma, J., *Soliton solutions in a diatomic lattice system*, Progr. Theoret. Phys. 62 (1979), 370–378.
279. Yosida, K., *Functional Analysis*, sixth edition, Springer, New York, 1995.
280. Zakharov, V.E., *Stability of periodic waves of finite amplitude on the surface of a deep fluid*, J. Appl. Mech. Tech. Phys. 9(1968), 86–94.
281. Zakharov, V.E., *Collapse of Langmuir waves*, Sov. Phys. JETP 35 (1972), 908–912.
282. Zakharov, V.E. and Shabat, A.B., *Interaction between solitons in a stable medium*, Sov. Phys. JETP 37 (1973), 823–828.

Index

Adjoint operator, 221, 222

Benjamin
  Equation, 19
  Solitary waves, 28
  Stability of solitary waves, 129
Benjamin-Ono
  Equation, 18
  Generalized, 18
  Local well-posedness, 51
  Periodic travelling wave solutions, 12, 39
  Regularized, 22
  Solitary waves, 27
  Stability of periodic waves, 192
  Stability of solitary waves, 98, 124
Benjamin-type equations, 127
Benney-Luke
  Cnoidal waves, 177
  Equation, 177
  Stability of cnoidal waves, 182
Blow up, 51, 70, 137, 195
Bourgain spaces, 52
Boussinesq
  Equation, 102
  Systems, 22

Class PF(2), 99
Class PF(2) discrete, 185, 192
Cnoidal wave, 10, 31, 35, 161, 164, 177, 188, 197, 209
Complementary modulus, 208
Complete
  Elliptic integral of the first kind, 208
  Elliptic integral of the second kind, 208
Concentration-Compactness method, 7, 106, 125, 127
Critical Korteweg-de Vries
  Equation, 37
  Instability of periodic waves, 193
  Periodic travelling wave solutions, 37
  Solitary waves, 27
  Stability of periodic waves, 193
Critical Nonlinear Schrödinger
  Equation, 137, 164, 194
  Instability of periodic waves, 195
  Periodic travelling wave solutions, 194
  Stability of periodic waves, 195
Critical solitary wave, 157

DispersiveWiki project, 49
Dnoidal wave, 10, 34, 41, 209
Duality, 206
Duhamel formula, 50

Embedding, 52

Floquet theory, 240
Fourier coefficient, 190
Fourier transform, 188, 201, 202, 203, 206

Gagliardo-Nirenberg-type inequality, 82, 125
Gear and Grimshaw system, 22, 157
Group, 50, 142

Hilbert transform, 203
Hypergeometric differential equation, 209

Instability
  Intervals, 241
  Linear, 153
  Orbital, 63
Instability for the generalized Benjamin equation, 139, 143
Instability for the generalized Korteweg-de Vries equation, 138, 150
Instability for weakly coupled KdV systems, 154
Intermediate long wave
  Equation, 18
  Solitary waves, 28
  Stability of solitary waves, 100

Jacobi form of Lamé's equation, 10, 162, 242
Jacobian elliptic functions, 210

Kadomtsev-Petviashvili-I
  Equation, 20
  Generalized, 20

Solitary waves, 30
Stability of solitary waves, 135
Kernel, 212
Korteweg-de Vries
Asymptotic stability of solitary waves, 88
Cnoidal wave solutions, 31, 164
Equation, 18
Generalized, 18
Local well-posedness, 51
Solitary waves, 27
Stability of cnoidal waves, 171, 188
Stability of solitary waves, 70

Lamé polynomials, 242
Lebesgue space, 201

Modified Korteweg-de Vries
Cnoidal wave solutions, 35
Dnoidal wave solutions, 34, 42
Equation, 33
Solitary waves, 27
Modulus, 208
Multiplication operator, 213, 226

Nonlinear Schrödinger
Cnoidal wave solutions, 196
Equation, 19
Ground state, 29, 137
Nonlocal equation, 43
Periodic standing wave solutions, 37
Normal elliptic integral
First kind, 208
Second kind, 208

Operator
Closed, 214
Fredholm, 235
Relatively bounded, 229
Semi-Fredholm, 235
Orbit, 63, 64

Parseval Theorem, 163, 191, 193
Periodic distribution, 206
Periodic Hilbert transform, 207
Periodic travelling wave, 26
Plancherel Theorem, 202
Poincaré, 174, 175, 193
Poisson Summation Theorem, 39, 43, 162, 163, 191

Range, 212
Regularized models for long waves, 21
Resolvent, 218
Riesz rearrangement inequality, 208

Schrödinger-Korteweg-de Vries
Bounded state solutions, 30
Systems, 20
Schwartz space, 203
Self-adjoint operator, 221, 222
Snoidal, 10
Sobolev spaces, 205, 206
Solitary waves, 26
Solitary waves of Benjamin-type equations, 127
Solitary waves of the GBO, 27, 107
Solitary waves of the GKdV, 27
Spectrum
Continuous, 218
Discrete, 232
Essential, 232
Point, 218
Residual, 218
Sub-additivity property, 113, 128, 131, 133
Stability
Asymptotic, 62
Closed operators, 229
Intervals, 241
Of the blow-up, 70, 81
Orbital, 63
Self-adjoint operators, 230
Semi-Fredholm operators, 235
Spectral, 65
Sturm-Liouville theory, 237
Sturm's Oscillation Theorem, 237
Symmetry decreasing rearrangement, 207
Symmetry operator, 224

Tempered distributions, 203
Threshold, 194
Travelling waves, 25

Well-posedness
Local, 49, 58
Global, 57

Titles in This Series

156 **Jaime Angulo Pava,** Nonlinear dispersive equations: Existence and stability of solitary and periodic travelling wave solutions, 2009

155 **Yiannis N. Moschovakis,** Descriptive set theory, 2009

154 **Andreas Čap and Jan Slovák,** Parabolic geometries I: Background and general theory, 2009

153 **Habib Ammari, Hyeonbae Kang, and Hyundae Lee,** Layer potential techniques in spectral analysis, 2009

152 **János Pach and Micha Sharir,** Combinatorial geometry and its algorithmic applications: The Alcála lectures, 2009

151 **Ernst Binz and Sonja Pods,** The geometry of Heisenberg groups: With applications in signal theory, optics, quantization, and field quantization, 2008

150 **Bangming Deng, Jie Du, Brian Parshall, and Jianpan Wang,** Finite dimensional algebras and quantum groups, 2008

149 **Gerald B. Folland,** Quantum field theory: A tourist guide for mathematicians, 2008

148 **Patrick Dehornoy with Ivan Dynnikov, Dale Rolfsen, and Bert Wiest,** Ordering braids, 2008

147 **David J. Benson and Stephen D. Smith,** Classifying spaces of sporadic groups, 2008

146 **Murray Marshall,** Positive polynomials and sums of squares, 2008

145 **Tuna Altinel, Alexandre V. Borovik, and Gregory Cherlin,** Simple groups of finite Morley rank, 2008

144 **Bennett Chow, Sun-Chin Chu, David Glickenstein, Christine Guenther, James Isenberg, Tom Ivey, Dan Knopf, Peng Lu, Feng Luo, and Lei Ni,** The Ricci flow: Techniques and applications, Part II: Analytic aspects, 2008

143 **Alexander Molev,** Yangians and classical Lie algebras, 2007

142 **Joseph A. Wolf,** Harmonic analysis on commutative spaces, 2007

141 **Vladimir Maz′ya and Gunther Schmidt,** Approximate approximations, 2007

140 **Elisabetta Barletta, Sorin Dragomir, and Krishan L. Duggal,** Foliations in Cauchy-Riemann geometry, 2007

139 **Michael Tsfasman, Serge Vlăduţ, and Dmitry Nogin,** Algebraic geometric codes: Basic notions, 2007

138 **Kehe Zhu,** Operator theory in function spaces, 2007

137 **Mikhail G. Katz,** Systolic geometry and topology, 2007

136 **Jean-Michel Coron,** Control and nonlinearity, 2007

135 **Bennett Chow, Sun-Chin Chu, David Glickenstein, Christine Guenther, James Isenberg, Tom Ivey, Dan Knopf, Peng Lu, Feng Luo, and Lei Ni,** The Ricci flow: Techniques and applications, Part I: Geometric aspects, 2007

134 **Dana P. Williams,** Crossed products of C^*-algebras, 2007

133 **Andrew Knightly and Charles Li,** Traces of Hecke operators, 2006

132 **J. P. May and J. Sigurdsson,** Parametrized homotopy theory, 2006

131 **Jin Feng and Thomas G. Kurtz,** Large deviations for stochastic processes, 2006

130 **Qing Han and Jia-Xing Hong,** Isometric embedding of Riemannian manifolds in Euclidean spaces, 2006

129 **William M. Singer,** Steenrod squares in spectral sequences, 2006

128 **Athanassios S. Fokas, Alexander R. Its, Andrei A. Kapaev, and Victor Yu. Novokshenov,** Painlevé transcendents, 2006

127 **Nikolai Chernov and Roberto Markarian,** Chaotic billiards, 2006

126 **Sen-Zhong Huang,** Gradient inequalities, 2006

125 **Joseph A. Cima, Alec L. Matheson, and William T. Ross,** The Cauchy Transform, 2006

124 **Ido Efrat, Editor,** Valuations, orderings, and Milnor K-Theory, 2006

123 **Barbara Fantechi, Lothar Göttsche, Luc Illusie, Steven L. Kleiman, Nitin Nitsure, and Angelo Vistoli,** Fundamental algebraic geometry: Grothendieck's FGA explained, 2005

122 **Antonio Giambruno and Mikhail Zaicev, Editors,** Polynomial identities and asymptotic methods, 2005

121 **Anton Zettl,** Sturm-Liouville theory, 2005

120 **Barry Simon,** Trace ideals and their applications, 2005

119 **Tian Ma and Shouhong Wang,** Geometric theory of incompressible flows with applications to fluid dynamics, 2005

118 **Alexandru Buium,** Arithmetic differential equations, 2005

117 **Volodymyr Nekrashevych,** Self-similar groups, 2005

116 **Alexander Koldobsky,** Fourier analysis in convex geometry, 2005

115 **Carlos Julio Moreno,** Advanced analytic number theory: L-functions, 2005

114 **Gregory F. Lawler,** Conformally invariant processes in the plane, 2005

113 **William G. Dwyer, Philip S. Hirschhorn, Daniel M. Kan, and Jeffrey H. Smith,** Homotopy limit functors on model categories and homotopical categories, 2004

112 **Michael Aschbacher and Stephen D. Smith,** The classification of quasithin groups II. Main theorems: The classification of simple QTKE-groups, 2004

111 **Michael Aschbacher and Stephen D. Smith,** The classification of quasithin groups I. Structure of strongly quasithin K-groups, 2004

110 **Bennett Chow and Dan Knopf,** The Ricci flow: An introduction, 2004

109 **Goro Shimura,** Arithmetic and analytic theories of quadratic forms and Clifford groups, 2004

108 **Michael Farber,** Topology of closed one-forms, 2004

107 **Jens Carsten Jantzen,** Representations of algebraic groups, 2003

106 **Hiroyuki Yoshida,** Absolute CM-periods, 2003

105 **Charalambos D. Aliprantis and Owen Burkinshaw,** Locally solid Riesz spaces with applications to economics, second edition, 2003

104 **Graham Everest, Alf van der Poorten, Igor Shparlinski, and Thomas Ward,** Recurrence sequences, 2003

103 **Octav Cornea, Gregory Lupton, John Oprea, and Daniel Tanré,** Lusternik-Schnirelmann category, 2003

102 **Linda Rass and John Radcliffe,** Spatial deterministic epidemics, 2003

101 **Eli Glasner,** Ergodic theory via joinings, 2003

100 **Peter Duren and Alexander Schuster,** Bergman spaces, 2004

99 **Philip S. Hirschhorn,** Model categories and their localizations, 2003

98 **Victor Guillemin, Viktor Ginzburg, and Yael Karshon,** Moment maps, cobordisms, and Hamiltonian group actions, 2002

97 **V. A. Vassiliev,** Applied Picard-Lefschetz theory, 2002

96 **Martin Markl, Steve Shnider, and Jim Stasheff,** Operads in algebra, topology and physics, 2002

95 **Seiichi Kamada,** Braid and knot theory in dimension four, 2002

94 **Mara D. Neusel and Larry Smith,** Invariant theory of finite groups, 2002